Wireless Sensor Networks: Signals and Communication Technology

Wireless Sensor Networks: Signals and Communication Technology

Edited by **Bob Tucker**

NY RESEARCH
P R E S S

New York

Published by NY Research Press,
23 West, 55th Street, Suite 816,
New York, NY 10019, USA
www.nyresearchpress.com

Wireless Sensor Networks: Signals and Communication Technology
Edited by Bob Tucker

International Standard Book Number: 978-1-63238-464-5 (Hardback)

Contents

Preface

I am honored to present to you this unique book which encompasses the most up-to-date data in the field. I was extremely pleased to get this opportunity of editing the work of experts from across the globe. I have also written papers in this field and researched the various aspects revolving around the progress of the discipline. I have tried to unify my knowledge along with that of stalwarts from every corner of the world, to produce a text which not only benefits the readers but also facilitates the growth of the field.

This book is a well-structured analysis of wireless sensor networks. It explores the most recent developments in wireless networks technology. It describes the essential concepts and realistic aspects of wireless sensor networks and addresses challenges faced in their design, examination and usage. This book will serve as a complete reference for experts and even for students who seek to learn about the latest advancements in wireless sensor networks.

Finally, I would like to thank all the contributing authors for their valuable time and contributions. This book would not have been possible without their efforts. I would also like to thank my friends and family for their constant support.

Editor

Basic Concepts &
Energy Efficient Design Principles

Overview of Wireless Sensor Network

M.A. Matin and M.M. Islam

Additional information is available at the end of the chapter

1. Introduction

Wireless Sensor Networks (WSNs) can be defined as a self-configured and infrastructure-less wireless networks to monitor physical or environmental conditions, such as temperature, sound, vibration, pressure, motion or pollutants and to cooperatively pass their data through the network to a main location or sink where the data can be observed and analysed. A sink or base station acts like an interface between users and the network. One can retrieve required information from the network by injecting queries and gathering results from the sink. Typically a wireless sensor network contains hundreds of thousands of sensor nodes. The sensor nodes can communicate among themselves using radio signals. A wireless sensor node is equipped with sensing and computing devices, radio transceivers and power components. The individual nodes in a wireless sensor network (WSN) are inherently resource constrained: they have limited processing speed, storage capacity, and communication bandwidth. After the sensor nodes are deployed, they are responsible for self-organizing an appropriate network infrastructure often with multi-hop communication with them. Then the onboard sensors start collecting information of interest. Wireless sensor devices also respond to queries sent from a "control site" to perform specific instructions or provide sensing samples. The working mode of the sensor nodes may be either continuous or event driven. Global Positioning System (GPS) and local positioning algorithms can be used to obtain location and positioning information. Wireless sensor devices can be equipped with actuators to "act" upon certain conditions. These networks are sometimes more specifically referred as Wireless Sensor and Actuator Networks as described in (Akkaya et al., 2005).

Wireless sensor networks (WSNs) enable new applications and require non-conventional paradigms for protocol design due to several constraints. Owing to the requirement for low device complexity together with low energy consumption (i.e. long network lifetime), a proper balance between communication and signal/data processing capabilities must be found. This motivates a huge effort in research activities, standardization process, and industrial investments on this field since the last decade (Chiara et. al. 2009). At present time, most of the research on WSNs has concentrated on the design of energy- and computationally efficient algorithms and protocols, and the application domain has been restricted to simple data-oriented monitoring and reporting applications (Labrador et. al. 2009). The authors in (Chen et al., 2011) propose a Cable Mode Transition (CMT) algorithm, which determines the minimal number of active sensors to maintain K-coverage of a terrain as well as K-connectivity of the network. Specifically, it allocates periods of inactivity for cable sensors without affecting the coverage and connectivity requirements of the network based only on local information. In (Cheng et al., 2011), a delay-aware data collection network structure for wireless sensor networks is proposed. The objective of the proposed network structure is to minimize delays in the data collection processes of wireless sensor networks which extends the lifetime of the network. In (Matin et al., 2011), the authors have considered relay nodes to mitigate the network geometric deficiencies and used Particle Swarm Optimization (PSO) based algorithms to locate the optimal sink location with respect to those relay nodes to overcome the lifetime challenge. Energy efficient communication has also been addressed in (Paul et al., 2011; Fabbri et al. 2009). In (Paul et al., 2011), the authors proposed a geometrical solution for locating the optimum sink placement for maximizing the network lifetime. Most of the time, the research on wireless sensor networks have considered homogeneous sensor nodes. But nowadays researchers have focused on heterogeneous sensor networks where the sensor nodes are unlike to each other in terms of their energy. In (Han et al., 2010), the authors addresses the problem of deploying relay nodes to provide fault tolerance with higher network connectivity in heterogeneous wireless sensor networks, where sensor nodes possess different transmission radii. New network architectures with heterogeneous devices and the recent advancement in this technology eliminate the current limitations and expand the spectrum of possible applications for WSNs considerably and all these are changing very rapidly.

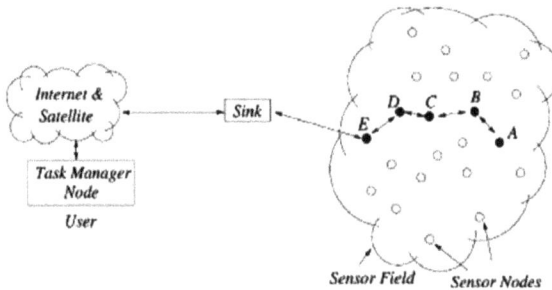

Figure 1. A typical Wireless Sensor Network

2. Applications of wireless sensor network

Wireless sensor networks have gained considerable popularity due to their flexibility in solving problems in different application domains and have the potential to change our lives in many different ways. WSNs have been successfully applied in various application domains (Akyildiz et al. 2002; Bharathidasan et al., 2001), (Yick et al., 2008; Boukerche, 2009), (Sohraby et al., 2007), and (Chiara et al., 2009;Verdone et al., 2008), such as:

Military applications: Wireless sensor networks be likely an integral part of military command, control, communications, computing, intelligence, battlefield surveillance, reconnaissance and targeting systems.

Area monitoring: In area monitoring, the sensor nodes are deployed over a region where some phenomenon is to be monitored. When the sensors detect the event being monitored (heat, pressure etc), the event is reported to one of the base stations, which then takes appropriate action.

Transportation: Real-time traffic information is being collected by WSNs to later feed transportation models and alert drivers of congestion and traffic problems.

Health applications: Some of the health applications for sensor networks are supporting interfaces for the disabled, integrated patient monitoring, diagnostics, and drug administration in hospitals, tele-monitoring of human physiological data, and tracking & monitoring doctors or patients inside a hospital.

Environmental sensing: The term Environmental Sensor Networks has developed to cover many applications of WSNs to earth science research. This includes sensing volcanoes, oceans, glaciers, forests etc. Some other major areas are listed below:

- Air pollution monitoring
- Forest fires detection
- Greenhouse monitoring
- Landslide detection

Structural monitoring: Wireless sensors can be utilized to monitor the movement within buildings and infrastructure such as bridges, flyovers, embankments, tunnels etc enabling Engineering practices to monitor assets remotely with out the need for costly site visits.

Industrial monitoring: Wireless sensor networks have been developed for machinery condition-based maintenance (CBM) as they offer significant cost savings and enable new functionalities. In wired systems, the installation of enough sensors is often limited by the cost of wiring.

Agricultural sector: using a wireless network frees the farmer from the maintenance of wiring in a difficult environment. Irrigation automation enables more efficient water use and reduces waste.

3. Design issues of a wireless sensor network

There are a lot of challenges placed by the deployment of sensor networks which are a superset of those found in wireless ad hoc networks. Sensor nodes communicate over wireless, lossy lines with no infrastructure. An additional challenge is related to the limited, usually non-renewable energy supply of the sensor nodes. In order to maximize the lifetime of the network, the protocols need to be designed from the beginning with the objective of efficient management of the energy resources (Akyildiz et al., 2002). Wireless Sensor Network Design issues are mentioned in (Akkaya et al., 2005), (Akyildizet al., 2002), (SensorSim; Tossim, Younis et al., 2004), (Pan et al., 2003) and different possible platforms for simulation and testing of routing protocols for WSNs are discussed in (NS-2, Zeng et al., 1998, SensorSim, Tossiim). Let us now discuss the individual design issues in greater detail.

Fault Tolerance: Sensor nodes are vulnerable and frequently deployed in dangerous environment. Nodes can fail due to hardware problems or physical damage or by exhausting their energy supply. We expect the node failures to be much higher than the one normally considered in wired or infrastructure-based wireless networks. The protocols deployed in a sensor network should be able to detect these failures as soon as possible and be robust enough to handle a relatively large number of failures while maintaining the overall functionality of the network. This is especially relevant to the routing protocol design, which has to ensure that alternate paths are available for rerouting of the packets. Different deployment environments pose different fault tolerance requirements.

Scalability: Sensor networks vary in scale from several nodes to potentially several hundred thousand. In addition, the deployment density is also variable. For collecting high-resolution data, the node density might reach the level where a node has several thousand neighbours in their transmission range. The protocols deployed in sensor networks need to be scalable to these levels and be able to maintain adequate performance.

Production Costs: Because many deployment models consider the sensor nodes to be disposable devices, sensor networks can compete with traditional information gathering approaches only if the individual sensor nodes can be produced very cheaply. The target price envisioned for a sensor node should ideally be less than $1.

Hardware Constraints: At minimum, every sensor node needs to have a sensing unit, a processing unit, a transmission unit, and a power supply. Optionally, the nodes may have several built-in sensors or additional devices such as a localization system to enable location-aware routing. However, every additional functionality comes with additional cost and increases the power consumption and physical size of the node. Thus, additional functionality needs to be always balanced against cost and low-power requirements.

Sensor Network Topology: Although WSNs have evolved in many aspects, they continue to be networks with constrained resources in terms of energy, computing power, memory, and communications capabilities. Of these constraints, energy consumption is of paramount importance, which is demonstrated by the large number of algorithms, techniques, and protocols that have been developed to save energy, and thereby extend the lifetime of the network. Topology Maintenance is one of the most important issues researched to reduce energy consumption in wireless sensor networks.

Transmission Media: The communication between the nodes is normally implemented using radio communication over the popular ISM bands. However, some sensor networks use optical or infrared communication, with the latter having the advantage of being robust and virtually interference free.

Power Consumption: As we have already seen, many of the challenges of sensor networks revolve around the limited power resources. The size of the nodes limits the size of the battery. The software and hardware design needs to carefully consider the issues of efficient energy use. For instance, data compression might reduce the amount of energy used for radio transmission, but uses additional energy for computation and/or filtering. The energy policy also depends on the application; in some applications, it might be acceptable to turn off a subset of nodes in order to conserve energy while other applications require all nodes operating simultaneously.

4. Structure of a wireless sensor network

Structure of a Wireless Sensor Network includes different topologies for radio communications networks. A short discussion of the network topologies that apply to wireless sensor networks are outlined below:

4.1. Star network (single point-to-multipoint) (Wilson, 2005)

A star network is a communications topology where a single base station can send and/or receive a message to a number of remote nodes. The remote nodes are not permitted to send messages to each other. The advantage of this type of network for wireless sensor networks includes simplicity, ability to keep the remote node's power consumption to a minimum. It also allows low latency communications between the remote node and the base station. The disadvantage of such a network is that the base station must be within radio transmission range of all the individual nodes and is not as robust as other networks due to its dependency on a single node to manage the network.

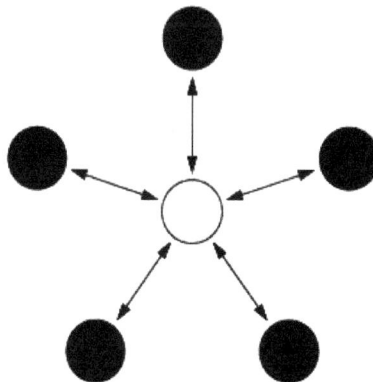

Figure 2. A Star network topology

4.2. Mesh network (Wilson, 2005)

A mesh network allows transmitting data to one node to other node in the network that is within its radio transmission range. This allows for what is known as multi-hop communications, that is, if a node wants to send a message to another node that is out of radio communications range, it can use an intermediate node to forward the message to the desired node. This network topology has the advantage of redundancy and scalability. If an individual node fails, a remote node still can communicate to any other node in its range, which in turn, can forward the message to the desired location. In addition, the range of the network is not necessarily limited by the range in between single nodes; it can simply be extended by adding more nodes to the system. The disadvantage of this type of network is in power consumption for the nodes that implement the multi-hop communications are generally higher than for the nodes that don't have this capability, often limiting the battery life. Additionally, as the number of communication hops to a destination increases, the time to deliver the message also increases, especially if low power operation of the nodes is a requirement.

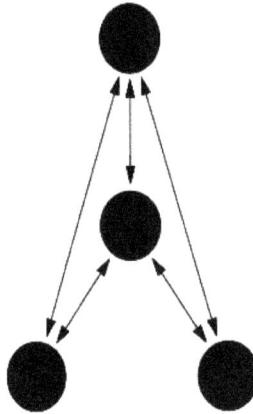

Figure 3. A Mesh network topology

4.3. Hybrid star – Mesh network (Wilson, 2005)

A hybrid between the star and mesh network provides a robust and versatile communications network, while maintaining the ability to keep the wireless sensor nodes power consumption to a minimum. In this network topology, the sensor nodes with lowest power are not enabled with the ability to forward messages. This allows for minimal power consumption to be maintained. However, other nodes on the network are enabled with multi-hop capability, allowing them to forward messages from the low power nodes to other nodes on the network. Generally, the nodes with the multi-hop capability are higher power, and if possible, are often plugged into the electrical mains line. This is the topology implemented by the up and coming mesh networking standard known as ZigBee.

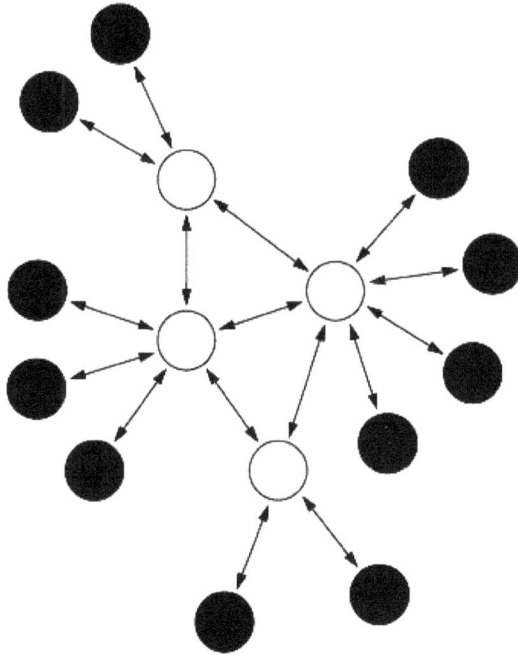

Figure 4. A Hybrid Star – Mesh network topology

5. Structure of a wireless sensor node

A sensor node is made up of four basic components such as sensing unit, processing unit, transceiver unit and a power unit which is shown in Fig. 5. It also has application dependent additional components such as a location finding system, a power generator and a mobilizer. Sensing units are usually composed of two subunits: sensors and analogue to digital converters (ADCs) (Akyildiz et al., 2002). The analogue signals produced by the sensors are converted to digital signals by the ADC, and then fed into the processing unit. The processing unit is generally associated with a small storage unit and it can manage the procedures that make the sensor node collaborate with the other nodes to carry out the assigned sensing tasks. A transceiver unit connects the node to the network. One of the most important components of a sensor node is the power unit. Power units can be supported by a power scavenging unit such as solar cells. The other subunits, of the node are application dependent.

A functional block diagram of a versatile wireless sensing node is provided in Fig. 6. Modular design approach provides a flexible and versatile platform to address the needs of a wide variety of applications. For example, depending on the sensors to be deployed, the signal conditioning block can be re-programmed or replaced. This allows for a wide variety of different sensors to be used with the wireless sensing node. Similarly, the radio link may be swapped out as required for a given applications' wireless range requirement and the need for bidirectional communications.

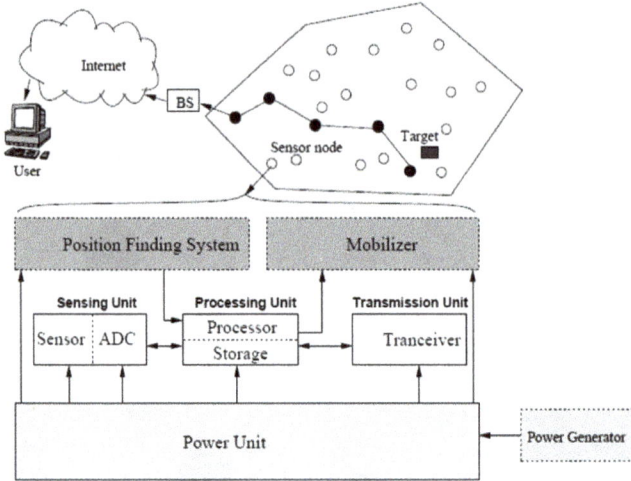

Figure 5. The components of a sensor node

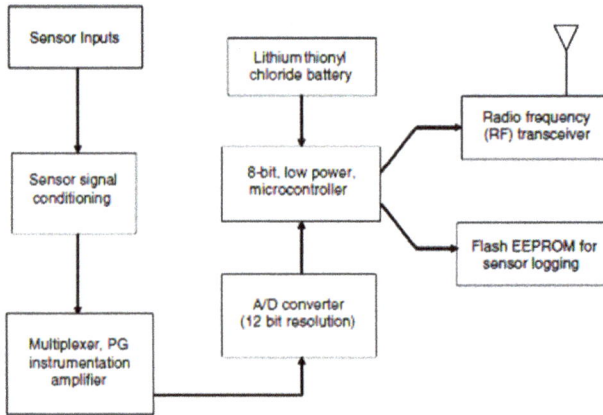

Figure 6. Functional block diagram of a sensor node

Using flash memory, the remote nodes acquire data on command from a base station, or by an event sensed by one or more inputs to the node. Moreover, the embedded firmware can be upgraded through the wireless network in the field.

The microprocessor has a number of functions including:

- Managing data collection from the sensors
- performing power management functions
- interfacing the sensor data to the physical radio layer
- managing the radio network protocol

A key aspect of any wireless sensing node is to minimize the power consumed by the system. Usually, the radio subsystem requires the largest amount of power. Therefore, data is sent over the radio network only when it is required. An algorithm is to be loaded into the node to determine when to send data based on the sensed event. Furthermore, it is important to minimize the power consumed by the sensor itself. Therefore, the hardware should be designed to allow the microprocessor to judiciously control power to the radio, sensor, and sensor signal conditioner (Akyildiz et al., 2002).

6. Communication structure of a wireless sensor network

The sensor nodes are usually scattered in a sensor field as shown in Fig. 1. Each of these scattered sensor nodes has the capabilities to collect data and route data back to the sink and the end users. Data are routed back to the end user by a multi-hop infrastructure-less architecture through the sink as shown in Fig. 1. The sink may communicate with the task manager node via Internet or Satellite.

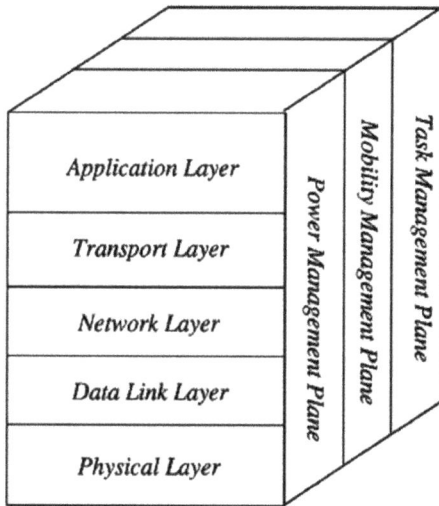

Figure 7. Wireless Sensor Network protocol stack

The protocol stack used by the sink and the sensor nodes is given in Fig. 7. This protocol stack combines power and routing awareness, integrates data with networking protocols, communicates power efficiently through the wireless medium and promotes cooperative efforts of sensor nodes. The protocol stack consists of the application layer, transport layer, network layer, data link layer, physical layer, power management plane, mobility management plane, and task management plane (Akyildiz et al., 2002). Different types of application software can be built and used on the application layer depending on the sensing tasks. This layer makes hardware and software of the lowest layer transparent to the end-user. The transport layer helps to maintain the flow of data if the sensor networks application requires it. The network layer takes care of routing the data supplied by the transport layer, specific multi-hop wireless routing protocols between sensor nodes and sink. The data link layer is responsible for multiplexing of data streams, frame detection, Media Access Control (MAC) and error control. Since the environment is noisy and sensor nodes can be mobile, the MAC protocol must be power aware and able to minimize collision with neighbours' broadcast. The physical layer addresses the needs of a simple but robust modulation, frequency selection, data encryption, transmission and receiving techniques.

In addition, the power, mobility, and task management planes monitor the power, movement, and task distribution among the sensor nodes. These planes help the sensor nodes coordinate the sensing task and lower the overall energy consumption.

7. Energy consumption issues in wireless sensor network

Energy consumption is the most important factor to determine the life of a sensor network because usually sensor nodes are driven by battery. Sometimes energy optimization is more complicated in sensor networks because it involved not only reduction of energy consumption but also prolonging the life of the network as much as possible. The optimization can be done by having energy awareness in every aspect of design and operation. This ensures that energy awareness is also incorporated into groups of communicating sensor nodes and the entire network and not only in the individual nodes (Bharathidasan et al. 2001).

A sensor node usually consists of four sub-systems (Bharathidasan et al. 2001):

- a computing subsystem : It consists of a microprocessor(microcontroller unit, MCU) which is responsible for the control of the sensors and implementation of communication protocols. MCUs usually operate under various modes for power management purposes. As these operating modes involves consumption of power, the energy consumption levels of the various modes should be considered while looking at the battery lifetime of each node.
- a communication subsystem: It consists of a short range radio which communicate with neighboring nodes and the outside world. Radios can operate under the different modes. It is important to completely shut down the radio rather than putting it in the Idle mode when it is not transmitting or receiving for saving power.
- a sensing subsystem : It consists of a group of sensors and actuators and link the node to the outside world. Energy consumption can be reduced by using low power components and saving power at the cost of performance which is not required.

- a power supply subsystem : It consists of a battery which supplies power to the node. It should be seen that the amount of power drawn from a battery is checked because if high current is drawn from a battery for a long time, the battery will die faster even though it could have gone on for a longer time. Usually the rated current capacity of a battery being used for a sensor node is less than the minimum energy consumption. The lifetime of a battery can be increased by reducing the current drastically or even turning it off often.

To minimize the overall energy consumption of the sensor network, different types of protocols and algorithms have been studied so far all over the world. The lifetime of a sensor network can be increased significantly if the operating system, the application layer and the network protocols are designed to be energy aware. These protocols and algorithms have to be aware of the hardware and able to use special features of the micro-processors and transceivers to minimize the sensor node's energy consumption. This may push toward a custom solution for different types of sensor node design. Different types of sensor nodes deployed also lead to different types of sensor networks. This may also lead to the different types of collaborative algorithms in wireless sensor networks arena.

8. Protocols & algorithms of wireless sensor network

In WSN, the main task of a sensor node is to sense data and sends it to the base station in multi hop environment for which routing path is essential. For computing the routing path from the source node to the base station there is huge numbers of proposed routing protocols exist (Sharma et al., 2011). The design of routing protocols for WSNs must consider the power and resource limitations of the network nodes, the time-varying quality of the wireless channel, and the possibility for packet loss and delay. To address these design requirements, several routing strategies for WSNs have been proposed in (Labrador et al., 2009), (Akkaya et al., 2005), (Akyildiz et al. 2002), (Boukerche, 2009, Al-karaki et al., 2004, Pan et al., 2003) and (Waharte et al., 2006).

The first class of routing protocols adopts a flat network architecture in which all nodes are considered peers. Flat network architecture has several advantages, including minimal overhead to maintain the infrastructure and the potential for the discovery of multiple routes between communicating nodes for fault tolerance.

A second class of routing protocols imposes a structure on the network to achieve energy efficiency, stability, and scalability. In this class of protocols, network nodes are organized in clusters in which a node with higher residual energy, for example, assumes the role of a cluster head. The cluster head is responsible for coordinating activities within the cluster and forwarding information between clusters. Clustering has potential to reduce energy consumption and extend the lifetime of the network.

A third class of routing protocols uses a data-centric approach to disseminate interest within the network. The approach uses attribute-based naming, whereby a source node queries an attribute for the phenomenon rather than an individual sensor node. The interest dissemination is achieved by assigning tasks to sensor nodes and expressing queries to relative to specific attributes. Different strategies can be used to communicate interests to the sensor nodes, including broadcasting, attribute-based multicasting, geo-casting, and any casting.

A fourth class of routing protocols uses location to address a sensor node. Location-based routing is useful in applications where the position of the node within the geographical coverage of the network is relevant to the query issued by the source node. Such a query may specify a specific area where a phenomenon of interest may occur or the vicinity to a specific point in the network environment.

In the rest of this section we discuss some of the major routing protocols and algorithms to deal with the energy conservation issue in the literatures.

1. Flooding: Flooding is a common technique frequently used for path discovery and information dissemination in wired and wireless ad hoc networks which has been discussed in (Akyildiz et al., 2002). The routing strategy of flooding is simple and does not rely on costly network topology maintenance and complex route discovery algorithms. Flooding uses a reactive approach whereby each node receiving a data or control packet sends the packet to all its neighbors. After transmission, a packet follows all possible paths. Unless the network is disconnected, the packet will eventually reach its destination. Furthermore, as the network topology changes, the packet transmitted follows the new routes. Fig. 8 illustrates the concept of flooding in data communications network. As shown in the figure, flooding in its simplest form may cause packets to be replicated indefinitely by network nodes.

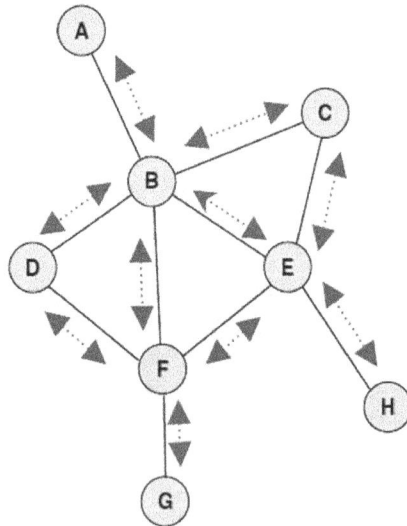

Figure 8. Flooding in data communication networks

1.   Gossiping:

To address the shortcomings of flooding, a derivative approach, referred to as gossiping, has been proposed in (Braginsky et al., 2002). Similar to flooding, gossiping uses a simple forwarding rule and does not require costly topology maintenance or complex route discovery algorithms. Contrary to flooding, where a data packet is broadcast to all neighbors, gossiping requires that each node sends the incoming packet to a randomly selected neighbor. Upon receiving the packet, the neighbor selected randomly chooses one of its own neighbors and forwards the packet to the neighbor chosen. This process continues iteratively until the packet reaches its intended destination or the maximum hop count is exceeded.

2.   Protocols for Information via Negotiation (SPIN):

Sensor protocols for information via negotiation (SPIN), is a data-centric negotiation-based family of information dissemination protocols for WSNs (Kulik et al., 2002). The main objective of these protocols is to efficiently disseminate observations gathered by individual sensor nodes to all the sensor nodes in the network. Simple protocols such as flooding and gossiping are commonly proposed to achieve information dissemination in WSNs. Flooding requires that each node sends a copy of the data packet to all its neighbors until the information reaches all nodes in the network. Gossiping, on the other hand, uses randomization to reduce the number of duplicate packets and requires only that a node receiving a data packet forward it to a randomly selected neighbor.

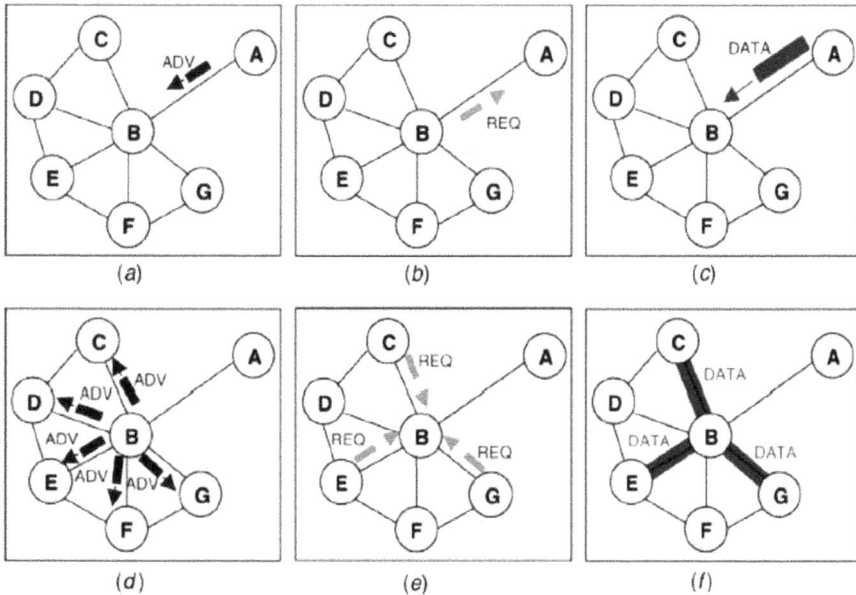

Figure 9. SPIN basic protocol operation

3.   Low-Energy Adaptive Clustering Hierarchy (LEACH)

Low-energy adaptive clustering hierarchy (LEACH) is a routing algorithm designed to collect and deliver data to the data sink, typically a base station (Heinzelman et. al. 2000). The main objectives of LEACH are:

- Extension of the network lifetime
- Reduced energy consumption by each network sensor node
- Use of data aggregation to reduce the number of communication messages

To achieve these objectives, LEACH adopts a hierarchical approach to organize the network into a set of clusters. Each cluster is managed by a selected cluster head. The cluster head assumes the responsibility to carry out multiple tasks. The first task consists of periodic collection of data from the members of the cluster. Upon gathering the data, the cluster head aggregates it in an effort to remove redundancy among correlated values. The second main task of a cluster head is to transmit the aggregated data directly to the base station over single hop. The third main task of the cluster head is to create a TDMA-based schedule whereby each node of the cluster is assigned a time slot that it can use for transmission. The cluster head announces the schedule to its cluster members through broadcasting. To reduce the likelihood of collisions among sensors within and outside the cluster, LEACH nodes use a code-division multiple access–based scheme for communication.

The basic operations of LEACH are organized in two distinct phases. The first phase, the setup phase, consists of two steps, cluster-head selection and cluster formation. The second phase, the steady-state phase, focuses on data collection, aggregation, and delivery to the base station. The duration of the setup is assumed to be relatively shorter than the steady-state phase to minimize the protocol overhead.

At the beginning of the setup phase, a round of cluster-head selection starts. To decide whether a node to become cluster head or not a threshold T(s) is addressed in (Heinzelman et. al. 2000) which is as follows:

$$T(s) = \begin{cases} \dfrac{P_{opt}}{1 - P_{opt}.(r.\bmod.\dfrac{1}{P_{opt}})}, & ifs \in G' \\ 0, & otherwise \end{cases} \tag{1}$$

Where r is the current round number and G is the set of nodes that have not become cluster head within the last $1/p_{opt}$ rounds. At the beginning of each round, each node which belongs to the set G selects a random number 0 or 1. If the random number is less than the threshold T(s) then the node becomes a cluster head in the current round.

4.   Threshold-sensitive Energy Efficient Protocols (TEEN and APTEEN):

Two hierarchical routing protocols called TEEN (Threshold-sensitive Energy Efficient sensor Network protocol), and APTEEN (Adaptive Periodic Threshold-sensitive Energy Efficient sensor Network protocol) are proposed in (Manjeshwar et al., 2001) and (Manjeshwar et al.,

2002) , respectively. These protocols were proposed for time-critical applications. In TEEN, sensor nodes sense the medium continuously, but the data transmission is done less frequently. A cluster head sensor sends its members a hard threshold, which is the threshold value of the sensed attribute and a soft threshold, which is a small change in the value of the sensed attribute that triggers the node to switch on its transmitter and transmit. Thus the hard threshold tries to reduce the number of transmissions by allowing the nodes to transmit only when the sensed attribute is in the range of interest. The soft threshold further reduces the number of transmissions that might have otherwise occurred when there is little or no change in the sensed attribute. A smaller value of the soft threshold gives a more accurate picture of the network, at the expense of increased energy consumption. Thus, the user can control the trade-off between energy efficiency and data accuracy. When cluster-heads are to change, new values for the above parameters are broadcast. The main drawback of this scheme is that, if the thresholds are not received, the nodes will never communicate, and the user will not get any data from the network at all.

5. Power-Efficient Gathering in Sensor Information Systems (PEGASIS):

Power-efficient gathering in sensor information systems (PEGASIS) (Lindsey et al., 2002) and its extension, hierarchical PEGASIS, are a family of routing and information-gathering protocols for WSNs. The main objectives of PEGASIS are twofold. First, the protocol aims at extending the lifetime of a network by achieving a high level of energy efficiency and uniform energy consumption across all network nodes. Second, the protocol strives to reduce the delay that data incur on their way to the sink.

The network model considered by PEGASIS assumes a homogeneous set of nodes deployed across a geographical area. Nodes are assumed to have global knowledge about other sensors' positions. Furthermore, they have the ability to control their power to cover arbitrary ranges. The nodes may also be equipped with CDMA-capable radio transceivers. The nodes' responsibility is to gather and deliver data to a sink, typically a wireless base station. The goal is to develop a routing structure and an aggregation scheme to reduce energy consumption and deliver the aggregated data to the base station with minimal delay while balancing energy consumption among the sensor nodes. Contrary to other protocols, which rely on a tree structure or a cluster-based hierarchical organization of the network for data gathering and dissemination, PEGASIS uses a chain structure.

6. Directed Diffusion:

Directed diffusion (Intanagonwiwat et al., 2000) is a data-centric routing protocol for information gathering and dissemination in WSNs. The main objective of the protocol is to achieve substantial energy savings in order to extend the lifetime of the network. To achieve this objective, directed diffusion keeps interactions between nodes, in terms of message exchanges, localized within limited network vicinity. Using localized interaction, direct diffusion can still realize robust multi-path delivery and adapt to a minimal subset of network paths. This unique feature of the protocol, combined with the ability of the nodes to aggregate response to queries, results into significant energy savings.

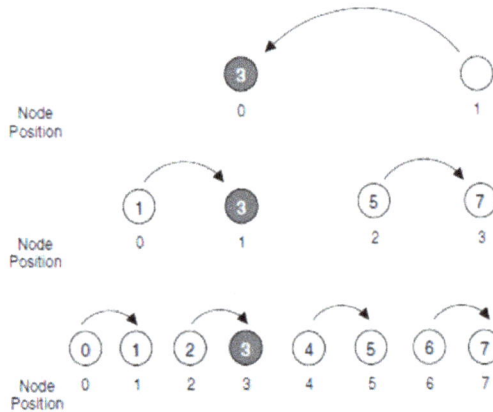

Figure 10. Chain-based data gathering and aggregation scheme

The main elements of direct diffusion include interests, data messages, gradients, and reinforcements. Directed diffusion uses a publish-and-subscribe information model in which an inquirer expresses an interest using attribute–value pairs. An interest can be viewed as a query or an interrogation that specifies what the inquirer wants.

7.   Geographic Adaptive Fidelity (GAF):

GAF (Xu et al., 2001) is an energy-aware location-based routing algorithm designed mainly for mobile ad hoc networks, but may be applicable to sensor networks as well. The network area is first divided into fixed zones and forms a virtual grid. Inside each zone, nodes collaborate with each other to play different roles. For example, nodes will elect one sensor node to stay awake for a certain period of time and then they go to sleep. This node is responsible for monitoring and reporting data to the BS on behalf of the nodes in the zone. Hence, GAF conserves energy by turning off unnecessary nodes in the network without affecting the level of routing fidelity.

9. Security issues in wireless sensor network

Security issues in sensor networks depend on the need to know what we are going to protect. In (Zia et al., 2006), the authors defined four security goals in sensor networks which are Confidentiality, Integrity, Authentication and Availability. Another security goal in sensor network is introduced in (Sharma et al., 2011).Confidentiality is the ability to conceal message from a passive attacker, where the message communicated on sensor networks remain confidential. Integrity refers to the ability to confirm the message has not been tampered, altered or changed while it was on the network. Authentication Need to know if the messages are from the node it claims to be from, determining the reliability of message's origin. Availability is to determine if a node has the ability to use the resources and the network is available for the messages to move on. Freshness implies that receiver receives the recent and fresh data and ensures that no adversary can replay the old data. This requirement is especially important when the WSN nodes use shared-keys for message communication, where a potential adversary can launch a replay attack using the old key as the new key is being refreshed and propagated to all the nodes in the WSN (Sen, 2009). To achieve the freshness the mechanism like nonce or time stamp should add to each data packet.

Having built a foundation of security goals in sensor network, the major possible security attacks in sensor networks are identified in (Undercoffer et al., 2002) . Routing loops attacks target the information exchanged between nodes. False error messages are generated when an attacker alters and replays the routing information. Routing loops attract or repel the network traffic and increases node to node latency. Selective forwarding attack influences the network traffic by believing that all the participating nodes in network are reliable to forward the message. In selective forwarding attack malicious nodes simply drop certain messages instead of forwarding every message. Once a malicious node cherry picks on the messages, it reduces the latency and deceives the neighboring nodes that they are on a shorter route. Effectiveness of this attack depends on two factors. First the location of the malicious node, the closer it is to the base stations the more traffic it will attract. Second is the percentage of messages it drops. When selective forwarder drops more messages and forwards less, it retains its energy level thus remaining powerful to trick the neighboring nodes. In sinkhole attacks, adversary attracts the traffic to a compromised node. The simplest way of creating sinkhole is to place a malicious node where it can attract most of the traffic, possibly closer to the base station or malicious node itself deceiving as a base station. One reason for sinkhole attacks is to make selective forwarding possible to attract the traffic towards a compromised node. The nature of sensor networks where all the traffic flows towards one base station makes this type of attacks more susceptible. Sybil attacks are a type of attacks where a node creates multiple illegitimate identities in sensor networks either by fabricating or stealing the identities of legitimate nodes. Sybil attacks can be used against routing algorithms and topology maintenance; it reduces the effectiveness of fault tolerant schemes such as distributed storage and disparity. Another malicious factor is geographic routing where a Sybil node can appear at more than one place simultaneously. In wormhole attacks an adversary positioned closer to the base station can completely disrupt the traffic by tunneling messages over a low latency link. Here an adversary convinces the nodes which are multi hop away that they are closer to the base station. This creates a sinkhole because adversary on the other side of the sinkhole provides a better route to the base station. In Hello flood attacks a Broadcasted message with stronger transmission power is pretending that the HELLO message is coming from the base station. Message receiving nodes assume that the HELLO message sending node is the closest one and they try to send all their messages through this node. In this type of attacks all nodes will be responding to HELLO floods and wasting the energies. The real base station will also be broadcasting the similar messages but will have only few nodes responding to it. Denial of service (DoS) attacks occur at physical level causing radio jamming, interfering with the network protocol, battery exhaustion etc. An specific type of DoS attack, Denial-of-service attack has been explored in (Raymond et al., 2009), in which a sensor node's power supply is targeted. Attacks of this type can reduce the sensor lifetime from years to days and have a devastating impact on a sensor network.

1. Layering based security approach:
*   Application layer

Data is collected and managed at application layer therefore it is important to ensure the reliability of data. Wagner (Wanger, 2004) has presented a resilient aggregation scheme which is applicable to a cluster based network where a cluster leader acts as an aggregator in sensor networks. However this technique is applicable if the aggregating node is in the range with all the source nodes and there is no intervening aggregator between the aggregator and source nodes. To prove the validity of the aggregation, cluster leaders use the cryptographic techniques to ensure the data reliability.

*   Network layer

Network layer is responsible for routing of messages from node to node, node to cluster leader, cluster leaders to cluster leaders, cluster leaders to the base station and vice versa.

*   Data link layer

Data link layer does the error detection and correction, and encoding of data. Link layer is vulnerable to jamming and DoS attacks. TinySec (Karlof et al., 2004) has introduced link layer encryption which depends on a key management scheme. However, an attacker having better energy efficiency can still rage an attack. Protocols like LMAC (Hoesel et al., 2004) have better anti-jamming properties which are viable countermeasure at this layer.

*   Physical Layer

The physical layer emphasizes on the transmission media between sending and receiving nodes, the data rate, signal strength, frequency types are also addressed in this layer. Ideally FHSS frequency hopping spread spectrum is used in sensor networks.

10. Conclusion & future work

The aim of this chapter is to discuss few important issues of WSNs, from the application, design and technology points of view. For designing a WSN, we need to consider different factors such as the flexibility, energy efficiency, fault tolerance, high sensing fidelity, low-cost and rapid deployment, above all the application requirements. We hope the wide range of application areas will make sensor networks an integral part of our lives in the future. However, realization of sensor networks needs to satisfy several constraints such as scalability, cost, hardware, topology change, environment and power consumption. Since these constraints are highly tight and specific for sensor networks, new wireless ad hoc networking protocols are required. To meet the requirements, many researchers are engaged in developing the technologies needed for different layers of the sensor networks protocol stack.

Future research on WSN will be directed towards maximizing area throughput in clustered Wireless Sensor Networks designed for temporal or spatial random process estimation, accounting for radio channel, PHY, MAC and NET protocol layers and data aggregation techniques, simulation and experimental verification of lifetime-aware routing, sensing spatial coverage and the enhancement of the desired sensing spatial coverage evaluation methods with practical sensor model.

The advances of wireless networking and sensor technology open up an interesting opportunity to manage human activities in a smart home environment. Real-life activities are often more complex than the case studies for both single and multi-user. Investigating such complex cases can be very challenging while we consider both single- and multi-user activities at the same time. Future work will focus on the fundamental problem of recognizing activities of multiple users using a wireless body sensor network.
Wireless Sensor Networks hold the promise of delivering a smart communication paradigm which enables setting up an intelligent network capable of handling applications that evolve from user requirements. We believe that in near future, WSN research will put a great impact on our daily life. For example, it will create a system for continual observation of physiological signals while the patients are at their homes. It will lower the cost involved with monitoring patients and increase the efficient exploitation of physiological data and the patients will have access to the highest quality medical care in their own homes. Thus, it will avoid the distress and disruption caused by a lengthy inpatient stay.

Author details

M.A. Matin
Institut Teknologi Brunei, Brunei Darussalam

M.M. Islam
North South University, Dhaka, Bangladesh

11. References

A. Boukerche. Algorithms and Protocols for Wireless, Mobile Ad Hoc Networks, John Wiley & Sons, Inc., 2009.

A. Manjeshwar and D. P. Agarwal, "APTEEN: A hybrid protocol for efficient routing and comprehensive information retrieval in wireless sensor networks," Parallel and Distributed Processing Symposium., Proceedings International, IPDPS 2002, pp. 195-202.

A. Manjeshwar and D. P. Agarwal, "TEEN: a routing protocol for enhanced efficiency in wireless sensor networks," In 1st International Workshop on Parallel and Distributed Computing Issues in Wireless Networks and Mobile Computing, April 2001.

B. Paul, M. A. Matin," Optimal Geometrical Sink Location Estimation for Two Tiered Wireless Sensor Networks" IET Wireless Sensor Systems, vol.1, no.2, pp.74-84, June 2011,doi: 10.1049/iet-wss.2010.0073, IET UK.

Bharathidasan, A., Anand, V., Ponduru, S. (2001), Sensor Networks: An Overview, Department of Computer Science, University of California, Davis 2001. Technical Report

C. Intanagonwiwat, R. Govindan, D. Estrin, "Directed Diffusion: A Scalable and Robust Communication Paradigm for Sensor Networks," Proceedings of the 6th ACM 226 ROUTING PROTOCOLS FOR WIRELESS SENSOR NETWORKS International Conference on Mobile Computing and Networking (MobiCom'00), Boston, MA, Aug. 2000, pp. 56–67.

C. karlof, N. Shastry and D. Wagner, TinySec: A link layer security architecture for wireless sensor networks, SenSys'04, November 3-5 2004, Baltimore, Maryland, USA

Chiara, B.; Andrea, C.; Davide, D.; Roberto, V. An Overview on Wireless Sensor Networks Technology and Evolution. Sensors 2009, 9, 6869-6896.

Chi-Tsun Cheng, Chi K. Tse, and Francis C. M. Lau, "A Delay-Aware Data Collection Network Structure for Wireless Sensor Networks", IEEE Sensors Journal, Vol. 11, No. 3, pp. 699-710, March 2011.

D. Braginsky, D. Estrin, "Rumor Routing Algorithm for Sensor Networks," Proceedings of the 1st Workshop on Sensor Networks and Applications (WSNA'02), Atlanta, GA, Oct. 2002.

D. Wagner, Resilient aggregation in sensor networks, In Proceedings of the 2nd ACM workshop on Security of ad hoc and sensor networks. ACM Press, 2004, pp. 78-87.

D.R. Raymond, R.C. Marchany, M.I. Brownfield, and S.F. Midkiff. Effects of Denial-of-Sleep Attacks on Wireless Sensor Network MAC Protocols. IEEE Transactions on Vehicular Technology, vol. 58, no. 1, pp. 367-380, 2009.

Fabbri, F.; Buratti, C.; Verdone, R.; Riihij arvi, J.; M ah onen, P. Area Throughput and Energy Consumption for Clustered Wireless Sensor Networks. In Proceedings of IEEE WCNC 2009, Budapest, Hungary, 2009

Heinzelman, W., Chandrakasan, A., Balakrishnan, H.: Energy-efficient communication protocol for wireless micro-sensor networks. In: Proceedings of the 33rd International Conference on System Sciences (HICSS), pp. 1–10 (2000)

I. F. Akyildiz, W. Su, Y. Sankarasubramaniam, and E. Cayirci. Wireless sensor networks: A survey. Computer Networks, 38(4):393–422, 2002.

I.F. Akyildiz, W. Su, Y. Sankarasubramaniam, E. Cayirci. A survey on sensor networks. IEEE Communications Magazine. 40 (8) (2002) 102–114.

J. Kulik, W. R. Heinzelman, H. Balakrishnan, "Negotiation-Based Protocols for Disseminating Information in Wireless Sensor Networks," Wireless Networks, Vol. 8, 2002, pp. 169–185.

J. Pan, Y. Hou, L. Cai, Y. Shi and S. X. Shen, 'Topology Control for Wireless Sensor Networks' Proc. 9th ACM Int. Conf. on Mobile Computing and Networking, San Diego, USA, September, 2003, pp. 286-29.

J. Sen. A survey on wireless sensor network security. International Journal of Communication Networks and Information Security (IJCNIS), 1(2):59–82, August 2009.

J. Undercoffer, S. Avancha, A. Joshi, and J. Pinkston, Security for sensor networks, 2002 CADIP Research Symposium.

J. Yick, B. Mukherjee, D. Ghosal. Wireless sensor network survey, Computer Networks 52 (12) (2008) 2292–2330.

J.N. Al-Karaki, A.E. Kamal, Routing techniques in wireless sensor networks: a survey, IEEE Wireless Communications (2004).

K. Akkaya, M. Younis, A survey on routing protocols for wireless sensor networks, Elsevier Journal of Ad Hoc Networks 3 (3) (2005) 325–349.

L.V. Hoesel and P. Havinga, A Lightweight Medium Access Protocol (LMAC) for wireless sensor networks: reducing preamble transmissions and transceiver state switches, in the proceedings of INSS, June 2004.

M A Matin, and Md. Nafees Rahman, "Lifetime improvement of Wireless Sensor Networks" 3rd IEEE International conference on Communication Software and Networks (ICCSN) 2011, Xi'an, China, May 27-29, 2011, pp.475-479.

M. A. Labrador, P. M. Wightman. Topology Control in Wireless Sensor Networks. Springer Science + Business Media B.V. 2009.

Ns-2 [Online]. Available: http://www.isi.edu/nsnam/ns/

S Sharma and S K Jena, "A Survey on Secure Hierarchical Routing Protocols in Wireless Sensor Networks", ICCCS'11 February 2011.

S. Lindsey, C. Raghavendra, "PEGASIS: Power-Efficient Gathering in Sensor Information Systems," IEEE Aerospace Conference Proceedings, 2002, Vol. 3, No. 9–16 pp. 1125–1130.

S. Waharte, R. Boutaba, Y. Iraqi, and B. Ishibashi, "Routing protocols in wireless mesh networks: challenges and design considerations," Multimedia Tools Appl., vol. 29, no. 3, pp. 285–303, 2006.

SensorSim [Online]: Available: http://nesl.ee.ucla.edu/projects/sensorsim/

Sohraby, K.; Minoli, D.; Znati, T. Wireless Sensor Networks: Technology, Protocols and Applications; John Wiley & Sons, Inc.: Hoboken, NJ, USA, 2007.

TOSSIM [Online]. Available: http://docs.tinyos.net/index.php/TOSSIM

Verdone, R.; Dardari, D.; Mazzini, G.; Conti, A. Wireless Sensor and Actuator Networks; Elsevier: London, UK, 2008.

Wilson, J. Sensor Technology Handbook; Elsevier/Newnes: Burlington, MA, USA, 2005.

X. Chen and N. Rowe, "An Energy-Efficient Communication Scheme in Wireless Cable Sensor Networks", Proc. of IEEE International Conference on Communications (IEEE ICC), June 2011

X. Han, X. Cao, E. L. Lloyd and C. Shen, "Fault-Tolerant Relay Node Placement in Heterogeneous Wireless Sensor Networks", IEEE Transaction on Mobile Computing, Vol. 9, No. 5, May 2010

X. Zeng, R. Bagrodia, and M. Gerla, "GloMoSim: A library for parallel simulation of large-scale wireless networks," SIGSIM Simulation Digest, vol. 28, no. 1, pp. 154-161, 1998.

Y. Xu, J. Heidemann, D. Estrin, "Geography-informed Energy Conservation for Ad-hoc Routing," In Proceedings of the Seventh Annual ACM/IEEE International Conference on Mobile Computing and Networking 2001, pp. 70-84.

Younis, O., Fahmy, S. HEED: a hybryd, energy-efficient, distributed clustering approach for ad hoc sensor networks. IEEE Trans. Mobile Comput. 3(4), 366–379 (2004)

Zia, T.; Zomaya, A., "Security Issues in Wireless Sensor Networks", Systems and Networks Communications (ICSNC) Page(s):40 – 40, year 2006.

Chapter 2

Maximum Lifetime Scheduling in Wireless Sensor Networks

Akshaye Dhawan

Additional information is available at the end of the chapter

1. Introduction

Wireless sensor networks (WSNs) have attracted a lot of recent research interest due to their applicability in security, monitoring, disaster relief and environmental applications. WSNs consist of a number of low-cost sensors scattered in a geographical area of interest and connected by a wireless RF interface. Sensors gather information about the monitored area and send this information to gateway nodes. The radio on board these sensor nodes has limited range and allows the node to transmit over short distances. In most deployment scenarios, it is not possible for each node to communicate directly to the sink and hence, the model of communication is to transmit over short distances to other peers in the direction of the sink nodes.

In order to keep their cost low, the sensors are equipped with limited energy and computational resources. The energy supply is typically in the form of a battery and once the battery is exhausted, the sensor is considered to be dead. The nodes also have limited memory and processing capabilities. Hence, harnessing the potential of these networks involves tackling a myriad of different issues from algorithms for network operation, programming models, architecture and hardware to more traditional networking issues. For a more detailed survey on the various computational research aspects of Wireless Sensor Networks, see the survey papers [2, 13, 24, 37, 39], or the more recent books [23, 28] and a special issue of the CACM [14].

This section focuses on the algorithmic aspects of Wireless Sensor Networks. Specifically, we look at the problem of covering a set of targets or an area for the longest duration possible. The next section focuses on a more detailed discussion of the problem and provides a formal statement for it. It is worth mentioning that there is an abundance of algorithmic research related to WSNs. A lot of this focuses on traditional distributed computing issues like localization, fault tolerance, robustness. This naturally raises the interesting question of how different are WSNs as a computational model than more traditional distributed computing environments or even ad-hoc networks? This question has been explored briefly in [43].

2. Coverage problems

Many intended applications of Wireless Sensor Networks involve having the network monitor a region or a set of targets. To ensure that the area or targets of interest can be covered, sensors are usually deployed in large numbers by randomly dropping them in this region. Deployment is usually done by flying an aircraft over the region and air dropping the sensors. Since the cost of deployment far exceeds the cost of individual sensors, many more sensors are dropped than needed to minimally cover the region. The leads to a very dense network and gives rise to an overlap in the monitoring regions of individual sensors.

A simplistic approach to meet the coverage objective would be to turn on all sensors after deployment. But this needlessly reduces the *lifetime* of the network since the overlap between monitoring regions implies that not all sensors need to be on at the same time. This can also lead to a very lossy network with several collisions happening in the medium access control (MAC) layer due to the density of nodes. In order to extend the lifetime of a sensor network while maintaining coverage, a minimal subset of the deployed sensors are kept active while the other sensors can sleep. Through some form of scheduling, this active subset changes over time until there are no more such subsets available to satisfy the coverage goal. In using such a scheme to extend the lifetime, the problem is two fold. First, we need to select these minimal subsets of sensors. Then there is the problem of *scheduling* them wherein, we need to determine how long to use a given set and which set to use next. For an arbitrarily large network, there are exponential number of possible subsets making the problem intractable and it has been shown to be NP-complete in [6, 20].

Centralized solutions like those in [6, 41] are based on assuming that the entire network structure is known at one node (typically the gateway node), which then computes the schedule for the network. The schedule is computed using *linear programming* based algorithms. Like any centralized scheme, it suffers from the problems of scalability, single point of failure and lack of robustness. The latter is particularly relevant in the context of sensor networks since sensor nodes are deployed in hostile environments and are prone to frequent failures.

Existing distributed solutions in [4, 5, 42] work by having a sensor exchange information with its neighbors (limited to k-hops). These algorithms use information like targets covered and battery available at each sensor to greedily decide which sensors remain on. Distributed algorithms are organized into rounds so that the set of active sensors is periodically reshuffled at the beginning of each round. The problem with these algorithms is that they use simple greedy criteria to make their decision on which sensors become active at each round and thus, do not efficiently take into account the problem structure.

3. Problem statement

The lifetime problem can be stated as follows. Given a monitored region R, a set of sensors S and a set of targets T, find a monitoring schedule for these sensors such that

- the total time of the schedule is maximized,
- all targets are constantly monitored, and
- no sensor is in the schedule for longer than its initially battery.

A related problem is that of monitoring an area of interest. In general, the area and target coverage problems have been shown to be equivalent. [3, 10, 41] provide ways to map an area to a set of points (targets) . In the work presented in the remainder of this dissertation, we focus on the target coverage problem with the implicit understanding that the algorithms and techniques presented can be translated to the area coverage problem by mapping the area to a set of points (virtial targets) with an appropriate granularity.

There are also several other variations of this basic problem. For example the $p\%$ coverage problem [30] requires only a certain percentage of all targets to be covered. The fault tolerant k-coverage version of this problem requires each target to be covered by at least k sensors [27, 47]. Also, the basic problem has been modified to include sensors that have adjustable sensing ranges, non uniform sensing shapes and othe heterogeneous sensor network models.

4. Related work

In this section, we briefly survey existing approaches to maximizing the lifetime of sensor networks, while meeting certain coverage objectives. [9] gives a more detailed survey on the various coverage problems and the scheduling mechanisms they use. [38] also surveys the coverage problem along with other algorithmic problems relevant to sensor networks. We end this section by focusing on two algorithms, LBP [4] and DEEPS [5], since we use them for comparisons against our algorithms.

A key application of wireless sensor networks is the collection of data for reporting. There are two types of data reporting scenarios: event-driven and on-demand [10]. Event-driven reporting occurs when one or more sensor nodes detect an event and report it to the sink. In on-demand reporting, the sink initiates a query and the nodes respond with data to this query.

Coverage problems essentially state how well a network observes its physical space. As pointed out in [32], coverage is a measure of the quality of service (QoS) for the WSN. The goal is to have each point of interest monitored by at least one sensor at all times. In some applications, it may be a requirement to have more than one sensor monitor a target for achieving fault tolerance. Typically, nodes are randomly deployed in the region of interest because sensor placement is infeasible. This means that more sensors are deployed than needed to compensate for the lack of exact positioning and to improve fault tolerance in harsh environments. The question of placing an optimal number of sensors in a deterministic deployment has been looked at in [17, 26, 34]. However, in this dissertation we focus on networks with very dense deployment of sensors so that there is significant overlap in the targets each sensor monitors. This overlap will be exploited to schedule sensors into a low power sleep state so as to improve the lifetime of these networks. Note that this definition of the network lifetime is different from some other definitions which measure this in terms of number of operations the network can perform [22].

The reason for wanting to schedule sensors into sense-sleep cycles that we talked about in Section 1, stems from the fact that sensor nodes have four states - *transmit, receive, idle and sleep*. As shown in [36] for the WINS Rockwell sensor, the transmit, receive and idle states all consume much more power than the sleep state - hence, it is more desirable for a sensor to enter a sleep state to conserve its energy. The goal behind sensor scheduling algorithms is to select the activity state of each sensor so as to allow the network as a whole to monitor its points of interest for as long as possible. For a more detailed look at power consumption

Name	Area/Target	Disjoint	Main Idea
Abrams, Goel [1]	Area	Yes	Greedy: Max uncovered area
Meguerdichian [33]	Area	No	Integer Linear Program
Cardei [6]	Target	Yes	Mixed Integer Programming
Shah [3]	Area	Yes	LP, Garg Könemann
Cardei [8]	Target	No	Integer Linear Program

Table 1. Centralized Algorithms

models for ad-hoc and sensor networks we refer the reader to [18, 19, 25]. We now look at coverage problems in more detail.

The maximum lifetime coverage problem has been shown to be NP-complete in [1, 6]. Initial approaches to the problem in [1, 6, 41] considered the problem of finding the maximum number of *disjoint* cover sets of sensors. This allowed each cover to be used independently of others. However, [3, 8] and others showed that using non-disjoint covers allows the lifetime to be extended further and this approach has been adopted since.

Broadly speaking, the existing work in this category can be classified into two parts - Centralized Algorithms and Distributed Algorithms. For centralized approaches, the assumption is that a single node (usually the base station) has access to the entire network information and can use this to compute a schedule that is then uploaded to individual nodes. Distributed Algorithms work on the premise that a sensor can exchange information with its neighbors within a fixed number of hops and use this to make scheduling decisions. We now look at the individual algorithms in both these areas.

A common approach taken with centralized algorithms is that of formulating the problem as an optimization problem and using linear programming (LP) to solve it [3, 6, 16, 33]. In [41], the authors develop a most-constrained least-constraining heuristic and demonstrated its effectiveness on variety of simulated scenarios. In this heuristic, the main idea is to minimize the coverage of sparsely covered areas within one cover. Such areas are identified using the notion of the critical element, defined as the element which is a member of the smallest number of sets. Their heuristic favors sets that cover a high number of uncovered elements, that cover more sparsely covered elements, that do not cover the area redundantly and that redundantly cover the elements that do not belong to sparsely covered areas. [33] is a followup work by the same authors in which they formulate the area coverage problem using a Integer LP and relax it to obtain a solution. They also presented several ILP based formulations and strategies to reduce overall energy consumption while maintaining guaranteed sensor coverage levels. Additionally, their work demonstrated the practicality and effectiveness of these formulations on a variety of examples and provided comparisons with several alternative strategies. They also show that the ILP based technique can scale to large and dense networks with hundreds of sensor nodes.

In order to solve the target coverage problem, [6] considers the disjoint cover set approach. Modeling their solution as a Mixed Integer Program shows an improvement over [41]. The authors define the disjoint set covers (DSC) problem and prove its NP-completeness. They also prove that any polynomial-time approximation algorithm for DSC problem has a lower bound of 2. They first transform DSC into a maximum-flow problem (MFP), which is then formulated as a mixed integer programming. Based on the solution of the MIP, the authors design a heuristic to compute the number of covers. They evaluate the performance by simulation, against the most constrainedï£¡minimally constraining heuristic proposed in [41] and found that their heuristics has a larger number of covers (larger lifetime) at the cost of a greater running time.

[3] formulates a packing LP for the coverage problem. Using the $(1 + \epsilon)$ Garg-Könemann approximation algorithm [21], they provide a $(1 + \epsilon)(1 + 2lnn)$ approximation of the problem. They also present an efficient data structureto represent the monitored area with at most n^2 points guaranteeing the full coverage which is superior to the previously used approach based on grid points in [41]. They also present distributed algorithms that tradeoff between monitoring and power consumption but these are improved upon by the authors in LBP and DEEPS.

A similar problem is solved by us for sensors with *adjustable* ranges in [16]. We present a linear programming based formulation that also uses the $(1 + \epsilon)$ Garg-Könemann approximation algorithm [21]. The main difference is the introduction of an adjustable range model that allows sensors to vary their sensing and communication ranges smoothly. This was the first model that allows sensors to vary their range to any value upto a maximum. The model is an accurate representation of physical sensors and allows significant power savings over the discreetly varying adjustable model.

A different algorithm to work with disjoint sets is given in [7]. Disjoint cover sets are constructed using a graph coloring based algorithm that has area coverage lapses of about 5%. The goal of their heuristic is to achieve energy savings by organizing the network into a maximum number of disjoint dominating sets that are activated successively. The heuristic to compute the disjoint dominating sets is based on graph coloring. Simulation studies are carried out for networks of large sizes.

[1] also gives a centralized greedy algorithm that picks sensors based the largest uncovered area. They have designed three approximation algorithms for a variation of the SET K-COVER problem, where the objective is to partition the sensors into covers such that the number of covers that include an area, summed over all areas, is maximized. The first algorithm is randomized and partitions the sensors within a fraction of the optimum. The other two algorithms are a distributed greedy algorithm and a centralized greedy algorithm. The approximation ratios are presented for each of these algorithms.

[8] also deal with the target coverage problems. Like similar algorithms, they also extend the sensor network life time by organizing the sensors into a maximal number of set covers that are activated successively. But they allow non-disjoint set covers. The authors model the solution as the maximum set covers problem and design two heuristics that efficiently compute the sets, using linear programming and a greedy approach. The greedy algorithm selects a critical target at each step. This is the least covered target. For the greedy selection step, the sensor with the greatest contribution to the critical target is selected.

The distributed algorithms in the literature can be further classified into greedy, randomized and other techniques. The greedy algorithms [1, 4, 5, 8, 29, 41] all share the common property of picking the set of active sensors greedily based on some criteria. [41] considers the area coverage problem and introduces the notion of a *field* as the set of points that are covered by the same set of sensors. The basic approach behind the picking of a sensor is to first pick the one that covers that largest number of previously uncovered fields and to then avoid including

Name	Area/Target	Disjoint	Main Idea
Sliepcivic [41]	Area	Yes	Greedy: Max uncovered fields
Tian [42]	Area	No	Geometric calculation of sponsored area
PEAS [45]	Area	No	Probing based determination of sponsored area
CCP [44]	Area	No	Random timers to evaluate coverage requirements
OGDC [46]	Area	No	Random back off node volunteering
Lu [29]	Area	No	Highest overall denomination sensor picks
Abrams [1]	Area	Yes	Randomized, Greedy picks max uncovered area
Cardei et al. [8]	Target	No	Sensor with highest contribution to bottleneck
LBP [4]	Target	No	Targets are covered by higher energy nodes
DEEPS [5]	Target	No	Minimize energy consumption for bottleneck target

Table 2. Distributed Algorithms

more than one sensor that covers a sparsely covered field. [1] builds on this work and presents three algorithms that solve variations of the set k-cover problem. The greedy heuristic they propose works by selecting the sensor that covers the largest uncovered area. [29] defines the sensing denomination (SD) of a sensor as its contribution, i.e., the area left uncovered when the sensor is removed. The authors assume that each sensor can probabilistically detect a nearby event, and build a probabilistic model of network coverage by considering the data correlation among neighboring sensors. The more the contribution of a sensor to the network coverage, the higher the sensori£¡s SD is. Based on the location information of neighboring sensors, each sensor can calculate its SD value in a distributed manner. Sensors with higher sensing denomination have a higher probability of remaining active.

[3] gives a distributed algorithm based on using the faces of the graph. If all the faces that a sensor covers are covered by other sensors with higher battery that are in an active or deciding state, then a sensor can switch off (sleep). Their work has been extended to target coverage in the load balancing protocol (LBP).

Some distributed algorithms use randomized techniques. Both OGDC [46] and CCP [44] deal with the problem of integrating coverage and connectivity. They show that if the communication range is at least twice the sensing range, a covered network is also connected. [46] uses a random back off for each node to make nodes volunteer to be the start node. OGDC addresses the issues of maintaining sensing coverage and connectivity by keeping a minimum number of sensor nodes in the active mode in wireless sensor networks. They investigate the relationship between coverage and connectivity. They also derive, under the ideal case in which node density is sufficiently high, a set of optimality conditions under which a subset of working sensor nodes can be chosen for complete coverage. OGDC algorithm is fully localized and can maintain coverage as well as connectivity, regardless of the relationship between the radio range and the sensing range. OGDC achieves similar coverage with an upto 50% improvement in the lifetime of the network. A drawback of OGDC is that it requires that each node knows its own location.

In [31] the authors combine computational geometry with graph theoretic techniques. The use Voronoi diagrams with graph search to design a polynomial time worst and average case algorithm for coverage calculation in homogeneous isotropic sensors. The also analyze and experiment with using these techniques as heuristics to improve coverage.

[44] sets a random timer for each node following which a node evaluates its current state based on the coverage by its neighbors. The authors present a Coverage Configuration Protocol (CCP) that can provide different degrees of coverage requested by applications. This flexibility allows the network to self-configure for a wide range of applications. They also integrate CCP to SPAN [11, 12] to provide both coverage and connectivity guarantees. [1] also present a randomized algorithm that assigns a sensor to a cover chosen uniformly at random.

A different approach has been taken in PEAS [42, 45]. PEAS is a distributed algorithm with a probing based off-duty rule is given in [45]. PEAS is localized and has a high resilience to node failure and topology changes. Here, every sensor broadcasts a probe PRB packet with a probing range γ. Any working node that hears this probe packet responds. If a sensor receives at least one reply, it can go to sleep. The range can be chosen based on several criteria. Note that this algorithm does not preserve coverage over the original area. The results for PEAS showed an increase in the network lifetime in linear proportion to the number of deployed nodes. In [42] the authors give a distributed and localized algorithm. Every sensor has an off-duty eligibility rule. They give an algorithm for a node to compute its sponsored area. To prevent the occurrence of blind-points by having two sensors switch off at the same time, a random back off is used. They show improved performance over PEAS.

To our knowledge, [44] was the first work to consider the k-coverage problem. [27] also addresses the k-coverage problem from the perspective of choosing enough sensors to ensure coverage. Authors consider different deployments with sensors given a probability of being active and obtain bounds for deployment. [47] solves the problem of picking minimum size connected k-covers. The authors state this as an optimization problem and design a centralized approximation algorithm that delivers a near-optimal solution. They also present a communication-efficient localized distributed algorithm for this problem.

Now, we look at the two protocols that we compare our heuristics against. The load balancing protocol (LBP) [4] is a simple 1-hop protocol which works by attempting to balance the load between sensors. Sensors can be in one of three states sense/on, sleep/off or vulnerable/undecided. Initially all sensors are vulnerable and broadcast their battery levels along with information on which targets they cover. Based on this, a sensor decides to switch to off state if its targets are covered by a higher energy sensor in either on or vulnerable state. On the other hand, it remains on if it is the sole sensor covering a target. This is an extension of the work in [3]. LBP is simplistic and attempts to share the load evenly between sensors instead of balancing the energy for sensors covering a specific target.

The other protocol we consider is DEEPS [5]. The maximum duration that a target can be covered is the sum of the batteries of all its nearby sensors that can cover it and is known as the life of a target. The main intuition behind DEEPS is to try to minimize the energy consumption rate around those targets with smaller lives. A sensor thus has several targets with varying lives. A target is defined as a *sink* if it is the shortest-life target for at least one sensor covering that target. Otherwise, it is a *hill*. To guard against leaving a target uncovered during a shuffle, each target is assigned an in-charge sensor. For each sink, its in-charge sensor is the one with the largest battery for which this is the shortest-life target. For a hill target, its in-charge is that neighboring sensor whose shortest-life target has the longest life. An in-charge sensor does not switch off unless its targets are covered by someone. Apart from this, the rules are identical as those in LBP protocol. DEEPS relies on two-hop information to make these decisions.

5. The lifetime dependency graph model

In this section, we introduce the Lifetime Dependency (LD) Graph as a model for the maximum lifetime coverage problem defined in Section 3. This model is a key contribution of this dissertation since the heuristics and algorithms that follow in subsequent sections rely heavily on the LD Graph.

Recall from Section 3 that given a sensor network and a set of static targets, the maximum lifetime sensor scheduling problem is to select a subset of sensors that covers all targets and then periodically shuffle the members of this subset so as to maximize the total time for which the network can cover all targets.

Since these sensors are powered by batteries, energy is a key constraint for these networks. Once the battery has been exhausted, the sensor is considered to be dead. The lifetime of the network is defined as the amount of time that the network can satisfy its coverage objective, i.e., the amount of time that the network can cover its *area* or *targets* of interest. Having all the sensors remain "on" would ensure coverage but this would also significantly reduce the lifetime of the network as the nodes would discharge quickly. A standard approach taken to maximizing the lifetime is to make use of the overlap in the sensing regions of individual sensors caused by the high densit y of deployment. Hence, only a subset of all sensors need to be in the "on" or "sense" state, while the other sensors can enter a low power "sleep" or "off" state. The members of this active set, also known as a *cover* set, are then periodically updated so as to keep the network alive for longer duration. In using such a scheduling scheme, there are two problems that need to be addressed. First, we need to determine how long to use a given cover set and then we need to decide which set to use next. This problem has been shown to be NP-complete [1, 6].

A key problem here is that since a sensor can be a part of multiple covers, these covers have an impact on each other, as using one cover set reduces the lifetime of another set that has sensors common with it. By making greedy choices, the impact of this dependency is not being considered, since none of the heuristics in the literature study this reduction in the lifetime of other cover sets caused by using a sensor that is a member of several such sets. The earlier disjoint formulations mentioned in the previous section, entirely avoided this problem by preventing it.

We capture this dependency between covers by introducing the concept of a local Lifetime Dependency (LD) Graph. This consists of the cover sets as nodes with any two nodes connected if the corresponding covers intersect. The graph is an example of an intersection graph since it represents the sensors common to different cover sets. By looking at the graph locally (fixed 1-2 hop neighbors), we are able to construct all the local covers for the local targets and then model their dependencies. Based on these dependencies, a sensor can then prioritize its covers and negotiate these with its neighbors. We also present some simple heuristics based on the graph. The material presented in this section was published in [35].

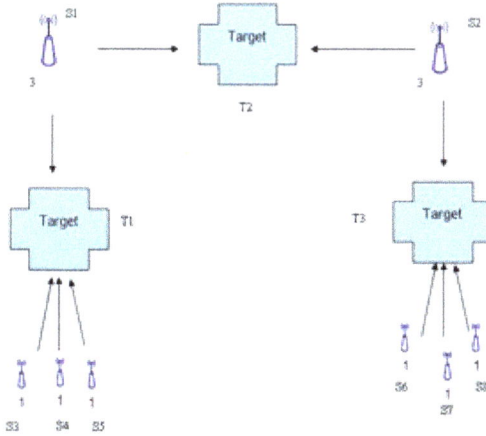

Figure 1. A sensor network

6. Symbols and definitions

Let us begin with a few basic conventions and definitions that will be used in the rest of this dissertation. Individual chapters will introduce additional notation as and when necessary. The notation presented here applies to the basic LD grapg model and will be utilized for all the chapters that follow.

We will use s_1, s_2, etc., to represent sensors, t_1, t_2, etc., to represent targets, and C, C', etc., to denote covers.

Let us assume we have n sensors and m targets, both stationary.

Consider the sensor network in Figure 1 with $n = 8, s = \{s_1, s_2, ..., s_8\}$ and $m = 3$ targets, t_1, t_2, and t_3.

We will employ the following definitions, illustrated using this network.

- $b(s)$: strength of the battery of sensor s; for example, $b(s_1) = 3$ while $b(s_3) = 1$.
- $T(s)$: set of targets that sensor s can sense; e.g., $T(s_1) = \{t_1, t_2\}$;
- $N(s, k)$: closed set of neighbors of sensor s at no more than k hops (i.e, those neighbors that s can communicate with using $\leq k$ hops) - this contains s itself; thus, $N(s_1, 1) = \{s_1, s_2, s_3, s_4, s_5\}$.
- Cover: C is a cover for targets in set T if
  (i) for each target $t \in T$ there is at least one sensor in C which can sense t and
  (ii) C is minimal. For example, the possible (minimal) covers for the two targets of s_1 are $\{s_1\}, \{s_2, s_3\}, \{s_2, s_4\}$ and $\{s_2, s_5\}$. There are other non-minimal covers as well such as s_1, s_2

which need to be avoided. Likewise, the possible covers for the only target of sensor s_3 are $\{s_1\}, \{s_3\}, \{s_4\}$ and $\{s_5\}$.

- $lt(C) = min_{s \in C} b(s)$, the maximum lifetime of a cover. The bottleneck sensor of the cover $\{s_2, s_3\}$ is s_3 with the weakest battery of 1. Therefore, $lt(\{s_2, s_3\}) = 1$.

An optimal lifetime schedule of length 6 for this network is $(\{s_1, s_6\}, 1)$, $(\{s_1, s_7\}, 1)$, $(\{s_1, s_8\}, 1)$, $(\{s_2, s_3\}, 1)$, $(\{s_2, s_4\}, 1)$, $(\{s_2, s_5\}, 1)$) where each tuple is a cover for the entire network followed by its duration.

7. Lifetime dependency (LD) graph

Let the local lifetime dependency graph be $G = (V, E)$ where nodes in V denote the local covers and edges in E exist between those pairs of nodes whose corresponding covers share one or more common sensors. For simplicity of reference, we will not distinguish between a cover C and the node representing it, and an edge e between two intersecting covers C and C' and the intersection set $C \cap C'$. Each sensor constructs its local LD graph considering its one- or two-hop neighbors and the corresponding targets. Figure 2 shows the local lifetime dependency graph of sensor s_1 in the example network of Figure 1, considering its one-hop neighbors $N(s_1, 1)$ and its targets $T(s_1)$.

In the LD graph, we will use the following two definitions:

- $w(e) = min_{s \in e} b(s)$, the weight of an edge e (if e does not exist, i.e., if e is empty, then $w(e)$ is zero).
- $d(C) = \sum_{e \in E \text{ and incident to } C} w(e)$, the degree of a cover C.

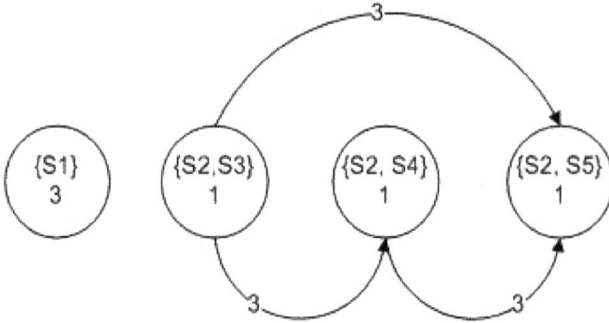

Figure 2. The local lifetime dependency graph of sensor s_1

In Figure 2, the two local covers $\{s_2, s_3\}$ and $\{s_2, s_4\}$ for the targets of sensor s_1 have s_2 in common, therefore the edge between the two covers is $\{s_2\}$ and $w(\{s_2\}) = 3$. Therefore, s_2's battery of 3 is an upper bound on the lifetime of the two covers collectively. It just so happens that the individual lifetimes of these covers are each 1 due to their bottleneck sensors and, therefore, a tighter upper bound on their total life is 2. In general, given two covers C and C', a tight upper bound on the life of two covers is $min(lt(C) + lt(C'), w(C \cap C'))$.

8. The basic algorithm

For the purpose of this explanation, without loss of generality, let us assume that the covers are constructed over one-hop neighbors. The algorithm consists of two phases. During the initial setup phase, each sensor calculates and prioritizes the covers. Then, for each reshuffle round of predetermined duration, each sensor decides its on/off status at the beginning, and then those chosen remain on for the rest of the duration.

Initial setup: Each sensor s communicates with each of its neighbor $s' \in N(s,1)$ exchanging mutual locations, battery levels $b(s)$ and $b(s')$, and the targets covered $T(s)$ and $T(s')$. Then it finds all the local covers using the sensors in $N(s,1)$ for the target set being considered. The latter can be solely $T(s)$ or could also include $T(s')$ for all $s' \in N(s,1)$. It then constructs the local LD graph $G = (V, E)$ over those covers, and calculates the degree $d(C)$ of each cover $C \in V$ in the graph G.

The "priority function" of a cover is based on its degree (lower the better). Ties among covers with same degree are broken first by preferring (i) those with longer lifetimes, then (ii) those which have fewer remaining sensors to be turned on, and finally (iii) by choosing the cover containing the smaller sensor id. A cover which has a sensor turned off becomes infeasible and falls out of contention. Also, a cover whose lifetime falls below the duration of a round is taken out of contention, unless it is the only cover remaining.

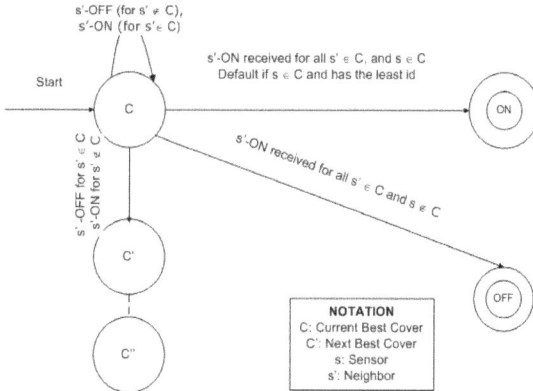

Figure 3. The state transitions to decide the On-Off Status

Reshuffle rounds: The automaton in Figure 3 captures the algorithm for this phase. A sensor s starts with its highest priority cover C as its most desirable configuration for its neighborhood. If successful, the end result would be switching on all the sensors in C, while others can sleep. Else, it transitions to the next best priority cover C', C'', etc., until a cover gets satisfied. The transitions are as follows.

- Continue with the best cover C: Sensor s continues with its current best cover C if its neighbor $s' \notin C$ goes off (thus not impacting the chances of ultimately satisfying C) or if neighbor $s' \in C$ becomes on (thus improving chances for C).

- To on/sense status: If all the neighboring sensors in cover C except s become on, s switches itself on satisfying the cover C for its neighborhood, and sends its on-status to its neighbors.

- To off/sleep status: If all the neighboring sensors in cover C become on thus satisfying C, and s itself is not in cover C, s switches itself off, and sends its off-status to its neighbors.

- Transition to the next best cover C': Sensor s transitions to the next best priority cover C', if (i) C becomes infeasible because a neighboring sensor $s' \in C$ has turned off, or (ii) priority of C is now lower because a sensor $s' \notin C$ has turned on causing another cover C', with same degree and lifetime as C, with fewer sensors remaining to be turned on.

The transitions from C' are analogous to that from C, with the possibility of even going back to C.

Correctness: We sketch a proof here that this algorithm ensures that, in each reshuffle round, all the targets are covered and the algorithm itself terminates enabling each sensor to decide and reach on/off status.

For contradiction, let us assume that in a given round a target t remains uncovered. This implies that either this target has no neighboring sensor within sensing range and thus network itself is dead, or else all the neighboring sensors which could have covered t have turned off. In the latter case, each of the sensor s whose $T(s)$ contains t has made the transition from its current best cover C to off status. However, s only does that if C covers all its targets in $T(s)$ and $s \notin C$. The last such sensor s to have turned off ensures that C is satisfied, which implies that all targets in $T(s)$ including t are covered, a contradiction. Next, for contradiction, let us assume that the algorithm does not terminate. This implies that there exists at least one sensor s which is unable to decide, i.e., make a transition to either on or off status. There are three possibilities: (i) all the covers of s have become infeasible, or

(ii) s is continually transitioning to the next best cover and none of them are getting satisfied, or

(iii) s is stuck at a cover C.

For case (i), for each cover C, at least one of its sensor $s' \in C$ has turned off. But the set of targets considered by sensor s is no larger than $T' = \bigcup_{s' \in N(s,1)} T(s')$. Since s itself can cover $T(s)$, there exist a target $t \in T' - T(s)$, from $T(s')$, that none of the cover sets at s are able to cover. This implies that s' is off, else $\{s, s'\}$ would have formed part of a cover at s covering t (given that s constructs all possible covers). This leads to the contradiction, as before turning off, s' ensures that $t \in T(s')$ is covered.

For case (ii), each transition implies that a neighbor sensor has decided its on/off status, thereby making some of the covers at s infeasible and increasingly satisfying portions of some other covers, thus reducing the choices from the finite number of its covers. Eventually, when the last neighbor decides, s will be able to decide as well becoming on if any target in $T(s)$ is still uncovered, else going off.

For case (iii), the possibility that all sensors are stuck at their best initial covers is conventionally broken by a sensor $s \in C$ with least id in its current best cover C pro actively becoming on, even though C may not be completely satisfied. This is similar to the start-up problem faced by others distributed algorithms such as DEEPS with similar deadlock breaking solutions. At a later stage, if s is stuck at C, it means that either all its neighbors have decided or one or more neighbors are all stuck. In the former case, there exists a cover C at s which will be satisfied with s becoming on (case i). The latter case is again resolved by the start-up deadlock breaking rule by either s or s' pro actively becoming on.

Message and time complexities: Let us assume that each sensor s constructs the covers over its one-hop neighbors to cover its targets in $T(s)$ only. Let $S = \{s_1, s_2, ..., s_n\}$ $\Delta =_{max s \in S} |N(s, 1)|$, the maximum number of neighbors a sensor can communicate with. The communication complexity of the initial setup phase is $O(\Delta)$, assuming that there are constant number of neighboring targets that each sensor can sense. Also, for each reshuffle round, a sensor receives $O(\Delta)$ status messages and sends out one. Assuming Δ is a constant practically implies that message complexity is also a constant. Let maximum number of targets a sensor considers is $\tau =_{max s \in S} |T(s)|$, a constant. The maximum number of covers constructed by sensor s during its setup phase is $O(\Delta^\tau)$, as each sensor in N(s,1) can potentially cover all its targets considered. Hence the time complexity of setup phase is $O((\Delta^\tau)2)$ to construct the LD graph over all covers and calculate the priorities. For example, if $\tau = 3$, the time complexity of the setup phase would be $O(\Delta^6)$. The reshuffle rounds transition through potentially all the covers, hence their time complexity is $O(\Delta^\tau)$.

9. Variants of the basic algorithm

We briefly discussed some of the properties of the LD graph earlier. For example, an edge e connecting two covers C and C' yields an upper bound on the cumulative lifetime of both the covers. However, if $w(e)$, which equals $b(s)$ for weakest sensor $s \in e$, is larger than the sum of the lifetimes of C and C', then the edge e no longer constrains the usage of C and C'. Therefore, even though C and C' are connected, they do not influence each other's lifetimes. This leads to our first variant algorithm.

Variant 1: Redefine the edge weight e as follows:

If $min_{s \in e} b(s) < lt(C) + lt(C')$, then $w(e) = min_{s \in e} b(s)$, else $w(e) = 0$.

Thus, when calculating the degree of a cover, this edge would not be counted when not constraining, thus elevating the cover's priority. Next, the basic framework is exploiting the degree of a cover to heuristically estimate how much it impacts other covers, and the overall intent is to minimize its impact. Therefore, we sum the edge weights emanating from a cover for its degree. However, if a cover C is connected to two covers C' and C'' such that both C' and C'' have the same bottleneck sensor s, s is depleted by burning either C' or C''. That is, in a sense, only one of C' and C'' can really be burned completely, and then the other is rendered unusable because s is completely depleted. Therefore, for all practical purposes, C' and C'' can be collectively seen as one cover. As such, the two edges connecting C to C' and C'' can be thought of as one as well. This yields our second variant algorithm.

Variant 2: Redefine the degree of a cover C in the LD graph as follows. Let a cover C be connected to a set of covers $V' = C_1, C_2, , C_q$ in graph G. If there are two covers C_i and C_j in V' sharing a bottleneck sensor s, then if $w(C, C_i) < w(C, C_j)$ then $V' = V' - C_j$ else $V' = V' - C_i$. With this reduced set of neighboring covers V', the degree of cover C is

$$d(C) = \sum_{C' \in V'} w(C, C')$$

In the basic algorithm, each sensors constructs cover sets using its one-hop neighbors to cover its direct targets $T(s)$. However, with the same message overheads and slightly increased time complexity, a sensor can also consider its neighbors' targets. This will enable it to explore the constraint space of its neighbors as well.

Variant 3: In this variant, each sensor s constructs LD graph over one-hop neighbors $N(s, 1)$ and targets in $\bigcup_{s' \in N(s,1)} T(s')$.

Variant 4: In the basic two-hop algorithm, each sensor s constructs LD graph over two-hop neighbors $N(s, 2)$ and targets in $\bigcup_{s' \in N(s,1)} T(s')$. In this variant, each sensor s constructs LD graph over two-hop neighbors $N(s, 2)$ and targets in $\bigcup_{s' \in N(s,2)} T(s')$.

10. Taming the exponential state space of the maximum lifetime sensor cover problem

If we consider the LD graph, it is quickly obvious that even creating this graph will take exponential time since there are 2^n cover sets to consider where, n is the number of sensors. However, the target coverage problem has a useful property - if the local targets for every sensor are covered, then globally, all targets are also covered. In [35], we make use of this property to look at the LD graph locally (fixed 1-2 hop neighbors), and are able to construct all the local covers for the local targets and then model their dependencies. Based on these dependencies, a sensor can then prioritize its covers and negotiate these with its neighbors. Simple heuristics based on properties of this graph were presented in [35] and showed a 10-15% improvement over comparable algorithms in the literature. [15] built on this work by examining how an optimal sequence would pick covers in the LD graph and designing heuristics that behave in a similar fashion. Though the proposed heuristics are efficient in practice, the running time is a function of the number of neighbors and the number of local targets. Both of these are relatively small for most graphs but theoretically are exponential in the number of targets and sensors.

A key issue that remains unresolved is the question of how to deal with this exponential space of cover sets. In this paper we present a reduction of this exponential space to a linear one based on grouping cover sets into *equivalence classes*. We use $[C_i]$ to denote the equivalence class of a cover C_i. The partition defined by the equivalence relation on the set of all sensor covers Given a set C and an equivalence relation $\Re$, the equivalence class of an element $C_i \in C$ is the subset of all elements in C which are equivalent to C_i. The notation used to represent the equivalence class of C_i is $[C_i]$. In the context of the problem being studied, C is the set of all sensor covers and for any single cover C_i, $[C_i]$ represents all other covers which are *equivalent* to C_i as given by the definition of some equivalence relation $\Re$. Our approach stems from the understanding that from the possible exponential number of sensor covers, several covers are very similar, being only minor variations of each other. In Section 11, we present the definition of the relation $\Re$, based on a grouping that considers cover sets equivalent if their lifetime is bounded by the same sensor. We then show the use of this relation to collapse the exponential LD Graph into an *Equivalence Class* (EC) Graph with linear number of nodes. This theoretical insight allows us to design a sampling scheme that selects a subset of all local covers based on their equivalence class properties and presents this as an input to our simple LD graph degree-based heuristic. Simulation results show that class based sampling cuts the running time of these heuristics by nearly half, while only resulting in a less than 10% loss in quality.

11. Dealing with the exponential space

In this section, we present our approach of dealing with the exponential solution space of possible cover sets. The next section utilizes these ideas to develop heuristics for maximizing the lifetime of the network. Even though the total number of cover sets for the network may be exponential in the number of sensors, for any given cover set, there are several other sets that are very *similar* to this set. We begin by attempting to define this notion of similarity by expressing it as an equivalence relation.

Definition 1: Let $\Re$ be an equivalence relation defined on the set of all sensor covers such that $C_i \Re C_j$ if and only if C_i and C_j share the same bottleneck sensor s_{bot}.

Theorem: $\Re$ is an equivalence relation
Proof: $\Re$ is reflexive, since $C_i \Re C_i$. $\Re$ is symmetric, since if $C_i \Re C_j$ then, $C_j \Re C_i$ since both covers C_i and C_j share the same bottleneck sensor. Finally, if $C_i \Re C_j$ and $C_j \Re C_k$, then $C_i \Re C_k$ and $\Re$ is transitive since if C_i shares the same bottleneck sensor with C_j and C_j shares the same bottleneck sensor with C_k then, clearly both C_i and C_k have the same bottleneck sensor in common. Therefore, $\Re$ is and equivalence relation. $\square$

Every equivalence relation defined on a set, specifies how to partition the set into subsets such that every element of the larger set is in exactly one of the subsets. Elements that are related to each other are by definition in the same partition. Each such partition is called an *equivalence class*. Hence, the relation $\Re$ partitions the set of all possible sensor covers into a number of *disjoint* equivalence classes.

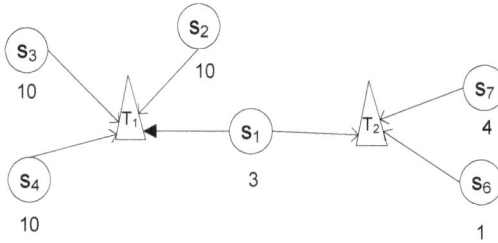

Figure 4. Example Sensor Network

Notation: Henceforth, we represent the equivalence class of covers sharing a bottleneck sensor s_i by $[s_i]$. Note that this is a slight abuse of notation since s_i is not a member of this class, but is instead the property that is common to all members of this class. Hence, $[s_i]$ can be read as the equivalence class for all covers having sensor s_i as their bottleneck sensor.

We now define what we would call the Equivalence Class (EC) Graph. Each node of this graph represents an equivalence class. Just as the LD graph models the dependency between sensor covers, the EC Graph models the dependency between classes of covers.

Definition 2: Equivalence Class Graph (EC). The Equivalence Class graph $EC = (V', E')$ where, V' is the set of all possible equivalence classes defined by $\Re$ and two classes $[s_i]$ and $[s_j]$ are joined by an edge for every cover in each class that share some sensor in common. Hence, the graph EC is a multi-edge graph.

The cardinality of the vertex set of the Equivalence Class Graph is at most n. This result follows from the observation that for any network of n sensors, there can be at most one equivalence class corresponding to each sensor, since every cover can have only one of the n sensors as its bottleneck (in case two or more sensors all have the same battery and are the bottleneck, sensor id's can be used to break ties).

To better understand these definitions, let us consider an example. Consider the sensor network shown in Figure 4. The network comprises of seven sensors, $s_1, ..., s_7$ and two targets, T_1, T_2. Observe that T_2 is the bottleneck target for the network since it is the least covered target (8 units of total coverage compared to 33 for T_1). Also note that only one sensor, s_1 can cover both targets.

For the given network, the set of all possible minimal sensor covers, S is,

$$S = \{\{s_2, \mathbf{s_6}\}, \{s_2, \mathbf{s_7}\}, \{s_3, \mathbf{s_6}\}, \{s_3, \mathbf{s_7}\}, \{s_4, \mathbf{s_6}\}, \{s_4, \mathbf{s_7}\}, \{\mathbf{s_1}\}\}$$

For each individual cover in this set, the bottleneck sensor is the sensor shown in bold face.

Figure 5 shows the Lifetime Dependency graph for these covers. As defined, an edge exists between any two covers that share at least one sensor in common and the weight of this edge is given by the lifetime of the common sensor having the smallest battery (the bottleneck). For example, an edge of weight 4 exists between C_1 and C_2 because they share the sensor s_2 having a battery of 4.

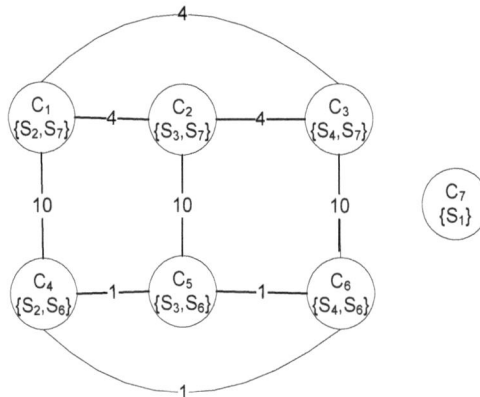

Figure 5. LD Graph for the example network

To obtain the EC Graph from this LD Graph, we add a node to represent the equivalence class for each sensor that is a bottleneck sensor for any cover. For the above example, given all sensor covers in the set S, there are three sensors s_1, s_6, s_7 that are each the bottleneck for one or more covers in S. Hence, the EC Graph is a three node graph. Figure 6 shows the complete EC Graph for the covers in S. There is a node corresponding to the equivalence class for each of the three sensors s_1, s_6, s_7 and for each cover in the class we retain edges to the class corresponding to the bottleneck sensor of the cover on which the edge terminated in the LD graph. Hence, we have three edges between the nodes s_6 and s_7.

It is key to realize that the EC graph is essentially an encapsulation of the LD Graph that can have at most n nodes. This view is presented in Figure 7, where we show the LD Graph that is embedded into the EC Graph. Each rectangular box shows the nodes in the LD graph that are in the same equivalence class. This figure also illustrates our next theorem.

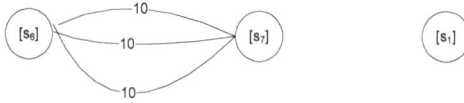

Figure 6. EC Graph for the example network

Theorem: For sensor covers in the same equivalence class, the induced subgraph on the LD Graph is a clique

Proof: This theorem states that for the nodes in the LD graph that belong to the same class, the induced subgraph is a clique. Since by definition, all sensor covers in a class $[s]$ share the sensor s as their bottleneck sensor, the induced subgraph will be a complete graph between these nodes. $\square$

Also, a subtle distinction has been made between inter-class edges and intra-class edges in going from the LD graph to the EC graph.

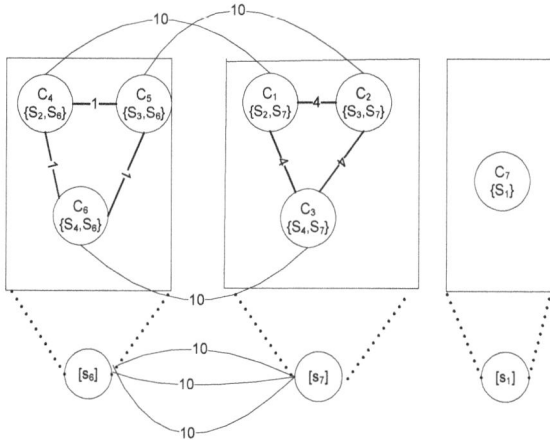

Figure 7. EC Graph for the example network along with the LD Graph embedded in it

12. Sampling based on the equivalence class graph

The previous section defined the concepts behind reducing the exponential space of covers in the LD Graph to the linear space of the EC Graph. In this section, we build on these concepts to discuss techniques for generating a limited number of covers for the LD Graph. Specifically, our goal is to improve the timing performance of the distributed algorithms we presented in [15, 35]. As presented, the EC Graph is not very useful since it still requires the exponential LD graph to be populated, before it can be constructed. However, by realizing that the exponential space of cover sets can be expressed in this linear space of equivalence classes, we can generate only a subset of the set of all covers.

Recall that even though the number of global sensor covers is exponential in the number of sensors, our heuristics presented in [15, 35] worked by constructing *local* covers. After exchanging one or two hop coverage information with neighboring sensors, a sensor can exhaustively construct all possible local covers. A local cover here is a sensor cover that covers all the local targets. The number of local covers is also exponential but is determined by the maximum degree of the graph and the number of local targets, typically much smaller values than the number of all sensors or targets. The heuristics then construct the LD graph over these local covers. The choice of which cover to use is determined by looking at properties of the LD graph such as the degree of each cover in the LD graph.

By making use of the idea of related covers in the same equivalence class, our goal is to use our existing heuristics from [15, 35] but to modify them to run over a subset of the local covers as opposed to all local covers. This should give considerable speedup and if the subset is selected carefully, it may only result in a slight reduction of the overall lifetime. We present such a local cover sampling scheme in Section 12.1 and then present the modified basic algorithm of [15, 35] to operate on this sample in Section 8. Finally, we evaluate the effectiveness of sampling in Section 13.

12.1. Local bottleneck target based generation of local cover sets

Understanding the underlying equivalence class structure, we now present one possible way of generating a subset of the local cover sets. Our approach is centered around the bottleneck target. For any target t_i, the total amount of time this target can be monitored by any schedule is given by:

$$lt(t_i) = \sum_{\{s \mid t_i \in T(s)\}} b(s)$$

Clearly there is one such target with the smallest $lt(t_i)$ value, and is hence a bottleneck for the entire network [40]. Without global information it is not possible for any sensor to determine if the global bottleneck is a target in its vicinity. However, for any sensor s, there is a least covered target in $T(s)$ that is the *local* bottleneck. A key thing to realize is the fact that the global bottleneck target is also the local bottleneck target for the sensors in its neighborhood. Hence, if every sensor optimizes for its local bottleneck target, then one of these local optimizations is also optimizing the global bottleneck target. We use t_{bot} to denote this local bottleneck target. Let C_{bot} be the set of sensors that can cover this local bottleneck target. That is,

$$C_{bot} = \{s \mid t_{bot} \in T(s)\}$$

Implementation: This understanding of bottleneck targets, along with our definition of equivalence classes, now gives us a simple means to generate local covers. Since no coverage schedule can do any better than the total amount of time that the global bottleneck can be covered, instead of trying to generate all local covers, what we really need are covers in the equivalence classes corresponding to each sensor $s_i \in C_{bot}$, such that each class can be completely exhausted. Also, to only select covers that conserve the battery of the sensors in C_{bot}, we want to ensure that the covers we generate are disjoint in C_{bot}. In terms of equivalence classes, for any two classes $[s_i]$ and $[s_j]$ such that $s_i, s_j \in C_{bot}$, we want to generate cover sets that are in these classes but do not include *both* s_i and s_j.

To generate such cover sets, we can start by picking only one sensor s'_{bot} in C_{bot}. This ensures that the local bottleneck target is covered. For each target t_i in the one/two-hop neighborhood being considered, we can then randomly pick a sensor s, giving preference to any $s \notin C_{bot}$. Note that this does not necessarily create a sensor cover in the class $[s'_{bot}]$, since any one of our randomly picked sensors could be the bottleneck for the cover generated. However, replacing that sensor with another randomly picked sensor that covers the same target ensures that the we finish by using a cover in $[s'_{bot}]$. Such a selection essentially ensures that we burn the entire battery of this sensor s'_{bot} in C_{bot} through different covers, while trying to avoid using other sensors in C_{bot}. This process is then repeated for every sensor in C_{bot}. Hence, instead of generating all local covers, we only generate a small sample (constant number) of these corresponding to the equivalence class for each sensor covering the bottleneck target and some related randomly picked covers. We already showed that there can be at most n equivalence classes for the network. Thus, the sampled graph generated has O(n) nodes. If we consider the maximum number of sensors covering any target as a constant for the network, sampling only takes cumulative time of $O(n\tau)$, where $\tau = max_{s \in S}|T(s)|$, since we do this for n sensors, each of which has a maximum of τ targets to cover, which are in turn covered by a constant number of sensors (as per our assumption). Even if this assumption is removed, in the worst case, all n sensors could be covering the same target making the time complexity $O(n^2\tau)$. Next, we run our basic heuristic from [35] on this sampled LD graph.

13. Performance evaluation

In this section, we evaluate the performance of the proposed sampling scheme and evaluate it against our degree based heuristics of [35]. By not constructing all local covers and instead constructing a few covers for key equivalence classes, we should achieve considerable speedup. But the effectiveness of sampling can only be evaluated by analyzing its tradeoff between faster running time for possible reduced performance. The objective of our simulations was to study this tradeoff. For completeness, we create both one-hop and two-hop versions of our sampling heuristic and also compare its performance to two other algorithms in the literature, the 1-hop algorithm LBP [4] and the 2-hop algorithm DEEPS [5].

In order to compare the equivalence class based sampling against our previous degree based heuristics, LBP, and DEEPS, we use the same experimental setup and parameters as employed in [4]. We carry out all the simulations using C++. For the simulation environment, a static wireless network of sensors and targets scattered randomly in $100m \times 100m$ area is considered. We conduct the simulation with 25 targets randomly deployed, and vary the number of sensors between 40 and 120 with an increment of 20 and each sensor with a fixed sensing range of $60m$. The communication range of each sensor assumed to be two times the sensing range [44, 46]. For these simulations, we use the linear energy model wherein the power required to sense a target at distance d is proportional to d. We also experimented with the quadratic energy model (power proportional to d^2). The results showed similar trends to those obtained for the linear model.

Figure 8 shows the Network Lifetime for the different algorithms. As can be seen from the figure, the sampling heuristics is only between 7-9% worse than the degree based heuristic. Sampling also outperforms the 1-hop LBP algorithm by about 10%. It is interesting to observe that for smaller network sizes, sampling is actually much closer to the degree-based heuristics in terms of performance.

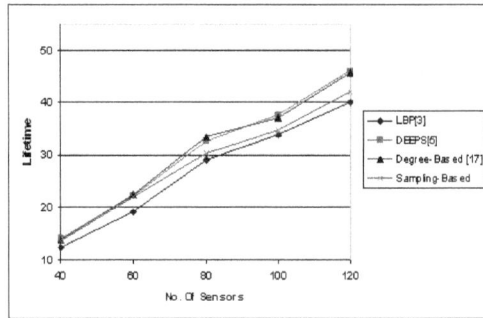

Figure 8. Comparison of Network Lifetime with 25 Targets

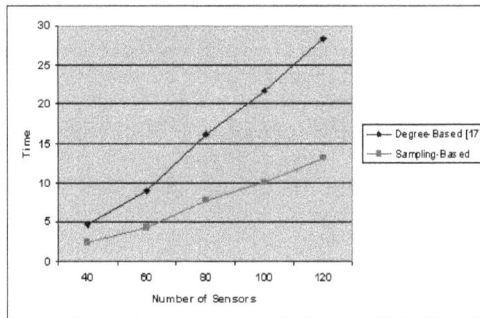

Figure 9. Comparison of Running Time with 25 Targets

Algorithm	$n=40$	$n=80$	$n=120$
LBP [4]	12.4	29.1	40.1
Degree-Based [35]	13.8	33.4	45.6
Sampling-Based	13.7	30.3	42.1
Randomized-Sampling	10.1	17.6	30.1

Table 3. Comparison of Network Lifetime for 1-hop algorithms

Now that we have seen that sampling works well when compared to the degree based heuristic, the question that remains to be answered is how much faster is the sampling algorithm? Figure 9 compares head-to-head the running time for the degree based heuristic (potentially exponential in m) and the linear time sampling algorithm. As can be seen from the figure the running time for the sampling algorithm is about half of the running time for the degree-based heuristic.

Finally, we individually study the 1-hop (Table 3) and 2-hop (Table 4) sampling heuristics with comparable algorithms. For the 1-hop algorithms, we also include a randomized-sampling algorithm that makes completely random picks for each target, without considering properties of the equivalence classes. The intention is to ensure that the performance of our sampling-heuristic can be attributed to the selection algorithm. For the 2-hop versions of

Algorithm	n=40	n=80	n=120
DEEPS [5]	14.1	32.7	46.1
Degree-Based (2-hop) [35]	15.2	36.2	49.6
Sampling-Based (2-hop)	14.4	33.4	47.5

Table 4. Comparison of Network Lifetime of 2-hop algorithms

our proposed sampling heuristic, the target set $T(s)$ of each sensor is expanded to include $\cup_{s' \in N(s,1)} T(s')$ and the neighbor set is expanded to all 2-hop neighbors, i.e., $N(s,2)$. Covers are now constructed over this set using the same process as before. As can be seen from the table, both the 1-hop and 2-hop version are under 10% worse than the comparable degree-based heuristics. Also, the 2-hop sampling slightly outperforms the DEEPS by a 5% improvement in network lifetime.

14. Conclusion

Despite a lot of recent research effort, creating real-world deployable sensor networks remains a difficult task. A key bottleneck is the limited battery life of sensor motes. Hence, energy conservation at every layer of the network stack is critical. Creating realistic theoretical models for problems in this domain that take this into account remains a challenge. Our work addresses energy efficiency at only point in the network stack. However, a holistic approach to energy efficiency design should not only account for energy concerns in each layer of the network stack for problems like routing, medium access etc., but also consider cross-layer issues and interactions.

In this chapter, we present innovative models and heuristics to address the coverage problem in Wireless Sensor Networks. Our work points to the potential of lifetime dependency graphs while serving to highlight the shortcomings of using standard distributed algorithms to this problem. In order to successfully bridge the gap between the theory and practice of wireless sensor networks, there is a clear need for algorithms that are designed keeping the unique constraints of these networks in mind. The improvements in network lifetime obtained by our approach using the dependency graph and heuristics that stem serve to underscore this point.

Acknowledgements

The author would like to thank his wife Kelli for her support while working on this chapter.

Author details

Akshaye Dhawan
Department of Mathematics and Computer Science, Ursinus College, USA.

15. References

[1] Abrams, Z., Goel, A. & Plotkin, S. [2004]. Set k-cover algorithms for energy efficient monitoring in wireless sensor networks, *Third International Symposium on Information Processing in Sensor Networks* pp. 424–432.

[2] Akyildiz, I., Su, W., Sankarasubramaniam, Y. & Cayirci, E. [2002]. A survey on sensor networks, *IEEE Commun. Mag.* pp. 102–114.

[3] Berman, P., Calinescu, G., Shah, C. & Zelikovsky, A. [2004]. Power efficient monitoring management in sensor networks, *Wireless Communications and Networking Conference (WCNC)* 4: 2329–2334 Vol.4.

[4] Berman, P., Calinescu, G., Shah, C. & Zelikovsky, A. [2005]. Efficient energy management in sensor networks, *In Ad Hoc and Sensor Networks, Wireless Networks and Mobile Computing* .

[5] Brinza, D. & Zelikovsky, A. [2006]. Deeps: Deterministic energy-efficient protocol for sensor networks, *Proceedings of the International Workshop on Self-Assembling Wireless Networks (SAWN)* pp. 261–266.

[6] Cardei, M. & Du, D.-Z. [2005]. Improving wireless sensor network lifetime through power aware organization, *Wireless Networks* 11: 333–340(8).

[7] Cardei, M., MacCallum, D., Cheng, M. X., Min, M., Jia, X., Li, D. & Du, D.-Z. [2002]. Wireless sensor networks with energy efficient organization, *Journal of Interconnection Networks* 3(3-4): 213–229.

[8] Cardei, M., Thai, M., Li, Y. & Wu, W. [2005]. Energy-efficient target coverage in wireless sensor networks, *INFOCOM 2005* 3.

[9] Cardei, M. & Wu, J. [2006]. Energy-efficient coverage problems in wireless ad hoc sensor networks, *Computer Communications* 29(4): 413–420.

[10] Carle, J. & Simplot-Ryl, D. [2004]. Energy-efficient area monitoring for sensor networks, *Computer* 37(2): 40–46.

[11] Chen, B., Jamieson, K., Balakrishnan, H. & Morris, R. [2001]. Span: An energy-efficient coordination algorithm for topology maintenance in ad hoc wireless networks, *Proceedings of the 7th ACM International Conference on Mobile Computing and Networking*, Rome, Italy, pp. 85–96.

[12] Chen, B., Jamieson, K., Balakrishnan, H. & Morris, R. [2002]. Span: An energy-efficient coordination algorithm for topology maintenance in ad hoc wireless networks, *ACM Wireless Networks* 8(5).

[13] Chong, C.-Y. & Kumar, S. [2003]. Sensor networks: evolution, opportunities, and challenges, *Proceedings of the IEEE* 91(8): 1247–1256.

[14] Culler, D. & Hong, W. [2004]. Wireless sensor networks, *Special Issue, CACM* .

[15] Dhawan, A. & Prasad, S. K. [2008]. Energy efficient distributed algorithms for sensor target coverage based on properties of an optimal schedule, *HiPC: 15th International Conference on High Performance Computing, LNCS 5374*.

[16] Dhawan, A., Vu, C. T., Zelikovsky, A., Li, Y. & Prasad, S. K. [2006]. Maximum lifetime of sensor networks with adjustable sensing range, *Proceedings of the International Workshop on Self-Assembling Wireless Networks (SAWN)* pp. 285–289.

[17] Dhillon, S. S. & Chakrabarty, K. [2003]. Sensor placement for effective coverage and surveillance in distributed sensor networks, Vol. 3, pp. 1609–1614 vol.3.

[18] Feeney, L. M. [2001]. An energy consumption model for performance analysis of routing protocols for mobile ad hoc networks, *Mob. Netw. Appl.* 6(3): 239–249.

[19] Feeney, L. & Nilsson, M. [2001]. Investigating the energy consumption of a wireless network interface in an ad hoc networking environment, *INFOCOM 2001. Twentieth Annual Joint Conference of the IEEE Computer and Communications Societies. Proceedings. IEEE*, Vol. 3, pp. 1548–1557 vol.3.

[20] Garey, M. R. & Johnson, D. S. [1979]. *Computers and Intractability: A Guide to the Theory of NP-Completeness*, W. H. Freeman & Co., New York, NY, USA.

[21] Garg, N. & Koenemann, J. [1998]. Faster and simpler algorithms for multicommodity flow and other fractional packing problems., *FOCS '98: Proceedings of the 39th Annual Symposium on Foundations of Computer Science*, IEEE Computer Society, Washington, DC, USA, p. 300.

[22] Giridhar, A. & Kumar, P. [2005]. Maximizing the functional lifetime of sensor networks, *Information Processing in Sensor Networks, 2005. IPSN 2005. Fourth International Symposium on*, pp. 5–12.

[23] Iyengar, S. & Brooks, R. [2005]. *Handbook of Distributed Sensor Networks*, Chapman and Hall/CRC.

[24] Iyengar, S. S. & Brooks, R. [2004]. Computing and communications in distributed sensor networks, *Special Issue, Jr. of Parallel and Distributed Computing* 64(7).

[25] Jung, E.-S. & Vaidya, N. H. [2005]. Power aware routing using power control in ad hoc networks, *SIGMOBILE Mob. Comput. Commun. Rev.* 9(3): 7–18.

[26] Kar, K. & Banerjee, S. [2003]. Node placement for connected coverage in sensor networks, *In Proc. of WiOpt 2003: Modeling and Optimization in Mobile, Ad-Hoc and Wirless Networks*.

[27] Kumar, S., Lai, T. H. & Balogh, J. [2004]. On k-coverage in a mostly sleeping sensor network, *MobiCom '04: Proceedings of the 10th annual international conference on Mobile computing and networking*, ACM, New York, NY, USA, pp. 144–158.

[28] Li, Y., Thai, M. T. & Wu, W. [2008]. *Wireless Sensor Networks and Applications*, Springer.

[29] Lu, J. & Suda, T. [2003]. Coverage-aware self-scheduling in sensor networks, *18th Annual Workshop on Computer Communications (CCW)* pp. 117–123.

[30] Mao, Y., Wang, Z. & Liang, Y. [21-25 Sept. 2007]. Energy aware partial coverage protocol in wireless sensor networks, *Wireless Communications, Networking and Mobile Computing, 2007. WiCom 2007. International Conference on* pp. 2535–2538.

[31] Megerian, S., Koushanfar, F. & Potkonjak, M. [2005]. Worst and best-case coverage in sensor networks, *IEEE Transactions on Mobile Computing* 4(1): 84–92. Senior Member-Srivastava, Mani B.

[32] Meguerdichian, S., Koushanfar, F., Potkonjak, M. & Srivastava, M. B. [2001]. Coverage problems in wireless ad-hoc sensor networks, *in IEEE INFOCOM*, pp. 1380–1387.

[33] Meguerdichian, S. & Potkonjak, M. [2003]. Low power 0/1 coverage and scheduling techniques in sensor networks, *UCLA Technical Reports 030001* .

[34] Patel, M., Chandrasekaran, R. & Venkatesan, S. [2005]. Energy efficient sensor, relay and base station placements for coverage, connectivity and routing, *Performance, Computing, and Communications Conference, 2005. IPCCC 2005. 24th IEEE International*, pp. 581–586.

[35] Prasad, S. K. & Dhawan, A. [2007]. Distributed algorithms for lifetime of wireless sensor networks based on dependencies among cover sets, *HiPC: 14th International Conference on High Performance Computing, LNCS 4873*, pp. 381–392.

[36] Raghunathan, V., Schurgers, C., Park, S., Srivastava, M. & Shaw, B. [2002]. Energy-aware wireless microsensor networks, *IEEE Signal Processing Magazine*, pp. 40–50.

[37] Römer, K. & Mattern, F. [2004]. The design space of wireless sensor networks, *IEEE Wireless Communications* 11(6): 54–61.

[38] Sahni, S. & Xu, X. [2004]. Algorithms for wireless sensor networks, *Intl. Jr. on Distr. Sensor Networks* 1.

[39] Schmid, S. & Wattenhofer, R. [2006]. Algorithmic models for sensor networks, *Parallel and Distributed Processing Symposium, 2006. IPDPS 2006. 20th International*, pp. 11 pp.–.

[40] Schmid, S. & Wattenhoffer, R. [n.d.]. Maximizing the lifetime of dominating sets.

[41] Slijepcevic, S. & Potkonjak, M. [2001]. Power efficient organization of wireless sensor networks, *IEEE International Conference on Communications (ICC)* pp. 472–476 vol.2.

[42] Tian, D. & Georganas, N. D. [2002]. A coverage-preserving node scheduling scheme for large wireless sensor networks, *WSNA: Proceedings of the 1st ACM international workshop on Wireless sensor networks and applications*, ACM, New York, NY, USA, pp. 32–41.

[43] Wattenhoffer, R. [n.d.]. Sensor networks: Distributed algorithms reloaded - or revolution?

[44] Xing, G., Wang, X., Zhang, Y., Lu, C., Pless, R. & Gill, C. [2005]. Integrated coverage and connectivity configuration for energy conservation in sensor networks, *ACM Trans. Sen. Netw.* 1(1): 36–72.

[45] Ye, F., Zhong, G., Lu, S. & Zhang, L. [2002]. Peas: A robust energy conserving protocol for long-lived sensor networks, *IEEE International Conference on Network Protocols (ICNP)* 00: 200.

[46] Zhang, H. & Hou, J. [2005]. Maintaining sensing coverage and connectivity in large sensor networks, *Ad Hoc and Sensor Wireless Networks (AHSWN)* .

[47] Zongheng Zhou; Das, S.; Gupta, H. [11-13 Oct. 2004]. Connected k-coverage problem in sensor networks, *Computer Communications and Networks, 2004. ICCCN 2004. Proceedings. 13th International Conference on* pp. 373–378.

Chapter 3

Tradeoffs Among Delay, Energy and Accuracy of Data Aggregation for Multi-View Multi-Robot Sensor Networks

Wuyungerile Li, Ziyuan Pan and Takashi Watanabe

Additional information is available at the end of the chapter

1. Introduction

Due to the recent development in micro mechanics, electronics and wireless communication technologies, Wireless Sensor Network (WSN) has been a hot issue for many applications like monitoring, detecting, remote control, life saving etc. However, along with the applications in different area, the ways of deploying the sensor nodes are different. In some harsh environmental detection, sensor nodes are always dropt into the target area by aircraft which may lead some unnecessary troubles, for example some nodes are out of communication range, some nodes are broken, and lack of flexible because of its immobile characters et al.

The Multi-Robot Sensor Network (MRSN) which is comprised of large numbers of small, simple and inexpensive wireless robots can solve the problems mentioned above. In MRSN, besides perceived sensors, the robot also can be set up digital camera or voice recording equipment, even video camera according to the applications' requirement. Hence, it can detect more detailed information like pictures, video or sound etc.

In this section, at first we introduce the MRSN and its applications. The open problems and one of the effective methods, data aggregation is presented. At last we show a preview of this chapter.

1.1. Wireless multi-robot sensor networks

In wireless MRSN, from a viewpoint of sensor network communication, each robot senses data and transmits data to the adjacent lower node. To collect all data at the sink, data are sent by relay nodes in a multi-hop manner. However, due to the mobility of the robot, sometimes robots are out of the network area so that break the network connectivity. Hence, how to keep the connectivity all the time is a crucial issue so that it already became a hot research topic (Mi et, al., 2010). Besides, the reliability and robustness as well as secure et al. are some other concerned research topics. With respect to communication techniques in MRSN, a network with the goal of search and rescue is described in (Reich & Sklar, 2006). In this paper, they proposed an entirely distributed gradient propagation (GP) algorithm. Each sensor in the paper independently executes the GP algorithm and broadcasts after some independent, randomly chosen time interval. The robots sensors estimate their target by "hot" values and "cold" values, where the "hot" values become the searching target. In (Sheng, et al. 2006), for reliability and robustness, a distributed biding algorithm was proposed for multiple robots in exploration tasks to address the problems caused by the limited communication range. In this algorithm, all the robots work asynchronously. There are three states for each robot that (1) sensing and mapping, (2) bidding and (3) traveling. A distributed algorithm that makes mobile robots in a multi-robot system aware of network connectivity was discussed in (Leyzx, et al., 2009). The basic idea is to take a "fixed" robot as the reference robot that keep in touch with at least one neighboring robot from which a communication path to the reference robot can be established.

1.2. Applications of multi-robot sensor networks

Due to its flexibility, operability, mobility and self-organization, the applications of MRSN has been increasing (Maxim & Gaurav, 2005), (Trigui S, et, al., 2012). Harsh environmental monitoring is the most popular application of MRSN, for example let wireless robots get into Amazon rainforests where it is very dangerous for human get inside or let them climb to Mount Everest where there is not enough Oxygen for human and covered by snow all over the mountain. In medical application, if the MRSN can help the nurses to do some simple task like checking body temperature and sending to a doctor, which would save much more labors in some countries those short of nurses. One of the most important utilizations of MRSN is that it can be used to detect nuclear radiation and to accomplish some other relevant tasks. The most recent example is the Fukushima nuclear leakage where if a MRSN was applied, it would have alleviated damage. Some other applications like outer space monitoring (space junk detection), industrial monitoring (quality control), disaster monitoring (forest fire detection), agriculture monitoring (soil moisture detection), traffic monitoring (intelligent transport system) etc have also much potential.

1.3. Open problems in multi-robot sensor networks

In MRSN, when an event occurs, multiple robots in the near area sense the event data and generate an abundance of sensed data; however, many of the data generated in the same area are highly redundant. Hence the transmission and relaying of all generated data caused a big waste of bandwidth and energy; it also causes data collision and congestion so that result in low efficiency of data gathering. On the other hand, similar to wireless sensor network, WRSN could not avoid the shortcoming of lack of continuous energy supplement.

One may say robot node can be equipped a large capacity battery, but its energy consumption is also large due to its big size, moving, detecting and transmitting etc. An energy harvesting algorithm (Eu et, al., 2010) is proposed for WSN. According to energy harvesting technique, a robot can absorbs solar energy from sunlight. However, how can the robot manage its task in the night? The vibrational energy get from the environment is too lees to trigger the robot. Therefore, saving energy is the most feasible way in WRSN.

For energy saving, decreasing the redundant data and sending a representative data of the detected area are the most considerable strategy. With a view of reducing the quantity of the transmitted data, the well-known scheme is data aggregation (Rajagopalan &Varshney, 2006). Since a sensor node in WSN waits for a period of time to collect extensive quantity of data to aggregate, data aggregation leads long transmission delay and low data accuracy. Some application, like medical and architectural utilization requires more accurate data while disaster relief requires receiving data as soon as possible. However, energy, delay and accuracy are trading off each other, one can not improve three of them at the same time. Hence, how to control the trade off of energy, delay and accuracy among different applications is the problem we will solve in our work.

1.4. Data aggregation in multi-robot sensor networks

We focus on data aggregation technology for collecting data in MRSN. Data aggregation (Rajagopalan & Varshney, 2006) is a process of aggregating the data from multiple robot sensors to eliminate redundant data and provide fused information to the base station. Considering from the point of data redundant, data aggregation can collect the most efficiency data. However, transmission delay and data accuracy are also important in many applications such as military application and architectural application. Hence trading off transmission delay, energy consumption and data accuracy is an important issue. There are several typical algorithms of data aggregations. PEGASIS (Lindsey & Raghavendra, 2001) is one of energy efficiency chain based data aggregation protocols that employs a greedy algorithm. The main idea of PEGASIS is forming a chain among the sensor nodes so that each node receives (or transmits) fused data from (to) the closest neighbors. The data gathered are sent from node to node, and all the sensor nodes take turns to be the leader for transmission to the Base Station. Data Funnelling (Petrović, et. al, 2003) is another scheme that sends a stream of data from a group of sensor readings to destination. Moreover, they proposed a compression method called "coding by ordering" to suppress some readings and encoding the values in the ordering of the remaining packets. On the other hand, LEACH (Heinzelman W., 2000) is one kind of energy saving schemes in which a small number of clusters are formed in a self-organized manner. A designated sensor node in each cluster collects and combines data from nodes in its cluster, then transmits the result to the BS. Directed Diffusion (Intanagonwiwat, et. al, 2000) is a kind of data centric routing protocols. The sink broadcasts an interest message to all the sensor nodes, and the nodes gather and transmit the sink-interested data to the sink. When the receiving data rate becomes low, the sink starts to attract other higher quality data.

Regarding the trade-offs, (Boulis, et. al, 2003) proposed an energy-accuracy tradeoffs algorithm for periodic data-aggregation which is a threshold-based scheme where the sensors compare their fused estimations to a threshold to make a decision of regarding transmission. Energy-latency tradeoffs algorithm (Yu at. el., 2004) is proposed for minimizing the overall energy consumption of the networks within a specific latency constraint where data aggregation is performed only after a node successfully collects data from all its children and its own local generated data. ADA (Adaptive Data Aggregation) (Chen et. al., 2008) is an adaptive data aggregation (ADA) for clustered wireless sensor networks. In ADA, sensed data are aggregated on two levels; one is aggregated at sensor nodes controlled by the reporting frequency (temporal reliability) of nodes; another is aggregated at cluster heads controlled by the aggregation ratio (spatial reliability). The reliability of observed data that is decided by the number of arrival data at sink node is compared with the reliability of desired data, which is decided by the application. According to comparison, nine characteristic regions and nine states are defined in which the eight states must change into the desired state through the calculating and adjusting of observed reliability.

Most of the previously mentioned works focus on energy saving and aggregate as much data as possible. As a result, they prolong the transmission delay. Many works aimed to achieve energy-delay trade off, however they still have shortcomings for example (Yu at, el., 2004) has long waiting time at nodes with less event data while the constant latency makes the networks very inflexible in (Galluccio L. & Palazzo S., 2009). A desired energy-delay tradeoff is achieved in (Ye Z. et al., 2008); however the algorithm ignored the issue of data accuracy. Energy-delay-accuracy tradeoffs in (Mirian F. & Sabaei M.) and (Chen et al., 2008) adapt to a situation that could be described by the following question: 'what is the average temperature of this area at this hour?' The algorithms did not consider delay and accuracy among nodes and data, which may lead to large data deviation as well as transmission delay in some other applications.

1.5. Preview of our work

In this paper, at first, we show the analyses of transmission delay, energy consumption and data accuracy of non-aggregation, full aggregation and partial data aggregation with Markovian chain. The analytical results show that non-aggregation consumes much energy and full aggregation causes long transmission delay; but the proposed partial aggregation can trade off total delay, energy consumption, and data accuracy between non-aggregation and full aggregation. Then we intensively discuss the tradeoffs among energy consumption, transmission delay and data accuracy with a Trade Off Index (TOI). We discuss the TOI under the different conditions of accuracy dominant, energy dominant, and delay dominant. By comparing the TOI value among non-aggregation, full aggregation and partial aggregation in different data generation rates, we obtain the best TOI. The results show that with small data generation rate, non-aggregation is the best TOI; with moderate data generation rate, the partial aggregation is the best TOI while the data generation rate is large, the full aggregation is the best TOI. At last a multi view multi robot sensor network is discussed and a User Dependent Multi-view Video Transmission (UDMVT) scheme is introduced.

2. Preliminary concepts

In this section, we will introduce network topology, network parameters and the definitions of network parameters, which will be helpful in understanding our work clearly.

2.1. Wireless MRSN network topology

Fig. 1 depicts tandem network topology of the MRSN, the most basic and simplest model, which enables us to make an analytic model. The results can be extensible to other topologies that are more complex. In such kind of network, all the robots deployed statically in a flat area and have same role. The robots are allocated omni-directional antenna for wireless communication and have the same transmission ranges. When a robot senses data, it transmits the data to the sink, if the data could not get to the sink by one hop; the robot sends the data to the sink by multi-hop way.

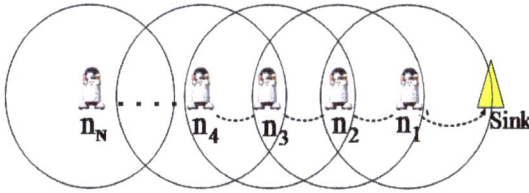

Figure 1. Tandem Multi-Robot Network

2.2. Definition

- n_i denotes the i-th node from the sink. N is a set of all nodes.
- n_{i+1} is called the adjacent upper node of , while n_{i-1} is the adjacent lower node of n_i.
- A set of nodes of $\{n_k \mid n_k \in \mid N \mid, k>i\}$ denotes the up-per nodes of n_i, while $\{n_k \mid n_k \in \mid N \mid, k <i\}$ denotes the lower nodes of n_i.
- Suffixes non, ful and par attached to terms mean non-aggregation, full aggregation and partial aggregation respectively.
- Arrival data denotes that the data come from adjacent upper nodes.
- Local generated data denotes that the data are generated at local nodes.
- Server: in our work, we assume that each node has a server to process data aggregation and data transmission.
- The MAC protocol used in this research is CSMA.
- The propagation delay between adjacent robots is negligible.

2.3. Aggregation factor

Here a robot aggregates its own generated data and received data from adjacent upper nodes before transmission. The sink does not participate in data aggregation. When data aggregation occurs at a robot node, the aggregation factor denotes the proportion of aggregated data size and local generated data size. It means that the aggregated data size is

AF times of the generated data size. AF=1 means that aggregated data have the same data size with generated data, and we assume there is one generated data at one time.

$$AF = \frac{\text{Aggregated data size}}{\text{Generated data size}} \tag{1}$$

2.4. Transmission delay

- Total delay $D(N)$ shows a time interval between the instance when event E_{ij} occurred at robot n_n and the instance when the sink receives D_{ij} in N hops networks.
- Data transmission time $\bar{c}'$ is defined as a time interval between the instance that data are transmitted from robot and the instance that the data are received at the adjacent lower robot.
- Channel waiting time $\tau^c(i)$: it is the time interval that data cannot utilize the channel.
- Event waiting time $\tau^e(i)$: In full aggregation, before a robot processes data aggregation, the arrival data have to wait for local generated data to be aggregated together, hence the waiting time of arrival data called event waiting time.

2.5. Energy consumption

Total energy consumption $E(N)$ is defined to be the sum of energy consumption of an event data that is generated at node n_n and finally received by sink node in N hops networks.

2.6. Data accuracy

We define the data accuracy as the proportion of collected data at sink and the amount of sensed data at all the robots.

3. Data aggregations

In this part, we analyze and evaluate the data aggregation simply in terms of non-aggregation, full aggregation and partial data aggregation.

3.1. Non-aggregation

The arrival data are transmitted to the adjacent lower node immediately after having been received; data neither wait for local generated data nor aggregate with any other data. The analytical model of non-aggregation is shown as follows in figure 2.

In the analytical model of node n_i in fig. 2, the average arrival rate from the upper node is approximates to Poisson distribution. Generated data rate at a node is assumed to be Poisson distribution. The generated data and arrival data join the service queue and wait for transmission. There is one server for data transmission at each node. All data in the queue will be sent based on first in first out. λ_i'' is the data rate upon exiting the server at node n_i.

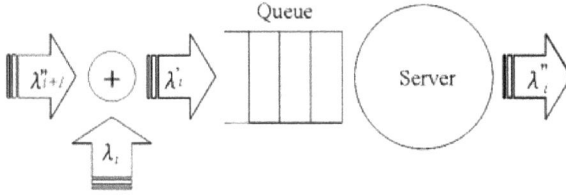

Figure 2. Analytical model of non-aggregation

- Arrival process to the queue

According to the analytic model, we find that the arrival rate to the queue is

$$\lambda_i' = \lambda_{i+1}'' + \lambda_i \qquad (2)$$

Strictly speaking, arrival data from the upper node is not Poisson distribution. However, for the purpose of simplicity, we approximate the process as Poisson distribution. Since the arrival data rate and local generated data rate are independent Poisson distributions, the sum of the two is also a new Poisson distribution.

- Service process

In our network model, each node has one server. The ACK packet transmission time is not considered. Data aggregating time is very short and negligible. Therefore the service time is one hop data transmission time. In our work, data transmission rate is v_c and local generated data size is S_i. Therefore the service time for each generated data is:

$$\tau^1 = \frac{S_i}{v_c} \qquad (3)$$

Since v_c and S_i are constant in non-aggregation, the service time for each data are fixed and constant.

From the above analysis we can determine that the queuing system approximates to M/D/1 model.

According to equation (4), the average data transmission time that we obtain at a node is:

$$\tau^1_{non}(i) = \frac{\tau^1}{(1 - \lambda_i' \tau^1)} \qquad (4)$$

According to queuing theory and equation (4), we determine the server waiting time as follows:

$$\tau^s_{non}(i) = \frac{\lambda_i'(\tau^1_{non}(i))^2}{2(1 - \lambda_i' \tau^1_{non}(i))} \qquad (5)$$

- Channel waiting time

Node n_i communicates with only one neighbor node at a time. If a neighbor node is transmitting data, node n_i has to wait until its neighbor node finishes transmission, due to the over hearing caused by the omni-directional antenna. This waiting time is defined as channel waiting time and is obtained the formulation as follows:

$$\tau_{non}^c(i) = 2\lambda_i' * \tau_{non}^1(i) * \tau_{non}^1(i) \tag{6}$$

3.1.1. Total delay

Total delay $D_{non}(N)$ is derived as follows where the number of hops from robot n_i to the sink is N.

$$D_{non}(N) = \sum_{i=1}^{N} (\tau_{non}^1(i) + \tau_{non}^s(i) + \tau_{non}^c(i)) \tag{7}$$

3.1.2. Energy consumption

Node n_i transmits data and relays arrival data from the upper nodes. Since the consumed energy is proportionate to the number of data transmissions, we can find the mean number of data $L^Q_{non}(i)$ in the service queue at node n_i according to Little's formula. The number of data in the queue waiting for data transmission is shown as follows:

$$L_{non}^Q(i) = \lambda_i' * T_{non}^s(i) \tag{8}$$

The λ'_i is arrival data rate at node n_i and $T_{non}^s(i)$ is time duration from data joining the queue to data having been received by the next neighbor node at node n_i, in case of non-aggregation, can be determined as follows:

$$T_{non}^s(i) = \tau_{non}^1(i) + \tau_{non}^s(i) + \tau_{non}^c(i) \tag{9}$$

According to equations (8) and (9), we obtain the whole energy consumption in N hops network as follows:

$$E_{non}(N) = \sum_{i=1}^{N} L_{non}^Q(i) * (P_t + 2P_r) \tag{10}$$

Here P_t and P_r denote the energy consumption for transmitting and receiving data.

3.1.3. Data accuracy

In non-aggregation, data are not aggregated and the packet drop occurs with the transmitting in real system. However, for simplicity, we assume there is no packet drop and retransmission, all the generated data will get to the sink, thus the data accuracy approaches to 100%.

3.2. Full aggregation

We define the full aggregation that the arrival data are sent to an adjacent lower node only after having been aggregated with local generated data at nodes. It means data transmission occurs only after a new local data generated a node. Hence, the waiting time for data aggregating at a node is decided by the data generation rate of the node. When there is local generated data, the node aggregates all the arrival data with generated data then waits for transmission at server. Data after aggregation undergo the same procedure as non-aggregation to detect the server and the channel for further transmission.

The analytical model of full aggregation is shown in figue3. Before explaining the model, we introduce queue A, queue B and "G." Queue A denotes the arrival data queue at a node that is waiting for local generated data for data aggregation. Data in Queue B are waiting for server; when the server is idle, data are transmitted to a neighbor node. The "G" is assumed as a virtual gate between queue A and queue B. Immediately after local generated data aggregate with the arrival data in queue A, the gate opens and lets the aggregated data join queue B.

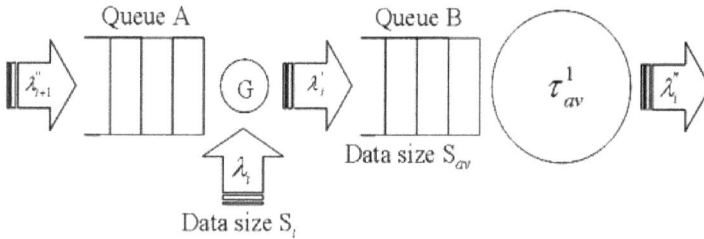

Figure 3. Analytical model of full aggregation

In full aggregation, the data join queue A with arrival rate of λ''_{i-1} and wait for new generated data. When an event occurs at local node, the node aggregates the generated data and all arrival data in queue A according to the aggregation factor Af. The size of aggregated data becomes S_{av} and the aggregated data join queue B with the rate of λ'_i to await further transmission. In full aggregation, the difference from non-aggregation is that we have to determine how long the arrival data wait for aggregation in queue A.

3.2.1. Event waiting time

To determine the event waiting time, we apply the state transition rate diagram. We describe the state transition rate diagram in fig. 4. The basic idea of the analysis is that data waiting in queue A for exponential distribution have an average of $1/2\lambda_i$. In the diagram, the state variable is the number of data waiting for an event.

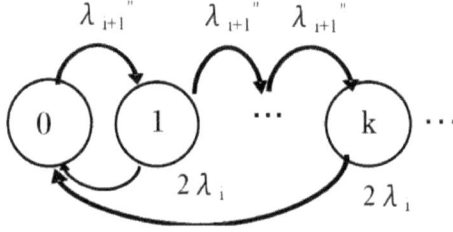

Figure 4. State transition rate diagrams of full aggregation

According to calculation of state probability distribution and Little's formula, we determine the event waiting time as shown below; more details please read (Li, et. al, 2010).

$$\tau_{ful}^{e}(i) = \frac{\lambda_{i+1}''}{2\lambda_i^2} \tag{11}$$

3.2.2. Total delay

From the definition of full aggregation we know that the arrival data join queue B only if there is new generated data at local node, hence the data arrival rate to queue B is equal to data generation rate at the node. The data generation rate abides by Poisson distribution; therefore the arrival data rate to queue B is Poisson distribution. Since the data arrival rate involves only one server and the data transmission time for server is fixed, according to queuing theory we model the queue by means of M/D/1 queue. Similar to non-aggregation, we determine the total delay D_{ful} (N) of the network in full aggregation is consists of event waiting time in queue A, server waiting time in queue B, channel waiting time and data transmission time at server.

$$D_{ful}(N) = \sum_{i=1}^{N} (\tau_{ful}^{e}(i) + \tau_{ful}^{s}(i) + \tau_{ful}^{c}(i) + \tau_{ful}^{1}(i)) \tag{12}$$

3.2.3. Energy consumption

The energy consumption is proportional to the number of data transmissions. $L_{ful}^{Q}(i)$ is the number of data at a robot n_i. The time duration from data joining Queue B to data having been received by the next neighbor node is $T_{ful}^{S}(i)$ and it can be obtained as follows:

$$T_{ful}^{s}(i) = \tau_{ful}^{s}(i) + \tau_{ful}^{c}(i) + \tau_{ful}^{1}(i) \tag{13}$$

According to Little's formula and equation (13), the amount of data in queue B is as follows:

$$L_{ful}^{Q}(i) = \lambda_i * T_{ful}^{s}(i) \tag{14}$$

Therefore, the whole energy consumption is determined as follows:

$$E_{ful}(N) = \sum_{i=1}^{N} L_{ful}^{Q}(i) * (p_t + 2p_r)$$
(15)

3.2.4. Data accuracy

In full aggregation, the aggregation factor Af=1. Thus, we can get the data accuracy in N hops transmission as follow:

$$A_c = \frac{1}{N}$$
(16)

3.3. Partial data aggregation

According to previous analyses of non-aggregation and full aggregation, we find that non-aggregation sends all the generated data to sink node which results in large energy consumption. In case of full aggregation, the arrival data must wait for local generated data to aggregate, which causes the prolonged transmission delay and low data accuracy for the data that come from nodes far away from sink.

To minimize these two shortcomings, we propose a partial data aggregation. The main idea of partial aggregation is that nodes process data aggregation and transmit data only if a) if there are new local generated data at a node or b) after waiting a holding time at a node; the inverse of the holding time we call random pushing rate λ^{D_i}. The analytical model of partial aggregation is shown as follows.

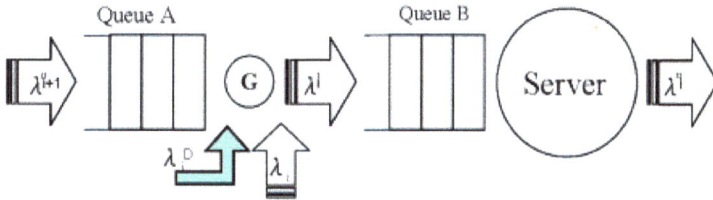

Figure 5. Analytical model of partial aggregation

For the purpose of simplifying our analytical model, we assume the arrival data rate from adjacent upper node is approximated to Poisson distribution and the arrival data join event waiting queue A in fig. 5. Data generation rate λ_i is assumed to be Poisson distribution. Random pushing rate is λ^{D_i} and assumed to be exponential distribution. If new generated data occur at a node or if holding time is over for arrival data, all the data are aggregated into one data, and the gate G opens and lets aggregated data join queue B. λ'_i is data arrival rate to queue B in which data are waiting for service (data transmission).

3.3.1. Event waiting time

Assume that a number of data are waiting for an event at robot n_i in queue; we describe the state transition diagram as shown in Fig.6.

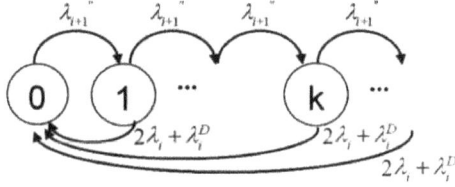

Figure 6. State transition rate diagram

Data are waiting in queue for the duration according to the exponential distribution of average $1/(2\lambda_i + \lambda_i^D)$. Similar to full aggregation, the event waiting time of partial aggregation can be determined as follows:

$$\tau_{par}^e = \frac{\lambda_{i+1}''}{(2\lambda_i + \lambda_i^D)^2} \tag{17}$$

3.3.2. Arrival process to Queue B

From the analytical process we find that arrival data rate λ_i is decided by the random pushing rate and data generation rate at a node. To determine the formulation of λ'_i, we calculate the property distribution of λ_i and λ^{D_i}. We define that λ^{D_i} and λ_i are the independent distribution X and Y. Through proofing of the property Y is bigger than X, we determine the arrival process to Queue B as follows; the proof can be found in (Li, et. al, 2010).

$$\lambda_i' = \lambda_i + \frac{\lambda_{i+1}''}{\lambda_{i+1}'' + \lambda_i^D} * \lambda_i^D \tag{18}$$

3.3.3. Total delay

Since the data generation rate is Poisson distribution and the random pushing rate abides by exponential distribution, the data arrival rate to queue B approximates to Poisson distribution. Therefore, we can confirm that the queuing system approximates to M/D/1 model. With the same way of full aggregation, the server waiting time and channel waiting time can be determined easily. Therefore, the total delay of partial aggregation is as follows:

$$D_{par}(N) = \sum_{i=1}^{N} (\tau_{par}^e(i) + \tau_{par}^c(i) + \tau_{par}^1(i) + \tau_{par}^s(i)) \tag{19}$$

3.3.4. Total energy consumption

In the N hops transmission in partial aggregation, total energy consumption $E_{par}(N)$ is the sum of transmission energy consumption, reception energy consumption and overhearing energy consumption. P_t and P_r are energy required for transmitting or receiving a data. The period of time that aggregated data wait in a queue for transmission can be determined as follows:

$$T_{par}^s(i) = \tau_{par}^s(i) + \tau_{par}^c(i) + \tau_{par}^1(i) \tag{20}$$

According to Little's formula and equation (25), we determine the amount of data in queue B at node n_i as follows:

$$L_{Par}^Q(i) = \lambda_i' * T_{par}^s(i) \tag{21}$$

Accordingly, we determine the total energy consumption for the network as follows:

$$E_{par}(N) = \sum_{i=1}^{N} L_{par}^Q(i) * (P_t + 2P_r) \tag{22}$$

3.3.5. Data accuracy

The total generated data $Lpar(N)$ in N hops network is obtained as follows:

$$L_{par}(N) = \sum_{i=1}^{N} \lambda_i * D_{par}(N) \tag{23}$$

The amount of data received by sink $Lpar(S)$ is as follows:

$$L_{par}(S) = \lambda_1' * D_{par}(N) \tag{24}$$

According to the definition and above equations, we determine the data accuracy as follows:

$$A_c = \frac{L_{par}(S)}{L_{par}(N)} * AF \tag{25}$$

3.4. Evaluation

Here we show the analytic results of the previous sections. The parameters are as below:

Transmission rate is 250[kbps], Data size is 4096 [bit], Energy consumption for data reception is 17.4 [mA] and for data transmission is 19.7 [mA]. In this section, we evaluate total delay, energy consumption and data accuracy when the aggregation factor is Af=1.

Fig. 7 to Fig. 9 show the total delay, energy consumption of whole network, robot energy consumption and data accuracy of five hops transmission where $\lambda_i=\lambda$. Partial-T1 and Partial-T2 are two sets of random pushing rate vectors in partial aggregation. We get the vectors randomly [1, 2, 3, 4, 5] and [5, 10, 15, 20, 25].

From figure 7, we find that when event generation rate is small, full aggregation has long transmission delay in comparison to non-aggregation. The reason for concaving up of delay of the full aggregation is that, when event generation rate is small, the received data has to wait for generated data longer duration. In addition at a robot near to the sink, the total delay increases because of the large waiting time due to the congestion around the sink. As long as total delay is concerned, non-aggregation is suitable for situation of small event generation rate. From the figure, we also find that the performances of partial-T1 and partial-T2 are between non-aggregation and full aggregation. If λ_i^D is infinite, it means fully non-aggregation, if λ_i^D is zero, it means fully aggregation.

Fig. 8 shows the energy consumption of the whole network. Obviously, non-aggregation consumes much more energy than full aggregation. Thus, full aggregation is suitable for energy consumption while non-aggregation is efficiency for transmission delay. The partial-T1 and partial-T2 has energy consumption between non-aggregation and full aggregation. In addition, the smaller random pushing rate vector set partial-T1 has less energy consumption than the set of partial-T2.

Fig. 9 shows the data accuracy of different data aggregation. From fig. 9, we find that the data accuracy of partial aggregation is between non-aggregation and full aggregation. The partial aggregation with the larger random pushing rate achieves higher data accuracy.

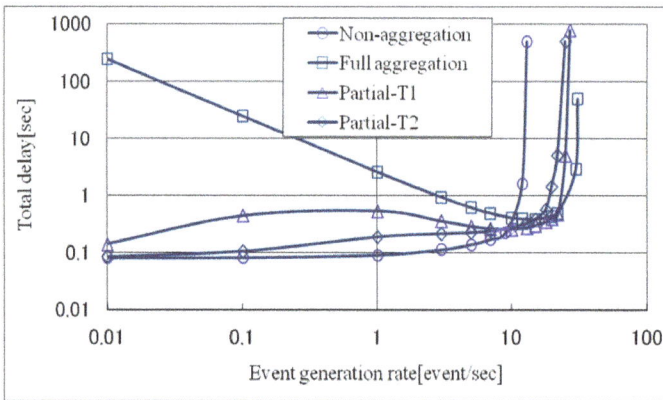

Figure 7. Total delay

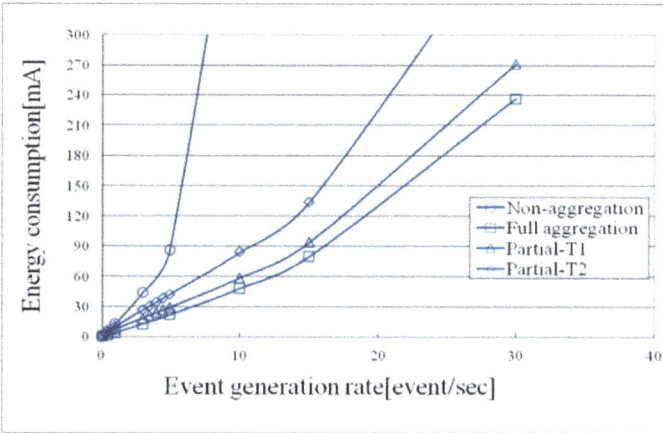

Figure 8. Energy consumption

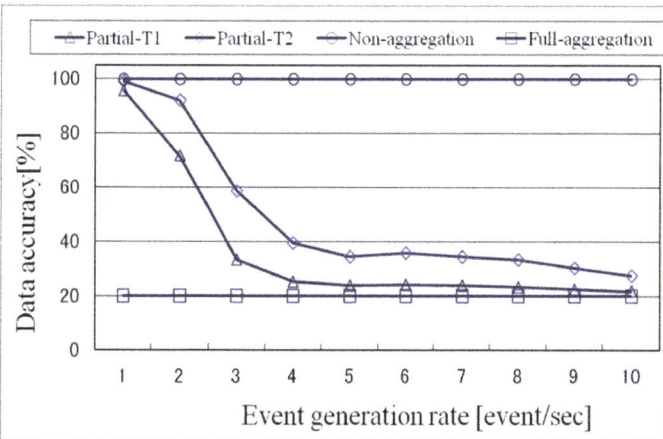

Figure 9. Data accuracy

From above evaluations we find that the partial aggregation with random pushing rate vectors can control the energy, delay and data accuracy between non-aggregation and full aggregation. Hence, one can achieve desired MSRN by controlling the random pushing rate.

4. Tradeoffs among accuracy, energy and delay

4.1. Trade off index TOI

Previous section clearly shows partial aggregation with random pushing rate λ_i^D can control the energy consumption, transmission delay and data accuracy. In MRSN, according to applications, delay taken to collect data, energy consumed by each sensor node for communication and data accuracy of the collected data are critical concerns and are in trade-off each other. Energy, delay and accuracy cannot reach full potential at the same time, but we can achieve the best possible tradeoff between them. To obtain the best trade-off value of practical application, we propose a Trade-Off Index (TOI). In the following subsections, we discuss energy, delay and accuracy of trade-offs in respect of TOI as criteria. Here E denotes total energy consumption, D denotes total delay, Ac denotes data accuracy. α, β, γ indicate the significance of accuracy, energy and delay and larger α, β, γ indicate more significance of energy, delay and accuracy. The smallest TOI value denotes the best data aggregation.

$$TOI = \frac{E^{\beta} * D^{\gamma}}{A_c^{\alpha}} \tag{26}$$

4.2. Applications of WSNs with different criteria

In MRSN, according to the different applications and objectives, we need different significances for transmission delay, energy consumption and data accuracy. Some application areas need to save energy because it is impossible to replace or recharge the battery. In some applications not only the energy is significant, but also the data freshness, such as in military monitoring and disaster monitoring; however data accuracy is most important in medical utilization and in quality control. According to real application, we formulate some of the applications according to the significances of energy, data accuracy and transmission delay in table 1.

Accuracy	Energy	Delay	Average event generation rate		
			month, day, hour	ten min, min ...	sec, ms, µs ...
L	S	S	Daily health care Medical application Survey		
			Home Automation Industrial Applications		
S	L	S	Agriculture Application Nature monitoring		
			River pollution detecting		
S	S	L	Outer space monitoring Architectural		
			Flood detection Meteorological system		

Table 1. Applications of MRSN

Here the "L" denotes large significance and "S" denotes small significance; the application is formed from left to right along a scale from smaller event generation rate to bigger generation rate. According to the table 1 we can decide the significant parameters of the application in order to perform our proposed TOI; we can achieve the best data aggregation corresponding to the applications.

4.3. Tradeoffs of different applications

In this section, we will investigate the tradeoffs among the applications of which data generation rate is in the range of 0.0001 to 100 events in per second, and here for corresponding to the event generation rate, we define the random pushing rate vectors as the same with event generation rate. We define the random pushing rate vectors as below:

As data generation rate of λ=0.0001, T= [0.0001, 0.0001, 0.0001, 0.0001, 0.0001],

As λ=0.001, set T= [0.001, 0.001, 0.001, 0.001, 0.001],
As λ= 0.01, set T= [0.01, 0.01, 0.01, 0.01, 0.01],
As λ= 0.1, set T= [0.1, 0.1, 0.1, 0.1, 0.1],
As λ= 1, set T= [1, 1, 1, 1, 1],
As λ= 10, set T= [10, 10, 10, 10, 10].

4.3.1. Accuracy significant networks

In accuracy significant utilization, we define α, β and γ as 2, 1, 1; however if the data accuracy is much more important than other two, we also can define α=3 or much larger. In this research, for simplify, we discuss none other but the case that significant parameter has the significance vector of 2 and the ordinary parameters are 1. According to TOI we can get the best result in fig. 10.

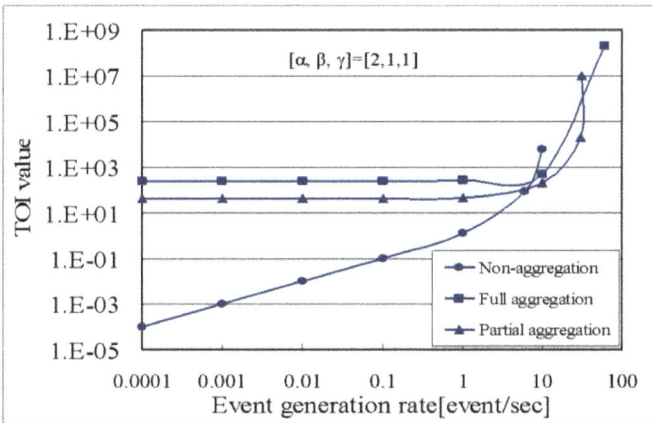

Figure 10. Tradeoffs of accuracy significant

From Fig. 10 we find that when the event generation rate is between 0.0001 and 4.0, non-aggregation is the best comparing with full and partial aggregation. When the data generation rate is between 4 and 30, the partial aggregation is the best, and the full aggregation is the best when data generation rate is larger than 30.

4.3.2. Energy significant networks

Here we discuss the case when energy is significant. The parameters are defined to be as below: $\alpha=1$, $\beta=2$ and $\gamma=1$. According to proposed TOI we can get the best TOI values when data generation rate is from 0.0001 to 100.

Fig. 11 shows the result. We find from the figure that in the region of data generation rate between 0.0001 and 4.0, the non-aggregation is the best TOI. When the data generation rate is about 4-10, the figure shows that the partial aggregation is the best; the full aggregation is the best when event generation rate is larger than 10.

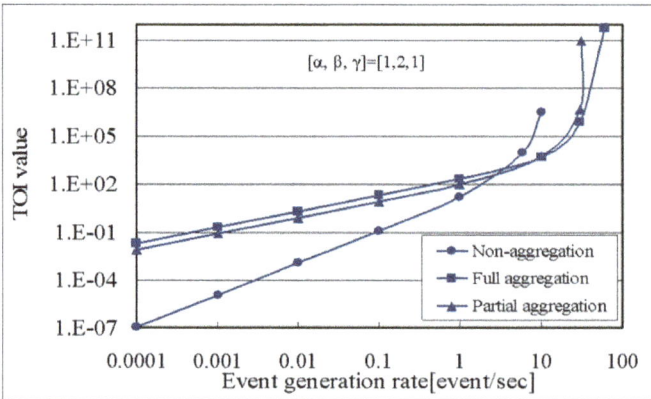

Figure 11. Tradeoffs of energy significant

4.3.3. Delay significant networks

In delay significant networks, α, β and γ is defined as 1, 1, 2, as shown in Fig. 12. From the figure we find that when event generation rate is between 0.0001 and 4.0, the non-aggregation is investigated to be the best TOI; and when the event generation rate is from 4 to 30, the partial aggregation is the best; the full aggregation is the best TOI when event generation rate is larger than 30.

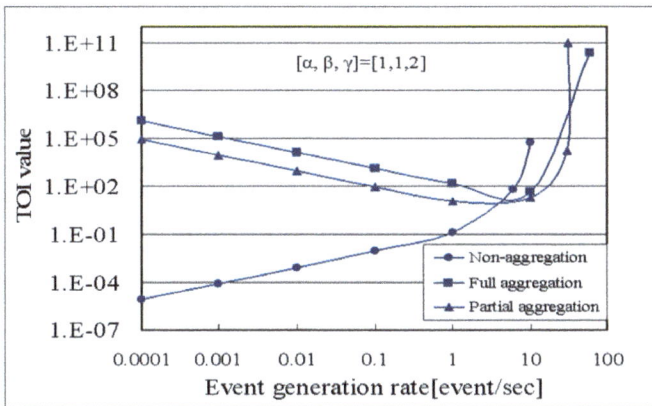

Figure 12. Tradeoffs of delay significant

4.3.4. Discussion

Let us summarize the data aggregation with best TOI according to different event generation rate in table 2. From the table we find that when event generation rate is small (0.0001-4.0, 6.0) the non-aggregation is the best TOI. Moreover, from the figures we find that in accuracy significant networks, the event data range of best TOI at non-aggregation is longer (0.0001-6.0) than any others. This is because, in non-aggregation, the data accuracy is 100%; and the other two have low data accuracy; when event generation rate is larger than 6, non-aggregation has very long delay because of the congestions around the sink node.

When event generation rate is moderate (4, 6-30), the partial aggregation is the best TOI except the case energy significance networks. In energy significance networks, the number of transmission in partial aggregation is much more than full aggregation, so the energy has great impact on partial aggregation with the significant of $\beta=2$. When data generation rate is large, due to the large number of transmission, the energy consumption is very high in non-aggregation and partial aggregation; therefore, the best TOI is the full aggregation in the networks with large event generation rate.

Best aggregation \ Significance	Non-aggregation	Partial aggregation	Full aggregation
$[\alpha, \beta, \gamma] = [1, 1, 1]$	0.0001-4.0 [event/sec]	4-30 [event/sec]	30-100 [event/sec]
$[\alpha, \beta, \gamma] = [2, 1, 1]$	0.0001-6.0 [event/sec]	6-30 [event/sec]	30-100 [event/sec]
$[\alpha, \beta, \gamma] = [1, 2, 1]$	0.0001-4.0 [event/sec]	4-10 [event/sec]	10-100 [event/sec]
$[\alpha, \beta, \gamma] = [1, 1, 2]$	0.0001-4.0 [event/sec]	4-30 [event/sec]	30-100 [event/sec]

Table 2. The best data aggregation

5. Multi-view multi-robot sensor networks

As we mentioned in the introduction section, applications of the MRSN will be more advanced if multi-cameras are equipped on the robot nodes. The reason is quite similar to human with more eyes. From the point of application, multi-view MRSN can be applied in security system that will not miss a corner. In addition, in the medical application, the multi-view MRSN can accomplish some complex and long time operations. Meanwhile it can achieve more accurate and small cut operation; besides, multi-view MRSN has quick reaction for the vary vital signs and other monitored parameters of the patient.

5.1. Introduction of multi-view video and open problem

The developments of camera and display technologies make recording a single scene with multiple video sequences possible. These multi-view video sequences are taken by closely spaced cameras from different angles. Each video sequence in the multi-view video presents a unique viewpoint of this scene. Therefore, user can switch the viewpoint by playing different video sequences. When a robot is equipped with multi-cameras, it will bring the user who controls the robot a broad perspective. The operator also can switch his viewpoints by playing different video sequences. However, since the multi-view video consists of the video sequences captured by multiple cameras, the traffic of multi-view video is several times larger than conventional multimedia, which brings the dramatic increase in the bandwidth requirement. However, as multi-view video is taken from the same scene, a large amount of inter-view correlation is contained in the video. Therefore, compression transmission technologies are especially important for multi-view video streaming.

The state of the art in multi-view representations includes Multi-View Video Plus Depth (Merkle et, al., 2007), Ray-Space (Smolic, et, al., 2006) and Multi-view Video Coding (MVC) (Vetro, et, al., 2008), (Mueller, et, al., 2006). However, the research on Multi-View Video Plus Depth sequences (Merkle et, al., 2007) suggests that with the addition of depth maps and other auxiliary information, the bandwidth requirements could increase. MVC is issued as an amendment to H.264/MPEG-4 AVC (Vetro, et, al., 2008), (Mueller, et, al., 2006). It was reported that MVC makes more significant compression gains than simulcast coding in which each view is compressed independently. However, even with the MVC, transmission bitrates for multi-view video are still high: about 5 Mbps for 704 × 480, 30fps, and 8 camera sequences with MVC encoding (Kurutepe, et, al. 2007).

5.2. User dependent multi-view video transmission

5.2.1. Switching models

In order to reduce traffic for multi-view video transmission, we have analyzed which frames should be displayed when the viewpoint is switched. Our work mainly focuses on the successive motion model (Pan, et, al., 2011, 2011). In the successive motion model as shown by Fig. 13, user is only able to switch to the neighboring views. In other words, if the multi-view video contains the views (1, 2… M), user is just able to switch from any view j to the view j', where max (1, j-1) ≤ j'≤ min(j+1, M). This kind of switching model is used in the applications such as free viewpoint TV and Remote Surgery System in which user's head is tracked to decide which views should be displayed.

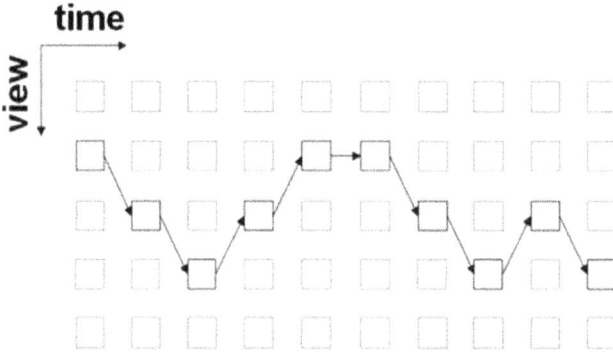

Figure 13. Switching models

5.2.2. User dependent multi-view video transmission (UDMVT)

In (Tanimoto, et. al., 2011), they developed two types of user interface for the Free Viewpoint TV. One showed one view according to the viewpoints given by user. With this type of user interface, the viewpoint of user can be switched by an eye/head-tracking system, moving the mouse of a PC or sliding the finger on the touch panel of a mobile player. In a real-time interactive multi-view video system (Lou, et, al., 2005), users can switch viewpoints by dragging the scroll bar to a different position. In the user interfaces of (Tanimoto, et. al., 2011) and (Lou, et, al., 2005), the changing of user's position, moving of mouse, sliding of finger and dragging of scroll bar are all successive motions. Since the switching models of these user interfaces are all successive motion models, it will take some time to switch from the current view to the neighboring view. For instance, in the head-tracking system, user needs to take some time to move from his current position to the next position for the new viewpoint. We call the speed with which user switch from one view to next view "switching speed." With different user and user interfaces, the switching speed is different. Even the same user may switch to a different switching speed each time.

In the successive motion model, which frames should be displayed when user starts to switch to the next view are decided by both the frame rate f (frame/s) of the multi-view video and the switching speed s (view/s) of user. Let k be the floor of the frame rate divided by switching speed: $k = \lfloor f/s \rfloor$. Fig. 14 presents the display of frames when k is 3, 2 and 1.

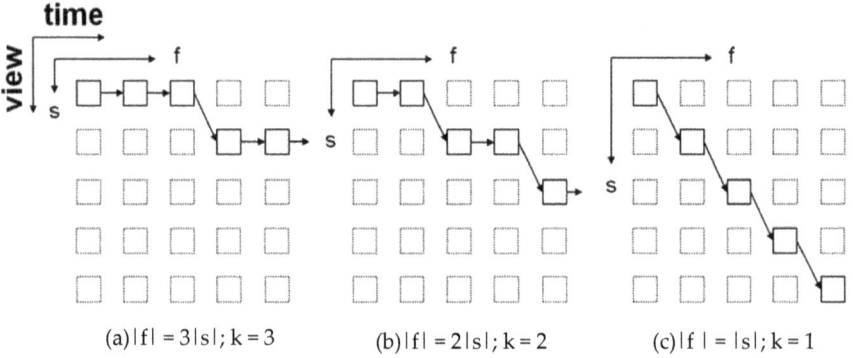

(a) $|f| = 3|s|; k = 3$ (b) $|f| = 2|s|; k = 2$ (c) $|f| = |s|; k = 1$

Figure 14. Multi-view video displays with different value of k.

Assuming the frame rates are the same, different frames should be displayed by these three different switching speeds, although they are switching to the same direction. If switching slows down, more frames of the current view should be displayed before the display changes to the next view. Otherwise, less frames of the current view should be displayed. Therefore, k denotes the number of the frames should be displayed in the current view after user starts to switch and before the user reaches the position where display should change to the next view. In practice, the frame rate is about 25~30 (frame/s). The value of switching speed depends on the density of the views and the speed of user interface. However, the switching speed is usually much slower than the frame rate. When the switching speed is about 2~5 (view/s), the k is about 5~15 (frame/view). For simplicity, k=1 and k = 2 are selected as the examples in this paper. Let $F_{i,j}$ denote the frame of view j at time instant i. By the three-tuples $N(p, f, s)$, it is able to predict a triangle area in which the frames are possible to be displayed in a subsequent period of time. p is the current position $F_{i0,j0}$. When the number of the views is M, R(t) is the set of frames that can be displayed at time instant t start from $F_{i0,j0}$. $F_{i,j'}$, R(t), in which:

$$i = i_0 + \lfloor f \times t \rfloor \tag{27}$$

$$j' \in \left[\max\left(1, j_0 - \lfloor s \times t \rfloor\right), \min\left(j_0 + \lfloor s \times t \rfloor, M\right) \right] \tag{28}$$

As the video continues to play, the frame at time instant i in (1) should be displayed starting from $F_{i0, j0}$. User can switch to the view $j_0 - \lfloor s \times t \rfloor$ or $j_0 + \lfloor s \times t \rfloor$ unless already at border view (view 1 or view M) during the period t. The user may also stop switching at any view before switching to view $j_0 - \lfloor s \times t \rfloor$ or $j_0 + \lfloor s \times t \rfloor$. Therefore, it is possible to display frames in view j' shown by (2). The triangles of frames are shown in Fig. 15 when k is 1 and 2, respectively. The frames in the triangle are called potential frames (PFs), which can be switched to and displayed. These frames should be encoded and transmitted. Those frames outside the triangle are called redundant frames (RFs). It is impossible to display RFs no matter how the user switches the viewpoint start from the current position. UDMVT reduces the transmission bitrate for multi-view video transmission by transmitting only the PFs without RFs.

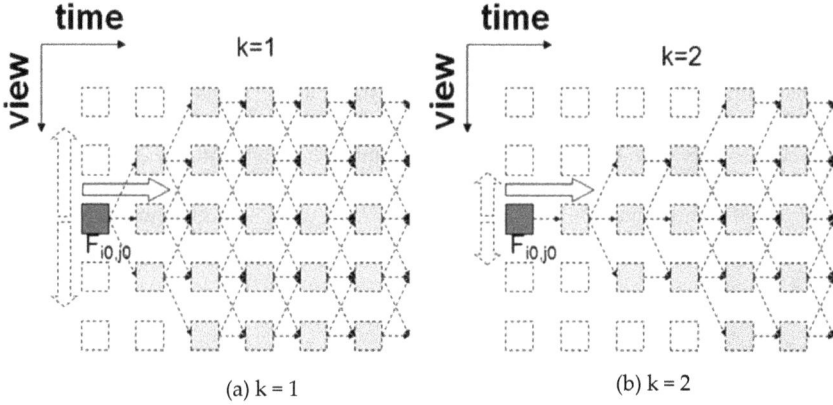

(a) k = 1   (b) k = 2

Figure 15. The triangles of the frames when k =1 (a) and k=2 (b). The M of the multi-view video is 5. Dotted line represents the possible display path.

When the length of the triangle is L, the number of RFs of the view j in the triangle is:

$$RFs(j) = \min(L, I(j))$$

I(j) is:

$$I(j) = |j - j_0| \times k = |j - j_0| \times \lfloor f / s \rfloor$$

In I(j), $| j - j_0 |$ is the distant between the view j and the current view j_0. The number of RFs in each triangle is:

$$\sum_{j=1}^{M} RFs(j) = \sum_{j=1}^{M} \min(L, I(j))$$

So the ratio of PFs to RFs is:

$$\frac{M \times L - \sum_{j=1}^{M} RFs(j)}{\sum_{j=1}^{M} RFs(j)} \tag{29}$$

From these expressions, it could be found that with the increase in the length L, the ratio of the PFs to RFs increases, which means that more frames should be encoded and transmitted. In other words, the triangle will be enlarged and finally all the frames at the same time instant are involved into the triangle, which is also shown in Fig. 15.

In order to overcome this problem, the N(p, f, s) should be fed back periodically, which is able to divide a large triangle into many smaller triangles as shown in Fig. 16. In the UDMVT, the N(p, f, s) is fed back periodically at the end of the triangle. The fed back N(p, f, s) from the end of the previous triangle is used to predict the next triangle. Therefore, only potential frames are transmitted each time and the transmission bitrate is reduced. N(f, p, s) should be detected at client and fed back periodically. At the server, N(p, f, s) is used to divide the frames into PFs and RFs. The transmission bitrate can be reduce by only transmitting the PFs and ignore the RFs. Although the transmission of RFs is unnecessary, encoding and transmitting the RFs can work as a kind of insurance against some special situations, such as the switching detection error.

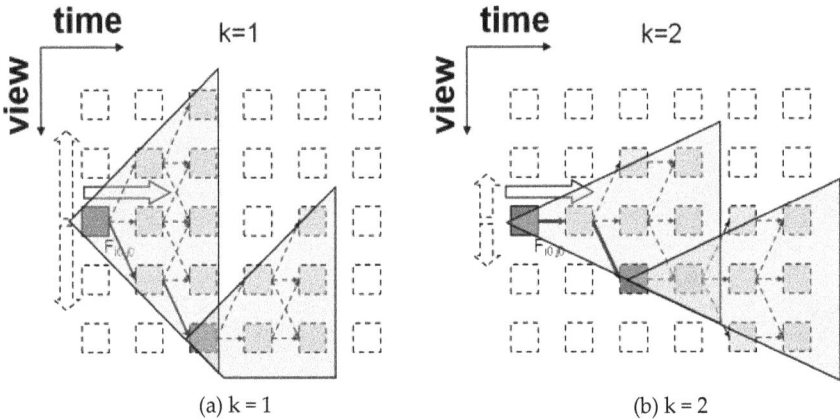

(a) k = 1   (b) k = 2

Figure 16. The triangles of the potential frames. Dotted line represents the possible display path while solid line represents the actual display.

6. Conclusions and future work

6.1. Conclusion

In this paper, at first, we analyzed the conventional non-aggregation, full aggregation, and our proposed partial aggregation with Markovian chain. The analytical result showed that, conventional method suffers large energy consumption with the highest accuracy, while full aggregation suffers long transmission delay, with the least accuracy. However, our proposed partial aggregation has the energy, delay and data accuracy between non-aggregation and full aggregation. When the random pushing rate becomes larger, the partial aggregation tends to non-aggregation and it tends to full aggregation with large random pushing rate. Hence, we find that the partial aggregation can trade off energy, delay and accuracy according to different applications. Secondly, we discussed the tradeoffs among data accuracy, transmission delay and energy consumption with different significances according to different applications by proposing tradeoff index (TOI). From the results, we find that non-aggregation has the best TOI for low event generation rate, that the partial aggregation does for moderate event generation rate, and that the full aggregation does for large event generation rate. At last, we discussed multi-view multi-robot sensor network from the viewpoint of potential applications, existing schemes and our proposed UDMVT.

6.2. Future work

For the future work, at first, we will discuss the random pushing rate to adapt the various changes of data generation rate and information content. For example, in an MRSN, when it is of the state of affairs, nodes generate much more event data than in normal case, that means data generation rate becomes larger. In this case we should decrease the random pushing rate to control the amount of data transmission. On the other hand, from the view point of information entropy, if the self information of generated data is high, it means the generated data are rare generating data. However, when a node applies the normal data aggregation and aggregates the data with normal data, the aggregated data cannot reflect the real situation which may lead bad result. In this case, we can increase the random pushing rate to send high self information data immediately without data aggregation. When the self information of generated data decreases, we decrease random pushing rate to control the quantity of data transmission. Secondly, in wireless sensor network, data are transmitted to sink node by multi-hopping way, which causes the uneven energy consumption on nodes at different locations. Hence, to keep all nodes in the network having the same energy consumption is our another future work.

Author details

Wuyungerile Li, Ziyuan Pan and Takashi Watanabe
Shizuoka University, Japan

7. References

Boulis, A., Ganeriwal, S. & Srivastava, M. (2003), Aggregation in sensor networks: an energy–accuracy trade-off. IEEE 1st Int'l WKsp. Sensor Network Protocols and Applications, USA, May 2003.

Chen H., Mineno H. & Mizuno H. (2008). Adaptive data aggregation scheme in clustered wireless sensor networks. Computer Communications, Vol 31, No. 12, (September 2008) pp. 3579–3585

Cheung G., Ortega A. & Cheung N. (2009). Bandwidth-Efficient Interactive Multiview Live Video Streaming using Redundant Frame Structures. Proceedings of 2009 APSIPA Annual Summit and Conference, Sapporo, Japan, October 2009.

Eu Z., H. Tan & Seah W. (2010). Opportunistic Routing in Wireless Sensor Networks Powered by Ambient Energy Harvesting. Computer Networks (Elsevier), Vol. 54, No. 17, (December 2010), pp. 2943-2966

Galluccio L. & Palazzo S. (2009) End-to-end delay and network lifetime analysis in a wireless sensor network performing data aggregation. Proceedings of the 28th IEEE conference on Global telecommunications, Honolulu, Hawaii, November 2009

Heinzelman W. (2000) Application-Specific Protocol Architectures for Wireless Networks. Massachusetts Institute of Technology, June 2000, Available from http://www-mtl.mit.edu/researchgroups/icsystems/pubs/theses/wendi_phd_2000.pdf

Intanagonwiwat, C., Govindan, R. & Estrin, D. (2000). Directed Diffusion: A Scalable and Robust Communication Paradigm for sensor networks. Proceedings of ACM Mobicom'2000, pp. 55-67, 2000.

Kurutepe E., Civanlar M. & Tekalp A. (2007) Client-driven selective streaming of multi-view video for interactive 3DTV, Circuits and Systems for Video Technology, IEEE Transactions on, Vol. 17, No. 11. (29 October 2007), pp. 1558-1565

Leyzx V., Bouraqadiy N., Stinckwichz S., Morarux V. & Doniecy A. (2009) Making Networked Robots Connectivity-Aware. Proceedings of the 2009 IEEE international conference on Robotics and Automation, Kobe, Japan,

Lindsey S. & Raghavendra C (2002) PEGASIS: Power Efficient GAthering in Sensor Information Systems. Aerospace Conference Proceedings, 2002. IEEE In Aerospace Conference Proceedings, IEEE, Vol. 3 (2002), pp. 3-1125-3-1130

Li W., Bandai M. & Watanabe T. (2010) Trade offs Among Delay, Energy and Accuracy of Partial Data Aggregation in Wireless Sensor Networks. Proceedings of the 2010 24th IEEE International Conference on Advanced Information Networking and Applications, Perth, Australia, pp. 917 – 924, April 2010

Lou J., Cai H. & Li J. (2005). A RealTime Interactive MultiView Video System. Proceedings of ACM international conference on Multimedia, 2005

Maxim A. & Gaurav S. (2005) Sensor network-mediated multi-robot task allocation. The Third International Multi-Robot Systems Workshop, pp. 27-38, March, 2005.

Mirian F. & Sabaei M. (2009) A Delay and Accuracy Sensitive Data Aggregation Structure in Wireless Sensor Networks. 2009 International Conference on In-formation Management and Engineering, Kuala Lumpur, Malaysia, 2009

Mueller K., Merkle P., Schwarz H., Hinz T., Smolic A., Oelbaum T. & Wiegand T. (2006). Multi-view video coding based on H.264/AVC using hierarchical B-frames. Proceedings of the Picture Coding Symposium 2006, CD-ROM: Beijing, April 2006

Merkle P., Smolic A., Müller K. & Wiegand T. (2007) Multi-view video plus depth representation and coding," IEEE International Conference on Image Processing 2007, San Antonio, TX, Sep. 2007

Mi Z., Yang Y. & Liu G. (2010) Connectivity control of mobile multi-robot networks. Proceeding of 2010 IEEE/ASME International conference on advanced intelligent mechatronics, Montreal, Canada, Jul. 2010.

Pan Z., Ikuta Y., Bandai M. & Watanabe T. (2011). A User Dependent System for Multi-view Video Transmission. IEEE International Conference on Advanced Information Networking and Applications (AINA), pp.732-739, Singapore, Mar. 2011

Pan Z., Ikuta Y., Bandai M. & Watanabe T. (2011). User Dependent Scheme for Multi-view Video Transmission. IEEE International Conference on Communications (ICC) 2011, Kyoto of Japan, Jun. 2011

Petrović D., Shah R., Ramchandran K. & Rabaey J. (2003). Data Funneling: Routing with Aggregation and Compression for Wireless Sensor Networks, Proceedings of the First IEEE. 2003 IEEE International Workshop on Sensor Network protocols, pp. 156 – 162, May 2003

Reich J. & Sklar B. (2006). Robot-sensor networks for search and rescue. In IEEE International Workshop on Safety, Security and Rescue Robotics, Gaithersburg, MD, August 2006

Rajagopalan R. & Varshney P. (2006). Data-Aggregation Techniques in Sensor Networks: A Survey. IEEE Communication Surveys and Tutorials, pp. 48-63, Fourth Quarter 2006

Smolic A., Mueller K., Merkle P., Fehn C., Kauff P., Eisert P. & Wiegand T. (2006) 3D video and free viewpoint video technologies, applications and MPEG standards. Proceedings of IEEE International Conference on Multimedia and Expo, July 2006

Sheng W, Yang Q, Tan J & Xi N. (2006) Distributed multi-robot coordination in area exploration. Robotics and Autonomous Systems (2006), Vol. 54 No.12 pp. 945–955

Tanimoto M, Tehrani M, Fujii T & Yendo T (2011). Free-Viewpoint TV. IEEE Signal Processing Magazine, vol. 28, Issue 1, pp. 67-76, Jan. 2011

Trigui S., Koubaa A, Jemaa M, Chaari I & Al-Shalfan K (2012) Coordination in a Multi-Robot Surveillance Application using Wireless Sensor Network. 16th IEEE Mediterranian Electrotechnical Conference, Hammamet (Tunisia), 25-28 March, 2012.

Vetro A, Pandit P, Kimata H, Smolic A & Wang Y. (2008). Joint Draft 8.0 on Multi-view Video Coding. Joint Video Team, Doc. JVT-AB204, Jul. 2008

Ye Z., Abouzeid A. & Ai J. (2009) Optimal stochastic policies for distributed data aggregation in wireless sensor networks (Oct. 2008). IEEE/ACM Trans. on Networking, Vol. 17, No. 5 pp. 1494 - 1507

Yu Y., Krishnamachari B. & Prasanna V. (2004). Energy-Latency Tradeoffs for Data Gathering in Wireless Sensor Networks. IEEE INFOCOM 2004. Twenty-third AnnualJoint Conference of the IEEE Computer and Communications Societies, HongKong, March 2004

Chapter 4

Calculation of an Optimum Mobile Sink Path in a Wireless Sensor Network

S. Chinnappen-Rimer and G. P. Hancke

Additional information is available at the end of the chapter

1. Introduction

A wireless sensor network (WSN) is a collection of spatially distributed, resource-constrained sensor nodes, deployed within an application area, to monitor a specific event or set of events. These sensor nodes are standalone devices without access to a continuous energy source and are located either within or close to the phenomena they are observing. The nodes communicate with one or more central control point(s), generally called a sink or base station. A typical sensor node comprises a sensing unit, a small processing unit to perform simple computations, a transceiver unit to connect nodes to the network and a power unit. Some nodes are also equipped with a location finding system [1]. A WSN application contains hundreds to thousands of sensor nodes. These sensor nodes are designed for unattended operation and are generally stationary after deployment.

One of the main criteria in designing a WSN application is prolonging network lifetime and preventing connectivity degradation through aggressive energy management. There is a trade-off between a node's energy, node range, size and cost. Due to the need to conserve battery lifetime, the sensor nodes operate with low duty cycles and communicate sporadically, over short distances with low data rates. In WSNs the flow of data is predominantly unidirectional, from nodes to sink [2]. The limited resources, non-renewable power supply and short radio propagation distances, (and hence large number required for deployment), of sensor nodes impose constraints on WSN applications not found in wired networks. A WSN differs from local area networks in the following key areas [3, 4]:

- Each sensor node communicates with one or more base stations (sinks). Traffic is mainly between individual sensor nodes and a base station.
- The network topology is a multi-hop star-tree that is either flat or hierarchical.
- They are used in diverse applications which may have different requirements for QoS and reliability.

- Most network applications require dense deployment and physical collocation of nodes.
- Individual sensor nodes have limited resources in terms of processing capability, memory and power.
- Power constraints result in small message sizes
- The placement of nodes in a WSN is application dependent and may not be pre-determined.

A WSN also differs from other wireless networks, such as cellular networks and mobile ad hoc networks (MANETS) because these networks are linked to a wired or renewable energy supply. In cellular networks and MANETS, the organising, routing and mobility management tasks focus on optimizing quality of service (QoS) and ensuring high bandwidth efficiency. There is a large amount of network traffic and the data rate is high to cater for the demand for multimedia rich data. These networks are designed to provide good throughput/delay characteristics under high mobility conditions [2]. Energy consumption is of secondary importance as the battery packs can be replaced or re-charged as needed.

As the term ``wireless'' implies, there is no fixed physical connection between sensors to provide continuous energy and an enclosed communication medium. This creates two problems, firstly, the sensor has a finite amount of energy, which once depleted, disables the sensor and hence reduces network lifetime. Secondly, all transmitted messages will be detected by any listening device within receiving range, which then has to decide whether to accept, forward or ignore the message. This signal transmission and reception has a power cost. In addition, many WSN applications do not have a pre-planned network topology and nodes are only aware of their immediate neighbours. When routing a message to a sink, the nodes closest to the sink receive a disproportionate amount of messages, resulting in their energy being consumed earlier.

Initial message routing protocols assumed the sink or destination node was in a fixed location, and that network nodes had no or limited knowledge of the network topology [5]. An area of active research for a number of years has been how to notify the central sink (or monitoring hub) about an event in real-time by utilising the minimum amount of power of sensor nodes. Strategies to improve node energy efficiency include using multiple sinks in the application area and the use of mobile sinks to collect data from stationary sensor nodes to prevent nodes close to a sink from having their energy depleted and hence decreasing network lifetime.

A model for optimum path movement of mobile sinks to reduce the number of messages transmitted and received by an individual sensor node is proposed. An investigation is conducted into the optimum route a mobile sink can travel that will reduce the number of messages transmitted within a network, allow equitable usage of all nodes to transfer an event message and still allow an event to be reported in real-time.

In the following sections a brief discussion of the use of mobile elements in WSNs as well as current research using mobile sinks and/or nodes to improve the energy efficiency of routing protocols is provided. The algorithm to transmit data from a sensor node to a mobile sink is discussed and the results analysed.

2. Mobile entities

The application and routing challenges presented by static nodes in a dense, multi-hop WSN has led to the investigation of the use of mobile elements in WSNs for data collection and/or dissemination. The advantages of using mobile entities in WSNs include [6, 7]:

1. Improved reliability as there is less contention and collisions within the wireless medium because data can now be collected directly through single or limited hop transmissions.
2. Reduced reliance on nodes located close to a static sink to route messages to the sink, resulting in increased energy efficiency and network lifetime.
3. Improved connectivity as mobile nodes can enable the retrieval of collected measurements from isolated regions of the sensor application area.
4. Sparse network architecture implies reduced application cost as fewer nodes are required and nodes can utilise mobile elements already present in the application area such as trains, cars, wildlife, and livestock etc.

The use of mobility in WSNs introduces complications not found in static WSN applications, such as detecting when nodes are within transmission range of a mobile sink, ensuring reliable data transfer as nodes may move as messages are exchanged, tracking sink location and design of a virtual backbone to store data reports so that the mobile sink can easily collect them, and managing sensor nodes to support sink mobility [6, 7].

Current strategies for data collection and dissemination using mobile elements include a rendezvous-based virtual infrastructure which uses limited and unlimited multi-hop relays to route data messages, or a backbone-based approach where mobile sinks only communicate with pre-defined cluster heads or gateways, or passive data collection where there is direct communication between the source and sink [7, 8].

The mobility patterns of mobile elements (sinks and relays) are dependent on the type of WSN application, its data collection requirements and the controllability of the mobile elements. Current mobility patterns can be classified into the following categories [6, 8]:

1. *Random mobility*: no network information required because communication does not occur regularly but with a distribution probability. This method does not provide optimal increases in network lifetime due to the need for continuous sink position updates and route reconstruction.
2. *Predictable or deterministic mobility*: mobile elements enter range of sensor nodes at regular, periodic times to collect data, and allow the sensor nodes to predict arrival of mobile entities.
3. *Controlled mobility*: the mobile elements movements are not predictable but are controlled by network parameters such as maximum and minimum residual energy of sensor nodes on a data route, event location, and the mobile elements trajectory and speed. In addition, the mobile entities can be instructed to visit individual nodes at specific times, and stop at nodes until they have collected all buffered data.

3. Related work

According to Akkaya, Younis and Bangad [9],, finding an optimal location for the sink in a multi-hop network is a complex problem, NP hard in nature. The complexity results mainly from two factors. The first factor is the potentially infinite possible positions that the gateway can be moved to. Second, for every interim solution considered during the search for an optimal location, a new multi-hop network topology needs to be established in order to qualify that interim solution in comparison to the current or previously picked location in the search. A mathematical formulation of the problem would involve a huge number of parameters including the positions of all deployed sensors, their state information such as energy level, transmission range, etc., and the sources of data in the networks.. The authors propose moving the sink to the top relay nodes location. The sink is assumed to know the geographical location of deployed sensors. In the solution proposed in this article, an optimum location is not sought but an optimum path for a mobile sink that will ensure equitable usage of all nodes to transport data messages to the mobile sink node.

Research undertaken by Somasundara et. al [10] shows that the energy consumption in a network using a mobile base station is significantly less than that of a static network. The authors propose moving the base station around the application area. When the base station is within range of sensor nodes, it collects event data. This is not an optimum real-time solution as the sensor nodes have to wait for the base station to arrive before transmitting event information but is feasible in delay-tolerant applications such as environmental monitoring. A key difference between this researcher's proposed ideas and the model presented here is that in the model presented here an optimum path within the application area, along which one or more mobile sinks travel is calculated.

Huang, Zhai and Fang [11] consider a wireless network where the sensors are mobile, (applications such as tracking free-ranging animals, both wild or farm livestock). The problem focused on in this paper is on improving the robustness of routing when there are path breakages in the communication channel due to node mobility. The suggested solution is the use of a cooperative, distributed routing protocol to combat path breakages. The writers assume that the intended path or route between the source and destination is already known and neighbouring nodes can be used if the communication channel on the intended path fails. Our primary research focus in this paper is the calculation of an optimum path for a mobile sink to reduce the number of messages required to be re-transmitted when sending a message to a sink in the WSN. However, we will have to take cognisance of possible path breakages that may occur during the development of optimal routes.

Vupputuri, Rachuri and Ram Murthy [12] use mobile data collectors to achieve energy efficient and reliable data communication. When an event occurs, sensor nodes inform the nearest data collector. The data collector aggregates the event information and with a specified reliability factor (R) informs the base station. The primary focus of the authors' investigation is determining a mobile strategy for the data collectors to ensure reliable and energy efficient event reporting. The mobility strategy does not consider how to optimise the changing locations of the data collectors. The authors focus on reducing the number of messages sent and received by nodes close to the base station to improve network lifetime as well as ensuring that multiple paths are used to improve network reliability.

Gu, Bozdag and Brewer [13] use a partitioning-based algorithm to schedule the movements of mobile sinks in order to reduce data loss due to buffer overflow while waiting for a sink to arrive. This aspect is ignored in our proposed solution. Other recent research activity in this field, include the work of Marta and Cardei [14] where mobile sinks change their location when the nearby sensors' energy becomes low, and determines the new location by searching for zones where sensors have more energy. Heinzelman, Chandrakasan and Balakrishnan [15] have proposed have proposed a combination hierarchical and cluster based scheme that groups sensors and appoints a cluster head to transmit messages to the sink, thus saving the surrounding nodes energy (LEACH). The small percentage of cluster heads are randomly re-selected to improve node longevity of nodes located close to cluster heads. Patel, Venkatesan and Chandrasekaran [16] propose a Lexicographic Maximum Lifetime Vector routing scheme to maximise the first, second and so forth set of nodes time until their battery energies are depleted.

The use of a mobile relay to route all traffic passing through a static node for a specified period of time, is discussed by Wang et. al. The mobile relay traverses a concentric circle that stays within a two-hop radius of the sink. The authors show that the use of a mobile relay can improve a WSN's lifetime by 130%. Additional experiments show that a mobile sink, moving around the perimeter of a large and dense network, can best optimise WSN lifetime compared to a mobile relay or using resource rich static relays located close to a static sink [17]. The results of this paper indicate that the mobile relay should be a maximum of two hops from a static sink and that only nodes within a maximum of 22 hops from the sink need to be aware of the location of the mobile relay. The use of both a mobile sink and a mobile relay prevent over-utilisation of static nodes located close to the sink to route messages to the sink and hence increase overall WSN lifetime. We do not consider the use of a mobile relay in the solution discussed in this chapter and focus exclusively on an optimum path for a mobile sink to follow within a WSN application area.

A multi-sink heuristic algorithm (HOP) is proposed by Ben Saad and Tourancheau to find the best way to move mobile sinks in order to improve the lifetime of large scale sensor networks. Sinks are relocated to nodes located the maximum number of hops from a sink as it is assumed that these node will have higher residual energy as the nodes will not be required to re-transmit messages destined for a sink [18]. The minimum amount of time a sink will spend at a specific location is 30 days. The proposed algorithm is compared against schemes using static sinks, sinks moving along the periphery of the network, sinks moving randomly and sinks moving according to an Integer Linear Programming algorithm, in terms of network lifetime and residual energy at each sensor node. The results of simulations indicate the HOP algorithm achieves significant improvement in network lifetime over the other algorithms and that there is more even distribution of residual energy per sensor node. The HOP algorithm differs from the solution proposed in this paper, because HOP assumes that the sinks are not continuously mobile but are moved after a specified number of days to different locations within the building.

4. Algorithm design

To reduce the number of messages received and re-transmitted by nodes closest to the sink, it is proposed that one or more mobile sinks follow a path in the application area based on the calculated number of hops from a sink. The path should (1) ensure reliable communication between nodes and sink(s), (2) ensure the even distribution of messages received and transmitted within the application area to reach a sink destination, and (3) enable real-time processing of event messages. Consider the following definitions in Table 1:

Variable	Description
X	Width of application area
Y	Length of application area
R	Node and mobile sink communication range
X_b	Minimum starting X point on the mobile path
X_e	Maximum ending X point on the mobile path
Y_b	Minimum starting Y point on the mobile path
Y_e	Maximum ending Y point on the mobile path
H	Number of hops to nearest node within communication range of the mobile sinks path.
d	Distance between each sink broadcast "hello" message.
N_{hello}	Number of times the "hello" message is broadcast to complete one loop around the calculated path
a	The constant acceleration of the mobile sink
d_{stop}	The constant deceleration of the mobile sink
v_i	Initial velocity of mobile sink
v_f	Final constant velocity of mobile sink
S_{av}	Distance the mobile sink has to traverse after accelerating from zero velocity to reach required velocity.
S_{dv}	Distance the sink has to traverse after decelerating from constant velocity to when the sink stops (zero velocity)
S_{cv}	Distance the sink has to traverse moving at constant velocity before next "hello" type message is broadcast
t_{av}	Time sink to accelerate from zero to constant velocity
t_{dv}	Time for sink to decelerate from constant velocity to zero
t_{cv}	Time for sink moving at constant velocity to traverse the required distance before next "hello" message is broadcast
T_{total}	Total time sink takes to complete one loop of its calculated path transmitting messages at required intervals
t_{stop}	Time a mobile sink will stop, broadcast a "hello" type message and wait for responses from surrounding nodes

Table 1. Definitions of variables used in calculations of mobile sink path

4.1. Calculation of optimum path for one mobile sink

For one sink, the optimum path must be equidistant from any furthest node in the application area. The maximum distance a message from a node on the perimeter of the application area travels before reaching a node within communication range of the mobile sink must be the same as the maximum distance from a node at the centre of the application area to a node within communication range of the mobile sink. If one sink is located in the centre of the application area, then for a square or rectangle shaped area, the number of hops a message from a node at the farthest end of a square or rectangular application area has to travel to reach the sink is approximately:

$$H_{sr\_num} = \left(\frac{X}{2*R}\right) + \left(\frac{Y}{2*R}\right). \tag{1}$$

For a circle shaped application area, nodes at the perimeter are distance r (where r is the radius) from the centre. Thus the number of hops for nodes on the perimeter is:

$$H_{circle\_num} = \frac{r}{R}. \tag{2}$$

To ensure equidistance between nodes at the centre of the application area and nodes at the perimeter of the application area, the number of hops should be almost the same, i.e. $\frac{H}{2}$.

Since the application area dimensions (X and Y for square and rectangular shapes or r for circular shapes), and the range of the nodes (R) is known, the maximum number of hops a message $\left(\frac{H}{2}\right)$ has to be re-transmitted before reaching a node that is within communication range of the mobile sink's path can be calculated as follows:

Square or rectangular shape

$$H_{sr} = \left(\frac{X}{4*R}\right) + \left(\frac{Y}{4*R}\right) \tag{3}$$

Circular shape

$$H_{circle} = \frac{r}{2*R} \tag{4}$$

$where, H_{(outer)} = low(H) = floor(H)$
$and, H_{(inner)} = high(H) = ceil(H)$

Once the number of hops has been calculated, the optimum path of a mobile sink can be calculated as shown below:

Calculation of optimum path:

Square or rectangular shape:

$$X_b = R * \left(\frac{H_{sr}}{2}\right) \tag{5}$$

$$X_e = X - \left(R * \left(\frac{H_{sr}}{2}\right)\right) \tag{6}$$

$$Y_b = R * \left(\frac{H_{sr}}{2}\right) \tag{7}$$

$$Y_e = Y - \left(R * \left(\frac{H_{sr}}{2}\right)\right) \tag{8}$$

Circular shape:

$$r_{path} = H_{circle} * R \tag{9}$$

Consider the nodes placed in a 300mx300m WSN as shown in Figure 1. The nodes' range is assumed to be 30m and thus each node is placed 30m from the previous node. The optimum path for the mobile sink calculated based on Equations [5], [6], [7] and [8]. Nodes within the immediate communication range of the mobile sink node act as temporary stores for any message destined for the sink. As the sink passes along the path, these nodes pass the message to the sink. This results in a short delay between the time an event occurs and the time the sink receives the message. If the sink needs to be notified immediately, the node can calculate where in the mobile path the sink currently is and re-route the message to the sink.

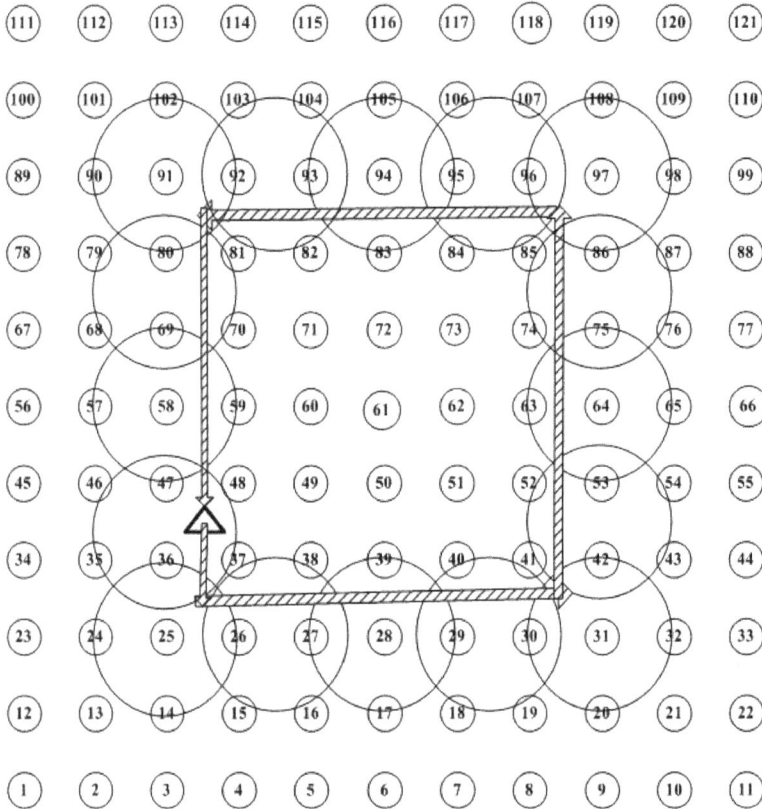

Figure 1. Path for a mobile node to follow in a 300mx300m application area.

4.2. Optimum path for multiple mobile sinks

For multiple sinks, the WSN application area should be sub-divided optimally. Thus the number of sinks and number of squares must be a square of a positive integer number, i.e. $N_{sink} = \{1^2, 2^2, 3^2, 4^2 \ldots\}$. The size of each square is calculated as follows:

$$x = \frac{X}{\sqrt{N_{sink}}} \; and \; y = \frac{Y}{\sqrt{N_{sink}}} \tag{10}$$

Using Equations [5], [6], [7], [8] and [10] the mobile path for each sink can be calculated. Figure 2 shows the number of sub-divisions and mobile paths calculated for the same 300mx300m WSN application area shown in Figure 1 for one mobile sink. The size of the application area is small, so for four square sub-divisions, each node in the WSN application area will be within communication range of the mobile path.

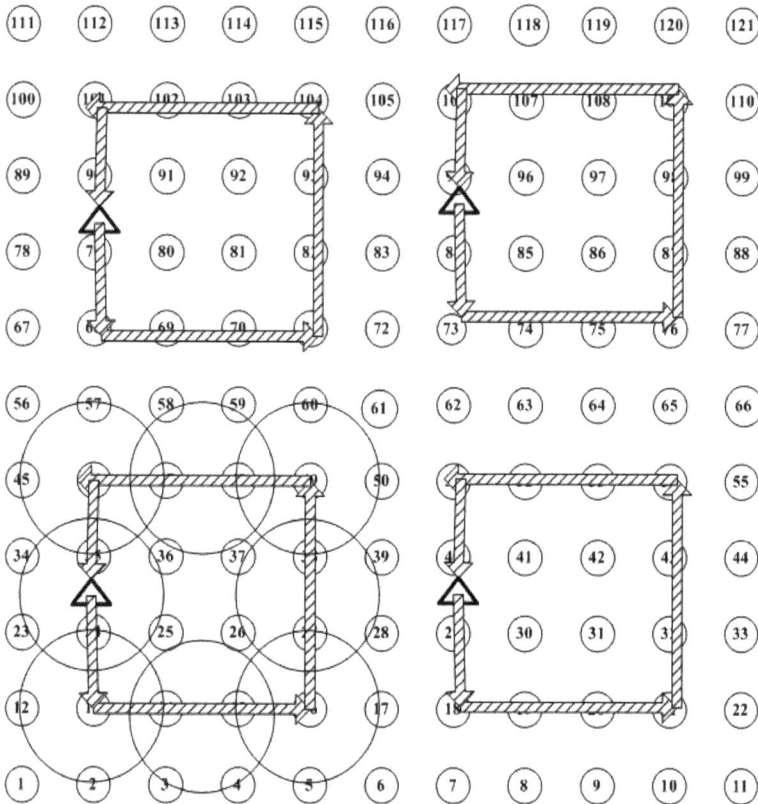

Figure 2. Four mobile sinks and each sinks path in a 300mx300m application area.

If there are multiple sinks, then the actual load is spread among more nodes. In Figure 2, the number of sinks and the optimum mobile path can be calculated to ensure that all nodes are within communication distance of a mobile sink's path, with the possible exception of nodes at the perimeter of the WSN application area. For example, nodes 1, 6, 11, 56, 61, 66, 111, 116 and 121 may require an intermediate node to pass the message on in Figure 2. To ensure connectivity, this set of nodes can be moved closer to the sink node's path, as shown in Figure 3. The path each sink has to travel is even shorter and hence the calculated time to complete one loop is less.

When an event occurs, the sensing nodes aggregate the data and elect a single node to forward the message to the sink. In Figure 2, as each node is one hop from the path of the mobile sink, the message will be stored by the elected node until the sink passes by and requests messages. In Figure 1, the message is stored by any node in direct communication range of the mobile sink as it moves along the path. Most nodes in the WSN application area of Figure 1 are two hops away from the path of the mobile sink. Nodes at each corner are at most three hops from the path of the mobile sink because it is assumed that the corner nodes are moved slightly into the application area as shown in Figure 3 to be within communication range of at least three nodes.

Only nodes which have a minimum of four immediate neighbours will re-transmit the event message. This ensures that nodes on the perimeter of the application area do not unnecessarily re-transmit the message. The event message is only re-broadcast until it is received by an intermediate node that is in direct communication range of the path of the mobile node. The message is stored and when the mobile node passes the intermediate node, all stored messages are transmitted to the mobile sink. Real-time event messages can be forwarded to nodes that will be closer to the sink's path based on the calculations described at the end of this section.

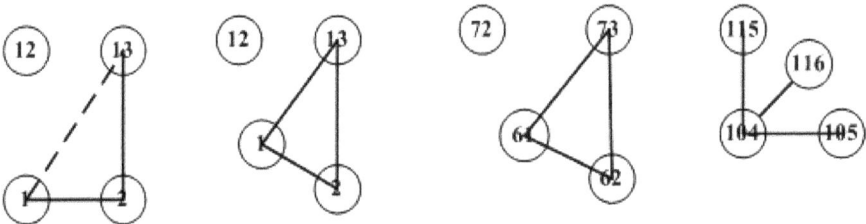

Figure 3. Moving corner nodes within communication range of mobile sink path

4.3. Calculation of distance between each "hello" broadcast message from mobile sink

To ensure complete coverage of all nodes neighbouring the path, the calculation of the distance between transmitting a "hello" broadcast message and waiting for responses from surrounding nodes is shown below:

$$d = R + \frac{R}{2}$$

$$d = \frac{3*R}{2} \tag{11}$$

The number of times the sink stops and broadcasts a "hello" type message is given by the following formula:

$$N_{hello} = \frac{2*(X_e - X_b) + 2*(Y_e - Y_b)}{d} \tag{12}$$

4.4. Time for a mobile sink to complete one loop around the path

The mobile sink first moves along the path and greets all nodes within communication range. The mobile sink transmits a "hello" greeting message at every d metres requesting any of the surrounding nodes to return any data messages they may have temporarily stored while waiting for the sink to return. The message contains the mobile sink's ID, velocity and acceleration, sink direction, its intended path, and when it calculates it will return to its current position as well as a list of all nodes that have responded to its greeting thus far. Initially, during the first loop of the mobile sink, the path list will be incomplete, as the sink is not yet aware of all nodes in its path range. When the mobile sink completes its first loop, it will have obtained a reasonably accurate network topology of all nodes within communication range of its path and their locations. The mobile node will re-broadcast this list as it continues to loop around its path, so that even if some nodes were asleep during previous cycles, these nodes can still obtain the list to update their records.

In the event that a real-time event message needs to be reported to the mobile sink, the initial node that is elected to receive the event message, as it is within communication range of the sink or the actual node that detected the event, can transmit the message to the sink, using this list and its knowledge of the mobile sink's velocity and intended path, to determine the optimum nodes to use to route the message to the sink. The messages transmitted between the nodes will travel faster than the mobile sink so the message will be delivered to the sink faster than waiting for the sink to pass by again.

4.5. Calculation of total time it takes a sink to complete one loop across the mobile path

4.5.1. Sink stop-start movement with non-uniform velocity

Initially the sink will have to move from a state of rest or initial velocity of zero to a constant, specified velocity. The time it takes a mobile sink to accelerate to a constant velocity can be calculated using the following equation:

$$a = \frac{v_f - v_i}{t_{av}}$$

The initial and final speed as well as the acceleration of the mobile sink is known. Thus, the time it takes the mobile sink to accelerate from zero and reach constant velocity is:

$$t_{av} = \frac{v_f - v_i}{a} \tag{13}$$

The distance the mobile sink has to traverse after accelerating from zero velocity to when the sink reaches the required constant velocity is:

$$s_{av} = \frac{1}{2} a t_{av}^2 \tag{14}$$

The sink will have to decelerate to stop before it broadcasts another "hello" type message. The time it takes a mobile sink to decelerate to a stop is similar to equation [13], i.e.

$$t_{dv} = \frac{v_{zero} - v_f}{d_{stop}} \tag{15}$$

The distance the mobile sink has to traverse after decelerating from constant velocity to zero velocity, i.e. to when the sink comes to a standstill, can be calculated as follows:

$$s_{dv} = \frac{1}{2} a t_{dv}^2 \tag{16}$$

Now, the time the mobile sink will spend at constant velocity can be calculated based on the distance between each time the sink node broadcasts a "hello" type message:

$$s_{cv} = d - (s_{av} + s_{dv}) = ut + \frac{1}{2} a t_{cv}^2$$

Since at constant velocity, a=0,

$$t_{cv} = \frac{s_{cv}}{v_f} \tag{17}$$

If t_{stop} is the time a mobile sink will stop, broadcast a "hello" type message and wait for responses from surrounding nodes, the total time for a node to complete one loop along the calculated path is given by the following formula:

$$T_{total} = N_{hello} * (t_{stop} + t_{av} + t_{cv} + t_{dv}) \tag{18}$$

Each node within communication range of the mobile sink's path must be able to perform the above calculations. When one of these nodes receives an event message that it has to re-transmit to the mobile sink, it can calculate the time delay before the sink will again pass by, based on the above equations. The node can then determine, based on the message status and urgency, whether to wait for the mobile sink to pass within communication range or whether to route the message to a node closer to the mobile sink. The sink path information contained in the previous "hello" message is used to determine which node to request to forward the message to the sink. As electromagnetic waves travel much faster than the mobile sink, this will ensure that the event message reaches the sink in real-time.

4.5.2. Sink movement with uniform velocity

The previous calculations are based on the mobile sink stopping before it broadcasts a "hello" type message. The stopping and re-starting by the mobile sink will increase the time it takes a mobile sink to complete a loop around the calculated mobile path. A variation on the above calculations is to assume that the mobile sink moves at constant velocity without stopping. When the mobile sink reaches a "hello" broadcast point it will transmit a "hello" type message to all nodes and continue moving at constant velocity. Because electromagnetic waves travel much faster than the mobile sink, the mobile sink should be able to send and receive all responses from surrounding nodes before it moves out of radio range. Then Equation [18] becomes:

$$T_{total} = \frac{2*(X_e - X_b) + 2*(Y_e - Y_b)}{v_f} \tag{19}$$

Of course the mobile node will have to decelerate when it approaches a corner to turn, but within the experimental simulation it is assumed that this time to turn is negligible.

5. Experimental simulation

The experimental setup used the Network Simulator (NS-2). In NS-2 the mobile nodes move at constant velocity. As this was a simulation environment, the mobile node did not require time to accelerate to a constant final velocity or to decelerate when turning a corner. Therefore, the time calculations are based on the node moving at constant velocity around a square path.

Changes were made to certain C++ programs in the NS-2.3.5 version to enable the node to move along the specified path and periodically send "hello" type messages. A Tcl script defined the parameters of the path the node travelled on and stored the event messages received by nodes along the mobile node's path. When a mobile node passed by a node with stored event messages, the node would pass these messages onto the mobile sink.

Experiments were run to determine the time it takes to complete one loop around the calculated path as shown in Figure 1. This time was verified with the calculated time, using Equation (15). Thereafter an event message was broadcast from a node on the perimeter of the application area, and the effect on surrounding nodes was analysed. The velocity of the mobile sink was set at 10 m/s. The range of the nodes was assumed to be 30m.

Using Equation [3], the number of hops was calculated to be:

$$H_{sr} = \left(\frac{300}{(4*30)}\right) + \left(\frac{300}{(4*30)}\right)$$

$$= \frac{20}{4}$$

$$= 5 \text{ hops}$$

Using Equations [5], [6], [7] and [8] the path can be calculated as follows:

$$Y_b = X_b = R * \left(\frac{H_{sr}}{2}\right)$$

(use the integer value, i.e. floor()) $$= 30 * \left(\frac{5}{2}\right)$$

$$= 60$$

$$Y_e = X_e = 300 - \left(R * \left(\frac{H_{ST}}{2}\right)\right)$$

(use the integer value, i.e floor()) $= 300 - \left(30 * \left(\frac{5}{2}\right)\right)$

$$= 240$$

The corner node (node 1 from Figure 1) is moved slightly into the application area (i.e. x = 10 and y = 10) to enable a message from node 1 to reach a node with at least 4 neighbours (in this case node 13). Figure 4 shows the experimental network topology for a 300 by 300 application area with a node range of approximately 30m.

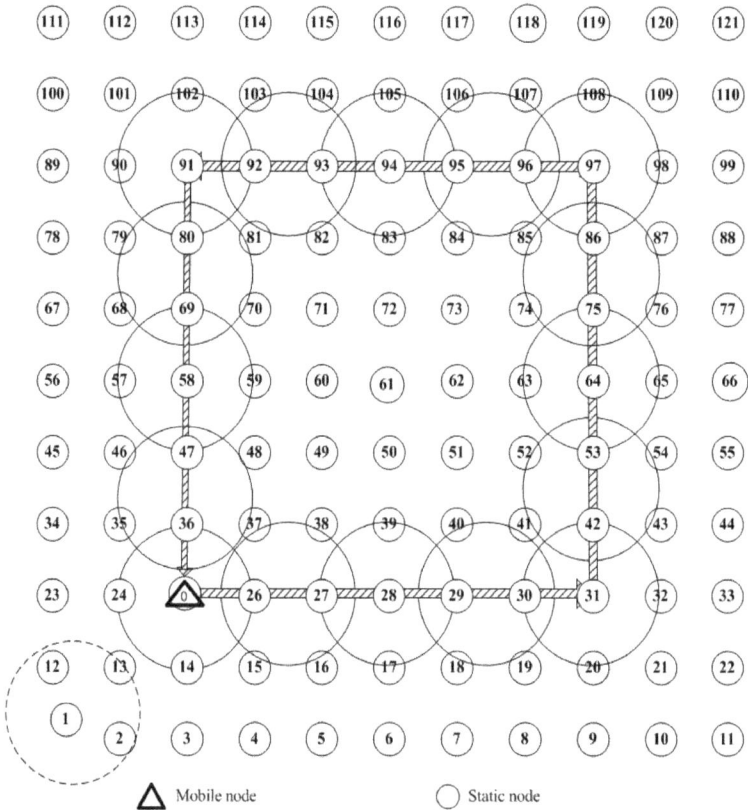

Figure 4. Experimental Setup with node 1 moved slightly into the application area.

6. Results and analysis

Using Equation [19] the time it will take a mobile node moving at a constant velocity of 10m/s to complete one loop along the calculated path is calculated.

$$T_{total} = \frac{2 * (240 - 60) + 2 * (240 - 60)}{10}$$

$$T_{total} = 72 \ seconds$$

The NS-2 Tcl script was run and the time taken for the mobile node to complete one loop around the calculated path as shown in Figure 1 is **72 seconds**.

An analysis of the number of messages received by nodes neighbouring the mobile node's path is shown in Figure 5. As can be seen, certain nodes receive more than one message. These nodes are on the perimeter of two intersecting "hello" type messages sent from the mobile sink as shown in Figure 1. Thus nodes 36, 37, 26, 27 etc. receive a "hello" message from the mobile node twice. To prevent this duplication of received messages from the mobile node, the researcher suggests that the neighbouring nodes go into sleep mode for a specified time period after receiving the first "hello" type message from the mobile node. This should ensure that all nodes neighbouring the mobile node's path only receive one message per complete loop of the circuit.

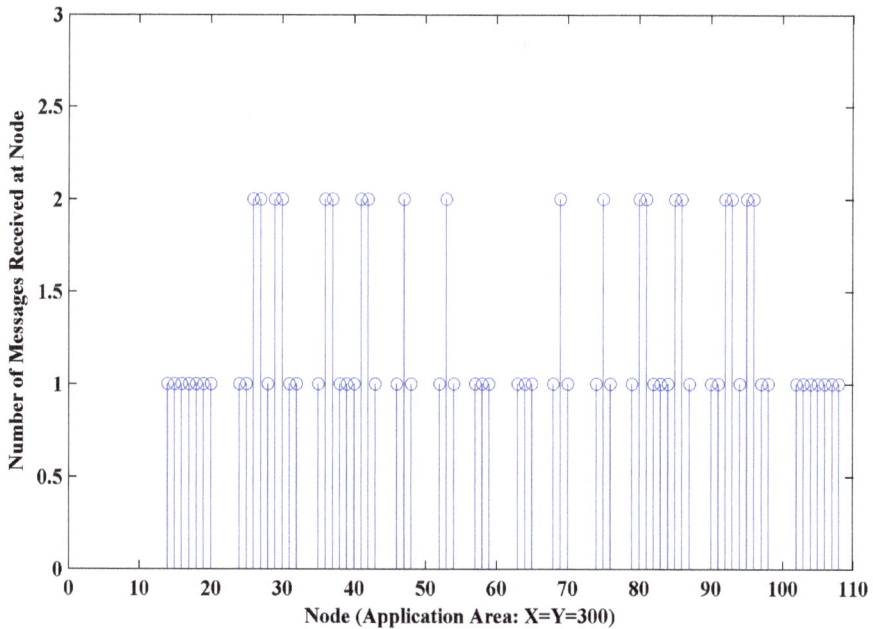

Figure 5. Number of messages received by nodes neighbouring mobile node's path

To conserve energy further, the number of times a mobile node will circumvent the path can be application-specific. For example, if sensor nodes are required to send updates to a sink periodically, the mobile node can traverse the path only during this time period. However, if the application requires the mobile node to monitor the area continuously for events and respond in real-time, the mobile sink has to move along the path and send "hello" type messages continuously.

However, the continuous sending of "hello" type messages at periodic intervals by the mobile node, does incur a cost. To reduce the number of messages transmitted within the application area further, (depending on the type of WSN application); a "hello" type message can be sent once at initialisation when the mobile sink first completes a loop along the path. All nodes along the perimeter will be able to calculate when the sink will pass by again and ensure that the node is awake during that time, if the node has event messages to relay. At the calculated time the perimeter node can proactively send a message to the mobile node informing it that it will begin transmitting event messages.

To determine the effect of the mobile node on reducing the total number of messages within the application area and received per individual node a series of tests were run for a message destined to the mobile node 0 (Figure 4) for both flooding and the mobile algorithms with various hop counts.

The effect of sending a message from a node on the perimeter of the application area to one of the nodes on the perimeter of the mobile node's path is analysed. For example, a message is sent from node *1* to nodes *14* and *24* (refer to Figure 4). As can be seen in Figure 6, only those nodes used to pass the event message receive more messages than those shown in Figure 5.

To further reduce the total number of messages sent and received when reporting an event message, the mobile node algorithm only allows nodes with four or more neighbours to re-broadcast the received event message. When a node that is within range of a mobile node receives a message (in this experiment, node 14 and node 24), it stores the message for collection by the mobile node 0. For flooding all nodes re-broadcast the message. The event message was sent at 3 seconds and the TCL script was set to end at 10 seconds, i.e., before the mobile node completes one full cycle around its predetermined path. The experiment was to compare flooding and using the mobile route algorithm in terms of the time to reach the destination and the total number of messages transmitted within the network as well as the number of messages received per node. Thus, the worst case scenario of assuming that the mobile node has just left the range of node 24 and node 14 and started on the path and will only return in approximately the time it takes to complete one full circumnavigation of the specified path is considered.

Figure 7 shows the number of messages received by node 14 and node 24 for the mobile algorithm and flooding. Node 14 and node 24 receive the same number of messages (2) for the mobile algorithm. In flooding the number of messages per node varies widely but node 14 tracks node 24 in terms of the number of messages received per hop count. In the mobile algorithm, node 14 and node 24 receive **two** messages, one from node 13 and one from the mobile node 0. As can be seen even for low hop counts, the number of messages when using flooding exceeds the number of messages for the mobile algorithm.

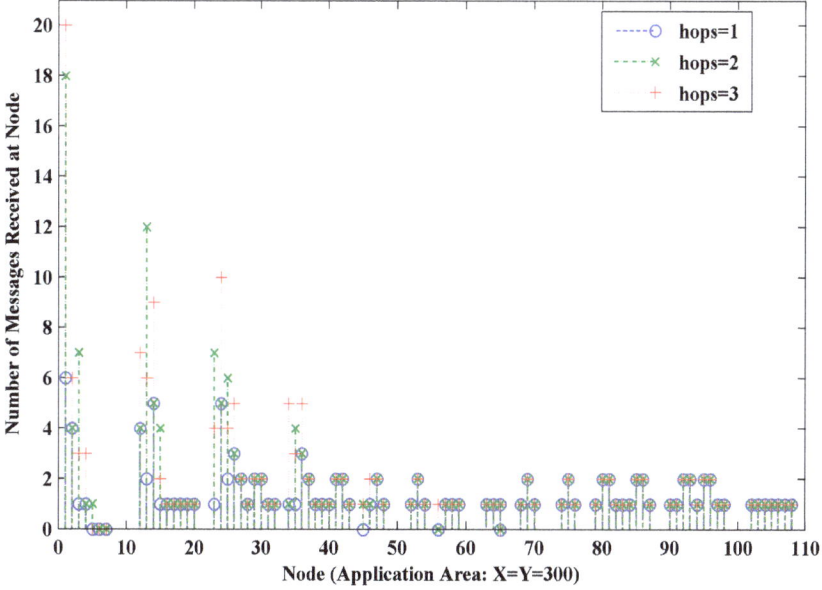

Figure 6. Number of messages per node when event message sent to node on mobile nodes perimeter

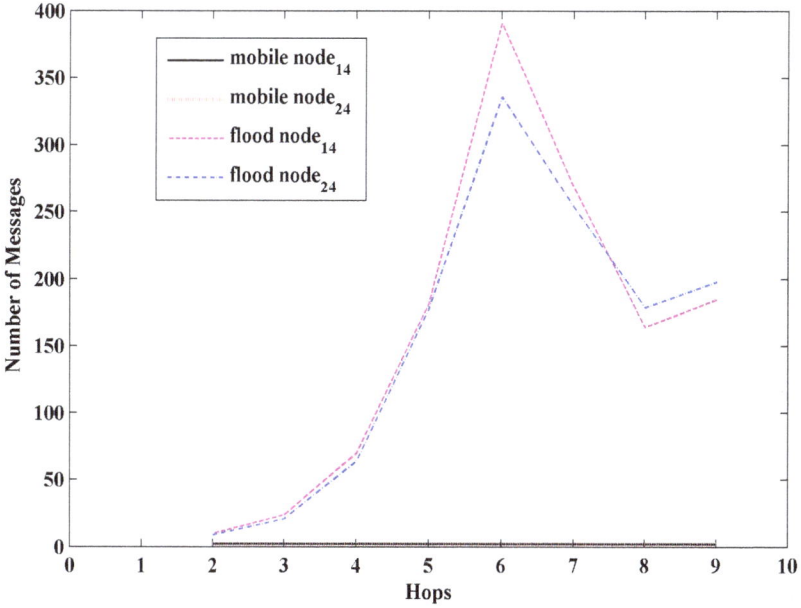

Figure 7. Number of messages for node 14 and node 24 for varying hop count

Figure 8 shows the number of messages per individual node when only nodes whose neighbours are equal to or greater than 4 re-transmit an event message. When compared to the number of messages received per node in Figure 6, it is obvious that by restricting which nodes re-transmit an event message, there is a significant decrease in the number of messages received or re-transmitted amongst individual nodes. Once the event message reaches a node that can convey the event message directly to the mobile node (in this case node 14 and node 24), the algorithm stops retransmitting the event message.

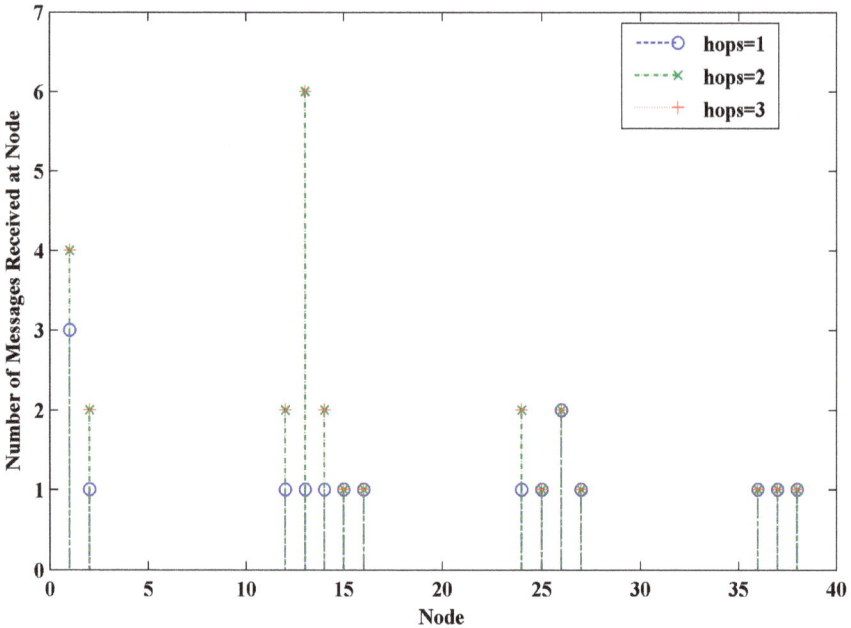

Figure 8. Restricting re-transmission of event messages to nodes with 4 or more neighbours

In Figure 6, the TCL script is run for 72 seconds whereas in Figure 8 the TCL script is only run for 10 seconds. Thus, not all the nodes along the perimeter of the mobile nodes path are shown receiving messages from the mobile sink in Figure 8. The number of messages for node 13 and node 24 are reduced by half when the number of neighbour nodes restriction rule applies.

Figure 9 shows the number of messages per individual node if a typical flooding message is sent from node 1 to a static node 0 located at the same X,Y coordinates as node 25. Flooding a message to the sink is the worst case scenario as more nodes receive (and possibly have to re-transmit) messages which depletes an individual node's limited energy and reduces node and network lifetime.

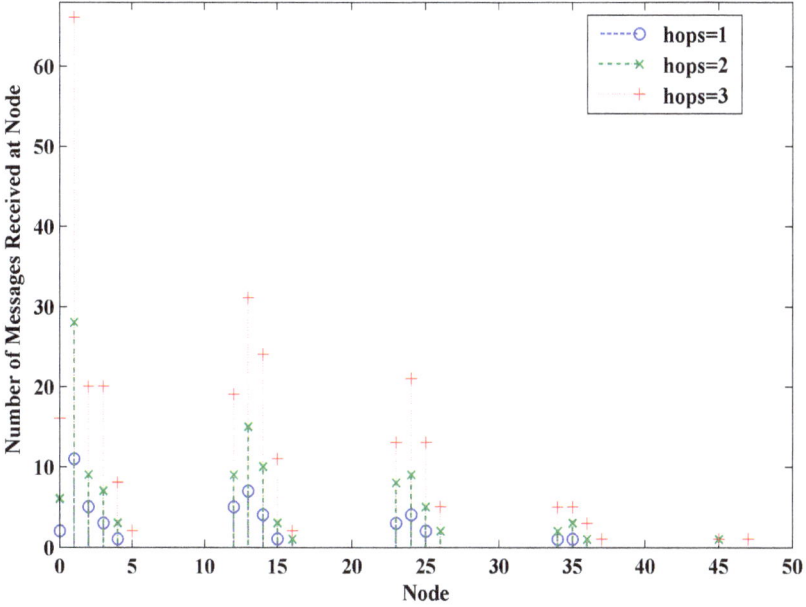

Figure 9. Number of messages per node for a flooding message sent to a static node 0 located at node 25.

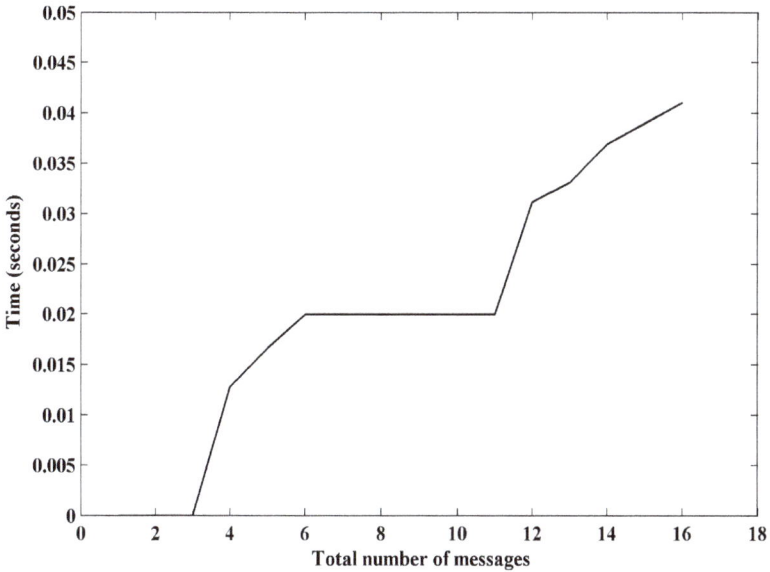

Figure 10. Time it takes for an event message sent from node 1 to reach node 14 or node 24

The time it takes for an event message to reach node 14 and node 24 using the mobile algorithm with re-transmissions of event messages restricted to nodes with four or more neighbours is shown in Figure 10. If the mobile node has just passed out of range of node 24 and node 14, then these nodes must wait for approximately 72 seconds before the mobile node is within range again. Alternatively, depending on the real-time requirement of the application, the perimeter nodes can retransmit the event message as electromagnetic waves travel faster than the mobile sink and the message should reach the sink in less than 72 seconds.

7. Conclusion

An optimum path for a mobile sink is calculated so that the number of hops that the message has to be re-transmitted is small. Because all neighbouring nodes can pass an event message to the sink, no specific set of nodes is overloaded with the task of routing event messages to the sink. This ensures more equitable usage of all sensor nodes in the network and hence increased node lifetime.

It has been shown that the number of messages received per node can be reduced by using a specific path for the mobile node/sink to move along. All neighbouring nodes can store messages when an event occurs, and if the sensor detecting the event is not an immediate neighbouring node along the path of the mobile sink, the number of hops that the message has to be re-propagated is small. By restricting the nodes that re-transmit a message to nodes with four or more neighbours, the number of messages received per individual node is further reduced.

Thus the use of a mobile sink moving along a calculated path around the application area can significantly reduce the number of messages received per individual node and hence increase node lifetime.

Author details

S. Chinnappen-Rimer
Department of Electrical and Electronic Engineering Science,
University of Johannesburg, Johannesburg, South Africa,

G. P. Hancke
Information Security Group, Royal Holloway, University of London, Egham,
United Kingdom and Department of Electrical, Electronic and Computer Engineering,
University of Pretoria, Pretoria, South Africa

8. References

[1] I. Akyildiz, W. Su, Y. Sankarasubramaniam and E. Cayirci, "Wireless sensor networks: A survey," *Computer Networks Journal*, vol. 38, no. 4, pp. 393-422, 2002.

[2] P. Rentala, R. Musunuri, S. Gandham and U. Saxena, "Survey of Sensor Networks.," 2002.

[3] C. Wang, K. Sohraby, B. Li, M. Daneshmand and Y. Hu, "A survey of transport protocols for wireless sensor networks," *IEEE Network*, vol. 20, no. 3, pp. 34-40, 2006.

[4] M. Katz and S. Shamai, "Transmitting to colocated users in wireless ad hoc and sensor networks," *IEEE Transactions on Information Theory*, vol. 51, no. 10, pp. 3540-3563, 2005.

[5] K. Akkaya and M. Younis, "A survey on routing protocols for wireless sensor networks," *Elsevier Journal of Ad Hoc Networks*, vol. 3, no. 3, pp. 325-349, 2003.

[6] M. D. Francesco, S. K. Das and G. Anastasi, "Data Collection in Wireless Sensor Networks with Mobile Elements: A Survey," *ACM Transactions on Sensor Networks*, vol. 8, no. 1, pp. 7:1-7:31, 2011.

[7] E. B. Hamida and G. Chelius, "Strategies for data dissemination to mobile sinks in wireless sensor networks," *IEEE Wireless Communications* , vol. 15, no. 6, pp. 31--37, 2008.

[8] Y. Faheem, S. Boudjit and K. Chen, "Data dissemination strategies in mobile sink Wireless Sensor Networks: A survey," in *Proceedings of the 2nd IFIP conference on Wireless days (WD'09)*, 2009.

[9] K. Akkaya, M. Younis and M. Bangad, "Sink repositioning for enhanced performance in wireless sensor networks," *Computer Networks*, vol. 49, pp. 512-534, 2005.

[10] A. Somasundara, A. Kansal, D. Jea and D. Estrin, "Controllably mobile infrastructure for low energy embedded networks," *IEEE Transactions on Mobile Computing*, vol. 5, no. 8, pp. 958-973, Aug. 2006.

[11] X. Huang, H. Zhai and Y. Fang, "Robust Cooperative Routing Protocol in Mobile Wireless Sensor Networks," *IEEE Transactions on Wireless Communications*, vol. 7, no. 12, pp. 5278- 5285, Dec. 2008.

[12] S. Vupputuri, K. Rachuri and C. S. Ram Murthy, "Using mobile data collectors to mprove network lifetime of wireless sensor networks with reliability constraints," *Journal of Parallel and Distributed Computing*, vol. 70, no. 7, pp. 767-778, July 2010.

[13] Y. Gu, D. Bozdag, R. Brewer and E. Ekici, "Data harvesting with mobile elements in wireless sensor networks," *Computer Networks*, vol. 50, no. 17, p. 3449–3465, 2006.

[14] M. Marta and M. Cardei, "Using sink mobility to increase wireless sensor networks lifetime," in *Proceedings of the 9th IEEE International Symposium on a World of Wireless, Mobile and Multimedia Networks (WoWMoM'08)*, 23-26 June 2008.

[15] W. Heinzelman, A. Chandrakasan and H. Balakrishnan, "Energy-efficient communication protocol for wireless sensor networks," in *In Proceedings of the Hawaii International Conference on System Sciences*, Hawaii, 2000.

[16] M. Patel, S. Venkatesan and R. Chandrasekara, "A Network-flow based Integral Optimal Algorithm for Lexicographic Maximum Lifetime Routing in Wireless Sensor Networks," in *Proceedings of 16th International Conference on Computer Communications and Networks, (ICCCN 2007)*, 2007 .

[17] W. Wang, V. Srinivasan and K.-C. Chua, "Using mobile relays to prolong the lifetime of wireless sensor networks.," in *Proceedings of the 11th annual international conference on Mobile computing and networking (MobiCom '05)*, Cologne, Germany, 2005.

[18] L. Ben Saad and B. Tourancheau, "Towards an Efficient Positioning of Mobile Sinks in Wireless Sensor Networks inside Buildings," in *3rd International Conference on New Technologies, Mobility and Security (NTMS)*, 2009.

Chapter 5

Spatial Communication Activity in Wireless Sensor Networks Based on Migrated Base Stations

Jan Nikodem, Marek Woda and Maciej Nikodem

Additional information is available at the end of the chapter

1. Introduction

There are many techniques used to conserve WSN energy, in order to prevent its premature dead. Longer distance transmission, involving a number of relaying nodes, increases energy consumption very fast. It is striven to receive a messages from nodes located as close as possible to a Base Station (BS). The nodes are deployed and we have no possibility to change its location. In order to achieve energy saver effect, more rational seems to be having mobile BS, especially that in real life there is usually only one. Typically, in WSN there are a lot of sources of messages. BS should be moved to the location where messages are flow evenly from all directions. If this condition is met, it prevents unnecessary BS movements in other directions. Furthermore, such BS location reduces consumption of energy spending for communication but, as a drawback, it reduces the WSN lifespan.

So, as it was assumed that in order to obtain the longer WSN lifespan, Base Station (BS) position can't be fixed, and it needs to be mobile. Having BS fixed to one position one agrees for quick nodes' energy depletion, since the messages routed along the same paths will drain energy to zero quite fast and render these nodes not operational, which ultimately would lead to network death.

There are advantages and disadvantages of moving BS closer to the origin of messages sent. The closer to the source of messages BS is, than less consumption of energy spending for communication in WSN is. However, if we move BS to close to a potential threat (e.g. source of fire, in case we monitor fire hazard in some area), this vital WSN element may be too exposed and ultimately damaged or even completely destroyed (which would render entire WSN no longer operational). Therefore special attention shall be brought to the idea how far BS shall be moved.

Another issue to consider is; how often BS should change its position? To minimize BS movement once the intensity of messages is neither changing rapidly nor area of these changes migrating too far, what would the threshold (or any other factor) that has influence on decision that BS won't be moved.

Since it is the common knowledge that migrating BS could help extend WSN lifespan, it is just a question; how this migration should be organized. There are several aspects to be investigated, among others: whether BS should change its position every time a message is received or not (if not how often should it be?), how far BS should move from its previous (original position), how the BS movement affects behavior of all nodes and the BS neighborhood. Another crucial aspect is how to notify the nodes from BS neighborhood that will soon become out of communication range with BS, when it is moving away from these nodes.

BS movement just a fracture of relay radio link range, seems to be energetically unreasonable, since just this kind of movement involves new distribution of nodes calculation and new relays designation, that consumes a lot of valuable energy resources.

2. Related works

There are a huge number of papers considered communication activity in WSN, related mainly to clustering and routing problems. On the one hand, scientists have discussed sensors' self-configuring [1], self-management [3, 1, 19], adaptive clustering [1, 9, 22] or concept of adjustable autonomy [5]. On the other hand, there are papers which discuss bio-inspired ideas and tend to extract some aspects of the natural world for computer emulation [6].

The WSN communication structure is crucial for BS migration. Authors [4] have shown that the communication topology of some biological, social and technological networks is neither completely regular nor completely random but stays somehow in between these two extreme cases. It is worth to mention papers [22, 19, 3] devoted to self-organizing protocols using both random and deterministic elements.

In order to effectively manage communication activities, one has to address the problems of sensor network organization and the subsequent reorganization and maintenance [22].

3. Communication, measurements and neighborhoods in WSN

Communication is one of two (along with measurements made by the nodes) primary forms of WSN network activity. As far measurements are made by the network nodes and can be carried out locally, completely independently, then a communication is a typical collective action in which, besides the transmitter and the receiver, relay nodes actively participate.

The active role of relay results from the limited range of radio communication. Awareness of energy preservation considerations causes that this communication range is much shorter than existing in WSN distances between nodes and BS. Then, in order to make sure that information (a packet) arrives to a destination (BS) from a source of information (a WSN node), an implementation of routing packet based on relays is requisite.

In order to describe mentioned above WSN activities, let us introduce concepts of actions and behavior. *Action* should be considered as the property of each network element such as: a sensor, a Base Station, a cluster head or a regular node. The *behavior*, on the other hand is an external attribute which can be considered either as an outcome of actions performed by the whole WSN or its subset (i.e. cluster, routing tree, group of nodes, neighborhood).

Action (Act) is a ternary relation which can be defined as follows:

$$Act: Nodes \times States \rightarrow States \tag{1}$$

Therefore, actions that can be taken by nodes of WSN can be represented as a Cartesian product over the sets of nodes (*Nodes*) and their possible states (*States*). Finally, new states are a result of every action taken.

Actions are executed individually by a single node of the network (e.g. measuring the environmental parameter) but some of them require that two or more neighboring nodes cooperate with each other to perform a particular action (e.g. during the message transmission, receiver interact with transmitter). Actions are taken depending on the actual state of the node (different actions will be taken during the network organization or normal operation phase) and lead to new state of the node. Actions may also change the state of the neighboring nodes (e.g. dual actions transmit - receive).

Since nodes are autonomous, each one can execute actions independently of others. Undoubtedly, this is an advantage since WSN as a whole can simultaneously execute a plenty of different tasks. On the other hand, some actions gain in importance only when two or more nodes cooperate with each other taking dual or related actions. For such actions nodes perform their actions in cooperation which means that these actions are related to each other. In such a case we say that actions are related. Routing in WSN is a good example of such related actions.

Let, *R* denotes, routing. We can construct the quotient set called *Behavior*, consist of elements which are called equivalence classes linked to the relation *R* and here denoted as:

$$Beh: Act/R = \{ act_x \in Act \mid act_x \, R \, x \} \tag{2}$$

So, routing activity is a *behavior* which draws on relations and describes dependencies between actions that are taken by nodes situated on a routing path. In other words, relations refer to actions that depend on each other and are taken together but not necessarily simultaneously – this is the relational way of thinking about the network activity. Detailed explanation of these concepts can be found in [12, 13, 16].

Concerning WSN structure, vicinity *V(k)* of a node *k* describes all what is placed in the radio link range of *k* node. This vicinity consists of various different components that belong to the WSN infrastructure and the other indirect elements that do not belong to WSN, although they play an important role in the behavior of the network. The set of objects from the first group can be called neighborhood *N(k)*, and a collection made of objects from the second group is defined as environment *E(k)*. The relationship between these three terms can be expressed as:

$$\mathcal{V}(k) = \mathcal{N}(k) \cup \mathcal{E}(k) \tag{3}$$

Coming back to mentioned above two crucial WSN activities, communication is a behavior which takes place within a neighborhood while measurements are actions related to environment. Further we will be working on communication aspects within WSN, so now let us come closer to this issue and begin from $Map(X; Y)$ expression that can be defined as a collection of mappings of set X onto set Y (surjection). Next, $Sub(X)$ is defined as a family of all X subsets and neighborhood $\mathcal{N}$ as a mapping

$$\mathcal{N} \in \big\{Map\big(Nodes; Sub(Nodes)\big)\big\} \tag{4}$$

Thus, $\mathcal{N}(k)$ denotes the the neighborhood of node k while, $\mathcal{N}(S)$ is the neighborhood of set of nodes S defined as:

$$\mathcal{N}(k)_{|k \in Nodes} = \{y \in Nodes|\ y\, \mathcal{R}_{\,\mathcal{N}}\, k\} \tag{5}$$

$$\mathcal{N}(S)_{|S \subset Nodes} = \{y \in Nodes|\ (\exists k \in S)(y\, \mathcal{R}_{\,\mathcal{N}}\, k)\} \tag{6}$$

where $y\, \mathcal{R}_{\,\mathcal{N}}\, k$ means that nodes y and k are in relation `to be neighbors'.

4. Spatial routing and routing chains in WSN

Getting back to the main WSN task, which is the monitoring of selected physical parameters of the given area, let's have look at how it is implemented. A packet containing measurement results is formed in the node that has made this measurement. The sources of packets are all nodes in WSN. We assume a regular frequency of measurements, forming packets and continuous uniform distribution of nodes within WSN area with probability density function

$$f(x) = \begin{cases} \dfrac{1}{Max_x}, & 0 \le x \le Max_x \\ 0, & 0 > x > Max_x \end{cases} \tag{7}$$

for both X and Y axes. Thus, we consider WSN as a collection of strongly homogenous elements (nodes). During WSN activity we do not affect either the place or the time of new packet creation.

Then a packet is transmitted to a base station via a routing path. Realization of this communication phase is based on the set of nodes cooperation that relay a packet. Short radio link communication range precludes (for many nodes) sending packet directly to the BS. Only a certain number of nodes can do this because only these are located within communication proximity (neighborhood) of the base station.

This node's communication phase with the base station has been described repeatedly in the literature [7, 20, 15]. Different criteria for assessing the effectiveness of the retransmission realization are being used. There are many different algorithms for packet routing. Some of these methods (proactive) determine the optimal routing path and exploit them as long as possible. Next, an algorithm strives for finding a new, an optimal path in new patch structure which yet again is exploited until an energy is depleted, etc. The other algorithms (proactive) determine the routing path, each time when it is needed. Transmission is then carried out closer to the current optimal routing path.

At this stage, we propose the following method of spatial route planning which is characterized by two important features:

- this method defines the area in which routing can be performed, while the existing methods were determined by a path,
- this method realizes inducing cooperation and at each node k on routing path, gives the choice of next subsequent relay $k+1$.

Using the spatial routing, nodes in the space S (Fig.1) can model the routing path collectively realizing inducing cooperation. These features give us a greater flexibility in modeling communication behaviors. Moreover, making a decision collectively (within the neighborhood) increases adaptability to a varied environmental conditions. Note that, if each relay node (in Fig.1) has only 5 choices, so on the way from s to BS made up of six relays we have $5^6 = 15625$ choices. It is an impressive number but we must remember that the routing path $(s, t_1, t_2, \ldots, t_6, BS)$ makes a chain, whose lifespan is determined by the weakest link.

In our case it is t_6 relay node. Why? Because in a sequence of relay nodes so many (all) choices is being created by $s, t_1, t_2, \ldots, t_5$ nodes. The last relay node t_6, as situated in the vicinity of BS has no choice. Since one possibility is not a choice. Selection starts with two or more possibilities. Consequently, t_6 has to send a packet to BS. A multitude of choices, and thus an ability to spread energy consumption on a certain subset of WSN nodes is not t_6 node merit. Moreover, this node represents a base station neighbors $\aleph(BS)$. Thus, whatever the route is, and how many choices for routing $(s, \ldots, BS)$ we have, each chosen route must end with one of these relay nodes which are neighbors of BS. What does this mean? We can spread an energy consumption for a routing path, forcing the nodes lying on its realization to work, but in the final retransmission phase, all packets, converge in the vicinity of the BS. So, we can offload nodes on a route, but we cannot relieve traffic going across nodes adjacent to the BS, because nothing can replace them.

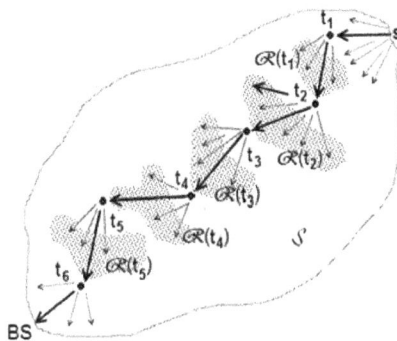

Figure 1. The map of choices during spatial routing from s to BS

Hence the idea, if we cannot distribute loads of the nodes from BS neighborhood and this results in depletion of energy resources, thereby shortening WSN lifespan, it should ensure a periodic exchange of BS neighbors on other nodes, which have so far not been exploited so intensively or simply have more energy. Such an exchange can take place in two ways, or we will shift nodes in WSN area, or location of BS will be subjected to shift. We prefer the second solution, as more practical in implementation. An octocopter - a flying autonomous agile aerial machine will be used for BS transportation.

5. The importance of base stations in terms of WSN maximum lifespan

The base station data acquisition absorbs the large amount of the network nodes energy resources. The largest losses occur in the nodes that are within the BS neighborhood. This is because they carry out the main burden of retransmission. A routing paths can (indeed be differently routed) by changing the relaying nodes, but the penultimate node of the path must always be one of the nodes within neighborhood $\mathcal{N}(BS)$.

The maximum life time of the network, expressed in number of packets, it might send to the BS, is:

$$LS^1(WSN) = \left[\sum_{k=1}^{l} LS(k) ;\ k \in \mathcal{N}(BS)\right] = LS(BS) \tag{8}$$

where $LS(k)$ is k node lifespan, and $l = Card\big(\mathcal{N}(BS)\big)$ is cardinality of $\mathcal{N}(BS)$ set.

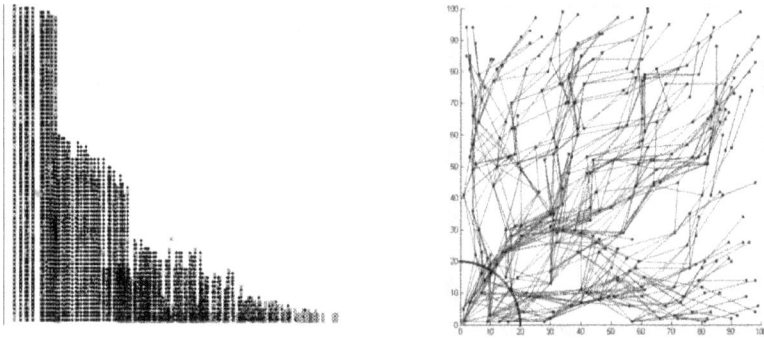

Figure 2. Single BS routing simulation – energy consumption of the WSN nodes

If all nodes in BS neighborhood $\mathcal{N}(BS)$ will lose their energy, WSN loses its consistency. This is a significant loss of consistency, which leads to BS isolation, and thus the loss of basic network function, which is to gather information from the specified area.

In order to extend a maximum lifespan of the network we can introduce more BS. Following the isolation of another BS, we lose contact with parts of the network served by this station, but the other part WSN still works. The maximum lifetime of the network (while maintaining its consistency) is obtained when there is a mutual isolation of the neighborhoods of all base stations. Then we use optimally the resources of all BS neighbors.

$$LS^b(WSN) = \sum_{j=1}^{b} LS(BS_j) = \sum_{j=1}^{b} \left[\sum_{k=1}^{l_j} LS(k); \ k \in \mathcal{N}(BS_j) \right] \qquad (9)$$

The number of BS neighbors also depends on the radio link range and node deployment density, but we have no impact on these parameters in the process of maximization of WSN lifespan. The longest lifespan can be achieved when we will make sure that the total number of neighbors of all BS was as large as possible. It should therefore deploy BS according to the following condition:

$$Card\left(\bigcup_{j=1}^{b} \mathcal{N}(BS_j) \right) \rightarrow \max \quad iff \ (\forall k, l \in \{1,2, \dots, b\}; k \neq l) \ (\mathcal{N}(BS_k) \cap \mathcal{N}(BS_l) = \emptyset) \quad (10)$$

The above condition (10) states that all partition of a set of WSN nodes (mutually exclusive and collectively exhaustive neighborhoods $\mathcal{N}(BS_j)$) provides maximal number of all BS neighbors.

Deployment of multiple BS in WSN area, optimal in the sense of network lifetime is a complex mixed optimization problem (known as mixed integer programming), even for a homogeneous network in terms of nodes and energy distribution, as well as density distribution of the messages occurrences. The relatively easier task is to define drainage areas, if we have established BS positions. The most frequently approach used in that case is the partition of the network into a clusters, in which BS serves a cluster head role. There are plenty of such partitions, but we have not met in the literature an algorithm that would guarantee that the partition into clusters meets the required optimality criteria. In addition to these drawbacks, the most important disadvantage is, that the real WSN networks are not giving up the theoretical assumptions (7). Nodes, even if they are homogeneous in terms of hardware, are randomly distributed and their distribution changes (nodes are dying during the WSN operation). Messages in the network are uniformly generated only during monitoring of the non-emergency situations (e.g. no fire in a monitored area). The presence of special circumstances significantly interferes with this distribution. A fire on some area of the network generates much more messages than when nothing special (unusual) happens. Hence also the diversification of energy consumption increases in this time. We need a smarter algorithm than one that finds an optimal multidimensional solution (several hundred to several thousand nodes) of the mixed programming with rare practical assumptions. Such an algorithm should take into account the dynamics of changes in the network and be run repeatedly, whenever there are changes in vital network parameters. As if that were not enough the algorithm should run in distributed mode, adjusting the solution to the local conditions. It should run adaptively in intensive monitoring area (areas under fire), and differently in areas of relatively stable monitored parameters.

6. WSN with one base station

6.1. The static base station issue

In order to comprehend the variety of interactions, a multitude of cases that may happen in WSN, let us begin our discussion from a case with one base station in WSN. What BS location will guarantee the longest lifetime of the network? According to (10) the best location should ensure the following condition:

$$Card(\mathcal{N}(BS)) \rightarrow max \qquad (11)$$

Assuming regular frequency of measurements, forming packets and continuous uniform distribution of nodes within WSN area with probability density function (7) (uniform distribution of homogenous nodes and messages) any, but not outlying location meets (11). Outlying (not fringe) location, means such a location for which BS radio link range falls within the ambit of the WSN area (**a**, **b**, **c** in Fig. 3.) Locations depicted as **e**, **d** (Fig.3) are inferior to the previous because for these locations the number of neighbors BS $(Card(\mathcal{N}(BS)))$ is lower.

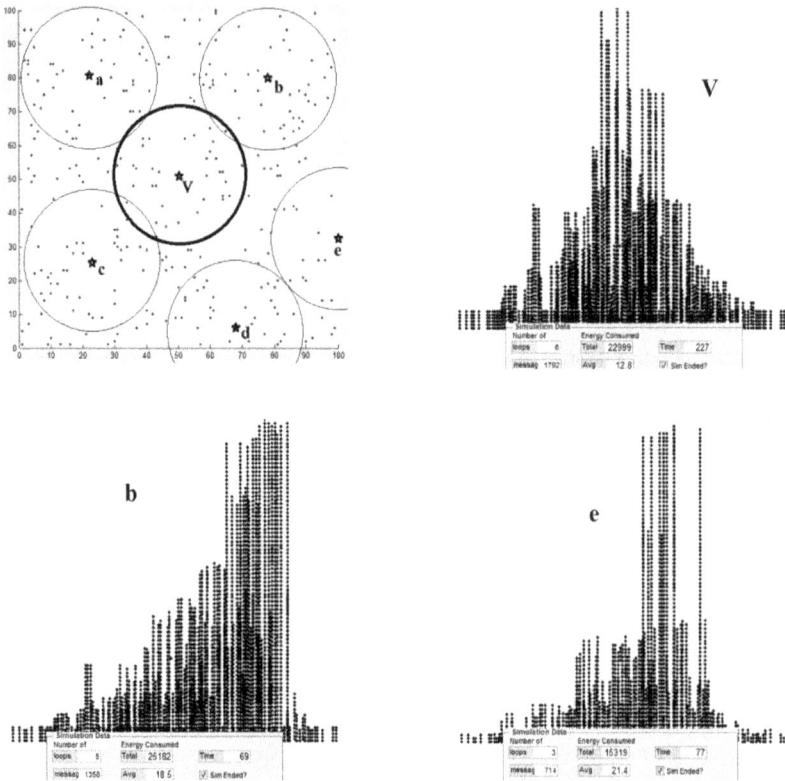

Figure 3. The base station locations and WSN lifespan simulation with energy consumption

We have a plenty of such sufficient locations in WSN as shown in Fig. 3 (for clarity there are marked only 3). In order to conform to the load uniformity postulate for each of BS neighbor, the center of area (**V** spot in Fig. 3) where network operates is the best place to locate BS. The obtained results show that **V** spot of is the best both in terms of the mean energy consumption spent for sending a single packet (only 12.8 energy units per packet), as well as in terms of WSN life expectancy (*1792* packets). In spot **b**, due to the greater distance between the subset of nodes (with coordinates *(x, y)* less than *(50, 50)*), the mean energy consumption for sending a single packet increases to *18.5* units. The network lifespan in this case is shorter (*1358* packets sent) because neighbors with coordinates above *(b, b)* were less intensively utilized for retransmission. So, these burdens were shifted on remaining neighbors, which resulted in faster BS isolation from the rest of the network, although some of its neighbors (those located above *(b, b)*) had left energy reserves. In **e** case, the situation was clearly the worst in terms of both an energy consumption and WSN lifespan. BS was using resources just only a half of neighbors that greatly shortened network lifespan, and much greater transmission distances increased mean energy consumption.

6.2. The migrated base station issue

BS placement in the **V** spot assures the longest WSN lifespan of all other possible static locations. But, whether a BS that migrates could not to assure longer WSN lifetime that being static (located all the time at the **V** spot)? Let us consider another BS position as a "new" base station in WSN, so analyzing (9), each nonzero element of the sum

$$LS^b(WSN) = \sum_{j=1}^{b} LS(BS_j),$$ (12)

increases the lifespan of the WSN.

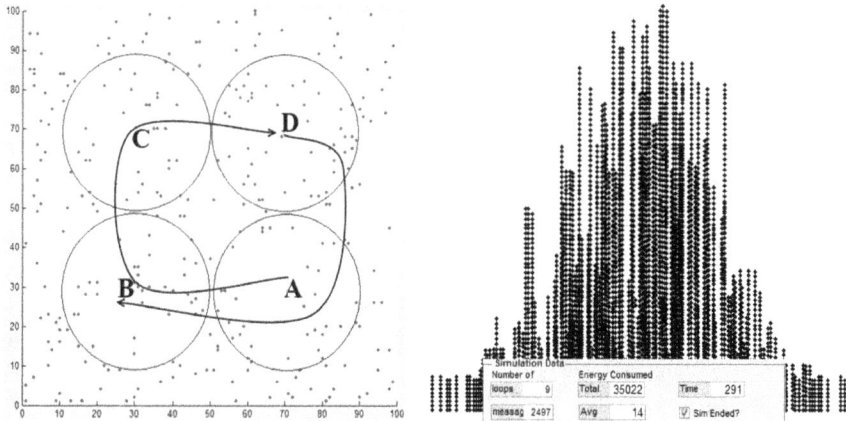

Figure 4. The base station migration and WSN lifespan simulation with energy consumption

The optimal deployment of *b* base stations is determined by formula (10) so, we assumed, in simulation, that the BS will travel in a way that its neighbors' sets in successive positions were disjunctive. As the number of nodes in the network was *N = 300*, so after receiving consecutive *300* packets, BS changed its position, moving clockwise (as shown in Fig.4). The new BS position was determined, so that a new set of BS neighbors did not have conjoint elements with all previous neighborhoods. After receiving *4x300 = 1200* packets, such a cycle was repeated until the energy of one of the nodes was drained out completely. The results are far better than those obtained when the BS was located in the best possible static position (V) and its location was fixed. Periodically migrated BS provides a larger number of neighbors increased the network lifespan but this issue reduce energetic efficiency (an average energy consumption per packet has increased noticeably). That was an obvious trade-off.

The number of neighbors (on average four times), we expected a commensurate increase in WSN lifespan. As a result we obtain a prolongation of WSN lifespan, but unfortunately it was not even doubled. Where we have lost so much potential energy resources (12)? Well, there are three reasons for this; firstly we do not know whether other BS migration path would not give better results. Secondly, the migration of the base station does not take into account changes in the WSN topology. Subsequent BS positions were determined before the WSN nodes start to be active. After another round, taking into consideration these nodes, which energy was almost drained, the new BS positions should always take this into account. Thirdly, a migrating base station is not equal to four ones still remaining in their initial locations. Each of these static BS supports only a part of the WSN and thus realizes communication more efficiently. One migrant BS serves the entire WSN and thus being in the **A** spot (see in Fig. 4) must receive packets sent from the vicinity of the nodes located in the **C** spot. So, we really know that BS at each position is working not optimally, generating such a significant loss of energy resources. Only a large number of neighbors make the total balance of such activity positive. In the case depicted in Fig. 4 a lot of energy is being simply wasted, hence far from the best, but yet better than previously had been achieved.

7. The adaptive migration of a base station

A static assignment of BS location takes into consideration anticipated (and what is more important static) sensors activity. It is a common knowledge that situation in WSN changes, some areas are more active some even dormant – it is very infrequent unlikely situation that entire WSN area is active. The routing activity entails substantial energy consumption and changes network communication conditions. A new situation requires changes and these involve BS location change – as per analogy to military tactical charts, no one will start re-positioning troops on a map from a scratch (deployment of a new map) but using runny movements of existing available units. Similarly with Base Station – smooth transition from one dynamic event to another entails migration of BS to follow resultant changes. WSN adaptation involves migration of the BS towards "hot" area, whereas the remaining region is covered cursorily.

Typically, at early life of WSN - its energy is distributed evenly across entire its area. Gradually with time, this changes. There are some nodes with no energy and WSN operation becomes problematic. Since the dynamic allocation of energy within the network is not (directly) possible, we propose the migration of BS that can greatly influence on energy distribution and consumption across the nodes. Since adaptive migration is a result of smart interaction between BS and its vicinity, now we consider how to determine a migration vector in the BS vicinity. In order to do so, a number of messages received by each node within BS neighborhood must be known. Having these numbers, for each node $n \in \mathcal{N}(BS)$ we calculate node's load quotient within BS neighborhood as

$$L_n(BS) = \frac{M_n}{\sum_{\mathcal{N}(BS)}^{i} M_i} \tag{13}$$

where: M_n is a number of messages received by n-th node,

$\sum_{\mathcal{N}(BS)}^{i} M_i$ - is a total number of all messages received by nodes within $\mathcal{N}(BS)$ neighborhood.

Once the load quotients of nodes are calculated, we take into account only few of BS neighbor nodes and treat $L_j(BS)$ values, as magnitude of vectors significant in determination of BS migration vector. Then using simple vectors addition of these significant $\left(\overrightarrow{L_j(BS)} ; \ j \in \mathbb{N} \right)$ vectors as components (Fig. 5), we shape BS migration vector $\overline{w}$ as follows:

$$\overline{w} = \sum_{\mathcal{N}^*(BS)}^{j} \overrightarrow{L_j(BS)} \tag{14}$$

where $\mathcal{N}^*(BS) \subset \mathcal{N}(BS)$ is a significant neighborhood of BS.

Now, in order to move BS we need to decide, how long this movement should be. It is being decided by α value, a movement distance factor that shapes BS movement distance from its original position.

$$k < \alpha \cdot \overline{w} < Range \tag{15}$$

where $Range$ is a BS radio link range parameter,

k is lower bounds parameter for BS movement distance.

The formula (15) provides some kind of neighborhood $\mathcal{N}(BS)$ continuity during the BS migration.

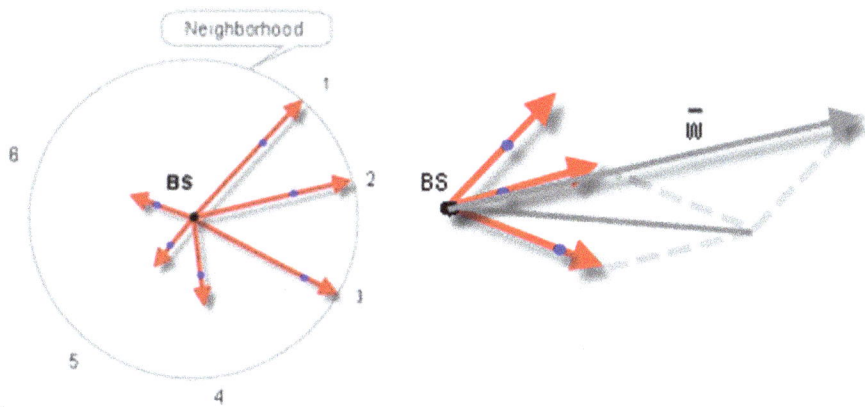

Figure 5. Shaping the BS migration vector

However there should be one more observation detected. We select only some of all calculated node's load quotients. One may wonder what and why such a criterion choice is? In our simulations, we were able to choose, building a BS migration vector $\overline{w}$, all the neighbors since we knew their locations. So there was no difficulty in defining direction and sense of all vectors. Selection of neighbors taking part in the shaping of the vector (14) allows for smart elimination of unwanted nodes in this process. For example, if a node transmits the parameter *temperature in my environment*, and this temperature is too high (potentially harmful), then BS should not migrate in that direction. Often in the real world, only some nodes locations are known to BS, then it is apparent to include only those nodes in the equation (14).

BS migration shall continue until such location is reached, in which a balanced number of messages reaches BS from all directions in its vicinity. Such a case, in the real world situation may never occur, so in order to stop redundant movements, to prevent further energy drain, we introduce indifference constant k (refer to (15)) that decides if any additional movement shall be done or not. If the left part of condition (15) is not fulfilled, BS remains on its previous position.

8. Accompanying issues

8.1. Hop zones distributions

All Wireless Sensor Networks has a circular shape of their close neighbor's communication range. If a Base Station is located in communication proximity of nodes, these nodes are in BS neighborhood, and if messages sent from a node cannot reach directly BS and this action requires relaying nodes - a distance from one node to another closer to BS is called hop, so a message having more than 1 relaying node on route to BS needs to travel through 2 hops. Hop zones gather nodes with the same communication distance to a BS. WSN rules enforcing messages send out to a node located directly within next hop. However, some nodes could potentially have more energy (or located on hop border) to breach the one hop limit and sent message out to another node in next hop. Communication between hop zones is primarily dependent on a node that initiates communication towards sink, because it shapes, by its communication range, entire traffic, deciding to which node in next hop a message is sent. This action is crucial, since a node starts a chain reaction in relaying node. A route will only change once energy of any node on this communication path is drained. Let's assume that we have capability to influence a node where and how far it sends a message. By this feature we may manipulate that messages are being sent as far as communication range extends which ultimately may lead to reduction of regular hops number on a long path. Whether this pays off, yet again, it results from WSN layout and participating nodes. As experiments shows [10] having variable number of hops (based on nodes arrangement) and feasibility to influence message distance send out, a noticeable amount of energy may be preserved.

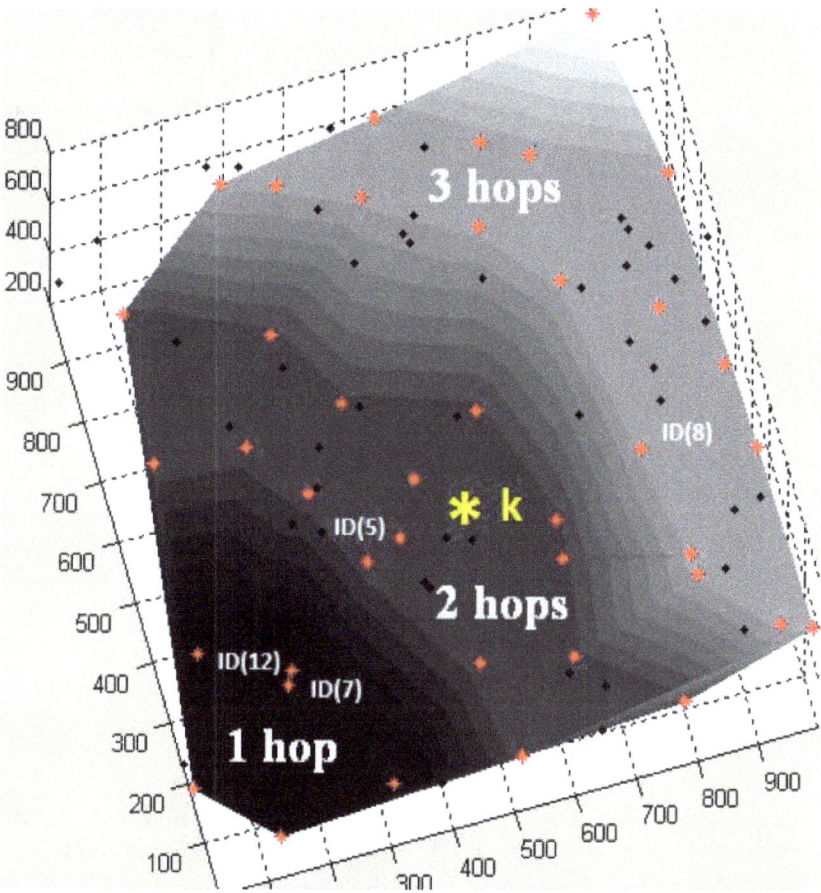

Figure 6. Nodes and hops distribution in a WSN

8.2. Location of nodes in WSN

Another vital aspect in WSN is mutual location of the nodes. Only BS can know its precise location among the other nodes (hence only BS is able to plot its migration vector). Not all WS node have capability to precisely define their position in space. Only a few of nodes have optional (built-in or add-in) GPS device, such nodes with GPS are exceptionally useful for remaining nodes without this capability, they can act as beacons – providing other nodes with their position at the same time enabling to set position on their own. Having three beacons in communicational range gives a node an opportunity to precisely set its position. After this operation such a node becomes an anchor which unfortunately cannot be used for any further nodes location designation. Node's (anchor) location could be determined either by Ad hoc On-Demand Distance Vector (AODV) Routing developed in Nokia Research Center, University of California, Santa Barbara and University of Cincinnati [17] or Angle of Arrival using Received Signal Strength [20, 21]. Node's position can be also relatively precisely determined based on loss of transmitted signal strength. When there are just two beacons and one anchor nodes location designation is still possible, but less precise than before. The least accurate location designation applies to a situation where there are just three anchors in node's proximity. In each case where anchor is used to location designation, it is being recognized as a classifier rather than a positioning element.

A constellation term, known from astronomy as a group of stars involved in a specific area of the celestial sphere that is shown in the relative position on the Earth night sky, could be also applied in WSN field, per analogy celestial bodies (nodes), sky (WSN), Earth (BS). Therefore constellation in WSN can also show only relative position, towards selected, known points. The advantage of constellation is that one need to know neither precise distance between nodes nor having any single beacon in a communicational proximity. One needs only select a reference point and then can calculate distance to it. Main disadvantage of this method is that it accumulates errors, the greater the determined distance the bigger inaccuracy. However these cumulative errors can be partially mitigated by introducing at least one beacon.

8.3. Flooding (new hop zones determination)

Another interesting, from energy saving standpoint, issue is how important (if at all) is sending information about zone number change (a hop zone, where a node is currently located). Normally, in a conventional WSN, when position of BS changes, zones and their numbering have to be designated again. Question is, whether we can avoid energetic cost on broadcasting info about all new zones? Let's concentrate on a situation, where we are focusing on a node located 15 hops from BS, does it make any sense to loose energy on broadcasting information that we are currently on 16th hop from BS, or maybe previous state of knowledge is good enough, in fact nothing important hasn't changed since the node still knows where to relay messages (set of nodes from preceding hop is still known). If this approach is taken *a priori* as reasonable solution, further investigation on where information about new zone is vital and where may be omitted. Shall a very simple rule (only nodes that directly can get s message from BS about its position change) or more sophisticated (per analogy to water wave flooding, where only nodes in a few first hop zones are being notified and then "notification wave" gradually fades) be applied. Having in mind these considerations only information about motion vector shall be propagated each time BS moves, and only every defined time slice forming new zones (only if required) and its new numbering (always) is being done. Described above scenario may resemble an attentive reader - MPEG compression algorithm, where in order to increase compression ratio (energy preserved), frame data is being deliberately lost (not each time full info about WSN state sent out), motion vectors are used, and I frames may be considered as information that is always being propagated across WSN and B / P frames is just partially sent data.

9. Conclusions

Two methods of WSN energy preservation (theoretically and simulation proven) are directed and spatial communication. Both in many cases may save energy expense of a certain nodes and ultimately the whole network, however infrequently these contribute to WSN lifespan prolongation. Introducing a migrated BS into a static WSN environment may bring further energy savings for whole network and assure that network lifespan will be much longer that in a static one. There is no need to underline how important WSN lifespan issue is; there is a common knowledge the longer the weakest WSN link (a node with the least energy left) is active the longer WSN lifespan is. Usually having this knowledge in mind, algorithms made attempt to distribute messages load (packets sent from source to a sink) evenly across neighboring nodes to prevent premature death of this node. Nonetheless it could remediate such situations, preventing BS isolation outside (the communication range) of other nodes, but in certain conditions (e.g. where BS was located in a corner of WSN area with only a few neighbors – Fig. 2) were ineffective. Such a case could be improved having a mechanism allowing migration of base station(s). BS migration could be fixed (based on defined criteria) or adaptive. Both are having their advantages and drawbacks. However their disadvantages may be neglected if the only criterion is WSN prolongation.

When the base station migrates in a WSN, every time it musts ensure that it does not lost contact with network nodes. Also, in order to ensure the best survival of relay nodes (which actively support it work), it should stay as close as possible to the most active regions of the network. Constantly changing its location the base station also changes its neighbors. This implies that the base station (BS) should match its velocity with currently neighboring nodes to keep abreast. Any collective behavior is solely based on observable phenomena within neighborhood. In the tested issue nodes within BS neighborhood helps it to calculate the velocity vector (14). However, change the location of BS must occur in accordance with the relay activity of the WSN nodes. The integration of these behaviors results in a stable BS location, where most active regions of the network are at least some minimum distance from BS. In terms of energy savings this is an optimal solution, however, because of the intense burdening neighborhood nodes is advantageous to periodically change the BS location.

During our experiments in both simulated and first real life environments we found some intriguing and at the same time interesting issues that for now are worth to be mentioned (irregular and not quite circularly shaped hop zones, BS location determination in WSN area, new hop zones determination after BS movement) and in a near future will a the subject of our research.

Author details

Jan Nikodem, Marek Woda and Maciej Nikodem
Institute of Computer Science,
Automatic Control, and Robotics, Wroclaw University of Technology, Wroclaw, Poland

Acknowledgement

This work was partially supported by E.U. Regional Development Fund and by Polish Government within the framework of the Operational Programme - Innovative Economy 2007-2013. Contract POIG.01.03.01-02-002/08-00, Detectors and sensors for measuring factors hazardous to environment - modeling and monitoring of threats.

10. References

[1] Cerpa A., Estrin D. ASCENT: Adaptive Self-Configuring Sensor Networks Topologies,IEEE Transactions On Mobile Computing, vol. 3, no. 3, Jul-Sep 2004

[2] Chaczko Z., Ahmad F.: Wireless Sensor Network Based System for Fire Endangered Areas, ICITA 2005, Sydney, 2005

[3] Chevallay C., Van Dyck R. E., Hall T. A.: Self-organization Protocols for Wireless Sensor Networks. In Thirty Sixth Conference on Information Sciences and Systems, March 2002

[4] Cohn, A.G., Bennett B., Gooday J.M., Gotts N.M., 1997. Representing and Reasoning with Qalitative Spatial Relations about Regions. In: Cohn, A.G., Bennett B., Gooday J.M., Gotts N.M, eds. Spatial and Temporal Reasoning, Dordrecht, Kulwer, 97-134

[5] Crandall J.W., Goodrich M.A.: Experiments in adjustable autonomy, in IEEE International Conference on Systems, Man, and Cybernetics, vol.3, Tucson, USA, 2001, 1624-1629

[6] Dressler F., Efficient and Scalable Communication in Autonomous Networking using Bio-inspired Mechanisms – An Overview, Informatica 29 (2005), pp. 183–188.

[7] Hunn N., Essentials of Short-Range Wireless, Cambridge University Press 2010, ISBN-978-0-521-76069-0

[8] Jaroń J.: Systemic Prolegomena to Theoretical Cybernetics, Scient. Papers of Inst. of Techn. Cybernetics, Wroclaw Techn. Univ., no. 45, Wroclaw, 1978

[9] Lin Ch.R., Gerla M.: Adaptive Clustering for Mobile Wireless Networks, IEEE Journal On Selected Areas In Communications, vol. 15, no. 7, Sep 1997

[10] Nikodem, J., 2008. Autonomy and Cooperation as Factors of Dependability in Wireless Sensor Network, Proceedings of the Conference in Dependability of Computer Systems, DepCoS - RELCOMEX 2008, 406-413. June 2008, Szklarska Poreba, Poland

[11] Nikodem J.,Klempous R., Chaczko Z., Modelling of immune functions in a wireless sensors network.W: The 20th European Modeling and Simulation Symposium. EMSS 2008, Campora S. Giovanni, Italy, 2008

[12] Nikodem J.,Nikodem M., Woda M., Klempous R., Chaczko Z.: Relation-based message routing in wireless sensor networks. in Smart wireless sensor networks / ed. by Hoang Duc Chinh, Yen Kheng Tan. Rijeka : InTech, 2010. s. 127-145

[13] Nikodem J., Klempous R., Nikodem M., Woda M., Chaczko Z.: Wireless sensors network simulator for modeling relation based communication activity. Fifth International Conference on Broadband and Biomedical Communications, IB2Com 2010, Malaga, Spain, 15-17 December 2010. [Piscataway, NJ] : IEEE, cop. 2010. [6] s.

[14] Mekkaoui K., Rahmoun A.: Short-hops vs. Long-hops - Energy efficiency analysis in Wireless Sensor Networks. Proceedings of the Third International Conference on Computer Science and its Applications (CIIA'11)

[15] Mills K. L, A brief survey of self-organization in wireless sensor networks, Wirel. Commun. Mob. Comput. 2007; 7, pp. 1–12.

[16] Nikodem J.: Relational approach towards feasibility performance for routing algorithms in wireless sensor network. Proceedings of International Conference on Dependability of Computer Systems, DepCoS - RELCOMEX 2009, Brunów, Poland, 30 June - 02 July, 2009 / eds Wojciech Zamojski. Los Alamito: IEEE Computer Society [Press], cop. 2009. s. 176-183.

[17] Perkins, C.; Belding-Royer, E.; Das, S. (July 2003). Ad hoc On-Demand Distance Vector (AODV) Routing. IETF. RFC 3561.

[18] Scerri P., Pynadath D., Tambe M.: Towards Adjustable Autonomy for the Real %World, Journal of Artificial Intelligence Research, vol.17, 2003

[19] Sohrabi K., Gao J., Ailawadhi V., Pottie G.J.: Protocols for Self-Organization of a Wireless Sensor Network, IEEE Personal Communications, Oct 2000

[20] Vollset E., Ezhilchelvan P., A survey of reliable broadcast protocols for mobile ad-hoc networks. Technical report, University of Newcastle, 2003

[21] Winfree S.: Angle of Arrival Estimation using Received Signal Strength with Directional Antennas, The Ohio State University. Department of Electrical and Computer Engineering Honors Theses; 2007

[22] Veyseh M., Wei B., Mir N.F.: An Information Management Protocol to Control Routing and Clustering in Sensor Networks, Journal of Computing and Information Technology - CIT 13 (1) 2005, 53-68

[23] Younis O., Fahmy S.: HEED: A Hybrid, Energy-Efficient, Distributed Clustering Approach for Ad Hoc Sensor Networks, IEEE Transactions On Mobile Computing, vol. 3, no. 4, Oct-Dec 2004

Chapter 6

Assessing the Vulnerabilities of Mission-Critical Wireless Sensor Networks

Mohamed M. A. Azim * and Aly M. Al-Semary †

Additional information is available at the end of the chapter

1. Introduction

Wireless Sensor networks (WSNs) have recently received increased attractiveness driven by many mission-critical applications such as battlefield reconnaissance and homeland security monitoring. Mission critical here refers to networking for application domains whose infrastructure and operations are absolutely necessary for an organization to carry out its mission [1]. Therefore, the main feature that must be guaranteed by all networks running mission critical applications is the network continuity.

However, due to the nature of the deployment field, these networks are vulnerable to natural disasters such as earthquakes, tornadoes or floods. Moreove, they are also subject to physical attacks such as an Electro-Magnetic Pulse (EMP) attack and security breaches such as sinkhole and selective forwarding attacks [2]. Such real world events may happen in particular geographical areas and disrupt specific parts of the network. Therefore, the geographical layout of the network topology determines the impact of such events on the network's connectivity.

Several contributions in the literature have addressed the failure modeling and survivability problems. The authors in [3, 4] tackled the single link failure problem in the logical topology. The authors in [5] focus on the dual link failure assumption. Most of these studies are based on a common assumption that failures are independent of their locations and randomly distributed across the network, which fails to reflect several real scenarios. Such real-world events have geographical nature, and therefore, the geographical structure of the network affects the impact of these events. Under such region failure scenarios, several network components within a geographically correlated region may be simultaneously destroyed,

* Author's permanent address: Faculty of Industrial Education, Beni-Suef University, Egypt
† Author's permanent address: Systems and Comp. Eng. Dept., Faculty of Eng., Al-Azhar University, Cairo, Egypt. This research is supported by the deanship of scientific research, Taibah University under Grant 717/431.

resulting in network holes, cuts (partitions) or even breakdown of the overall network connectivity as shown in Fig. 1. Therefore, it is essential to assess the vulnerabilities of mission critical networks to such region-failures.

Some research has been conducted to understand the impact of region failures on wired backbone networks such as [11–18]. On the other hand, the cut detection problem has been investigated by [19] and [20].

Recently, few studies have tackled the region-failure problem in wireless networks. The authors in [7–9] investigated the region- based connectivity issue in wireless networks and demonstrated the effect of the transmitting power on maintaining a region-based connectivity in the presence of single and multiple region failures. The authors in [10] proposed a more general Probabilistic Region Failure (PRF) Model to capture the key features of geographically correlated region failures. They also developed a framework to apply the PRF model for the reliability assessment of wireless mesh networks.

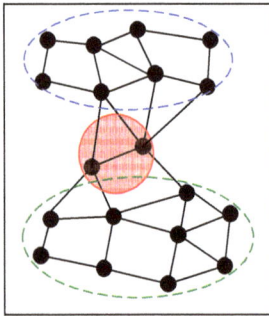

(a) Partitions due to single region-failure

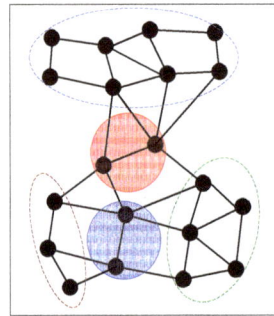

(b) Partitions due to dual region-failure

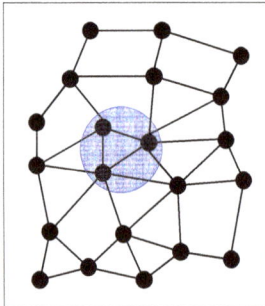

(c) A hole due to a single region-failure

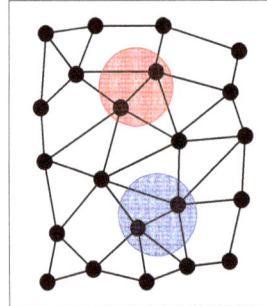

(d) Two holes due to dual region-failure

Figure 1. Example of Network Partitions and holes

All of the aforementioned studies about regional-failures consider a worst-case cut as the cut which maximizes or minimizes certain performance metric (such as capacity) of the intersected links. However, such a definition is inadequate to capture many realistic situations where the faulty nodes may influence larger number of nodes rather than larger number of link-cuts. we claim that, using the number of failed links as the main criteria for defining the worst-case region cut underestimates the impact of a region failure on the overall network performance.

Therefore, in this chapter, we first introduce a new definition for for a worst-case cut (partition) due to failure regions. Then, we identify the location of a disaster that would have the maximum impact on a test network using both definitions. Finally, we conduct a deeper analysis to understand the behavior of a single and dual region-failures on the network performance with mission-critical nodes. Our simulation results indicate that, current studies in regional failures under estimate the impact of the worst-case cut due to their dependence on a relaxed definition for the worst-case region-cut.

The rest of this chapter is organized as follows. Section 2 presents the problem investigation. Section 3 demonstrates our proposed scheme for identifying the worst-case cut under single and dual region-failures. Section 4 presents our experimental results. Finally, Section 5 concludes our chapter.

2. Problem investigation

Most of the available studies [21–23] in the literature consider that the failure probability of a node is independent of its location in the deployment area. Few studies [6–10] addressed the region failure problem of spatially correlated network nodes in the physical topology. Available studies focusing on region failures consider link-cuts due to a region-failure as the main criteria for identifying the worst-case region-failure and can be interpreted in terms of the network capacity and throughput. However, such fault scenario is inadequate to capture many realistic situations where the failure region may influence larger number of nodes rather than larger number of link-cuts as shown in Fig.2.

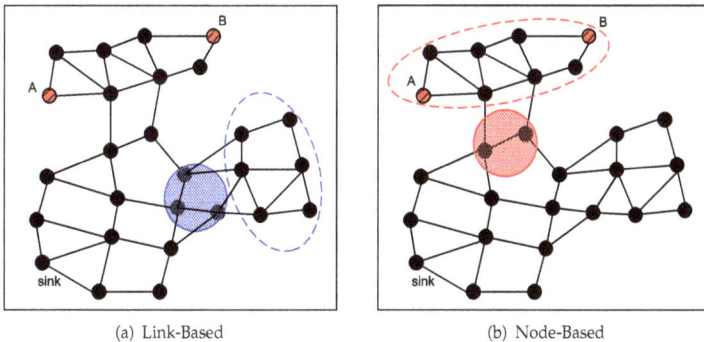

(a) Link-Based   (b) Node-Based

Figure 2. Selection of the Worst-case region cut

For the network shown in Fig. 2, suppose that all links have the same capacity. Then, it can be easily seen that the region failure in Fig. 2a leads to 10 link-cuts while the region failure shown in Fig. 2b leads to only 6 link-cuts. Based on the current studies, the region with the maximum number of link-cuts is identified as the worst-case cut (i.e the region in Fig. 2a). On the other hand, the failure region depicted in Fig. 2b disconnects the dashed region (8 nodes) from the rest of the network which has a greater impact on the network than the case shown in Fig. 2a which has only 6 disconnected nodes.

Therefore, we first propose a new model for the worst-case region-cut considering disconnected nodes (Node-Based) due to a region-failure as the major criteria for identifying the worst-case region-cut. In our proposed model, we may select a region with less number of link-cuts as the worst-case cut if it isolates a larger number of nodes. Moreover, in our proposed model mission-critical nodes are given more weight during the worst-case analysis process to reflect the importance of its service continuity.

3. Proposed model

The network model we consider here is composed of a base station, a sink node, and a set of distributed wireless sensor nodes. The base station can be inserted in any suitable place whether in the field or somewhere else. It is directly connected to the sink node through a wire or wireless link. The sink node is a wireless sensor with high capability in memory, processing, power, and wireless coverage. It works as an intermediate node between the base station and the other sensors. It receives commands from the base station and then conveys them to the deployed sensors. In addition, it collects data from the sensors and sends it to the base station. The other sensor nodes are categorized into mission-critical (MC) sensor nodes and regular sensor nodes. A mission-critical sensor node is a node that is responsible for sensing or reading mission-critical information such as a sensor node in the battle field. while the regular sensor node is any other node. These sensors which have limited capabilities in their battery-powers, memory, and processing are distributed all over the area of interest in such a way that any deployed node has at least one path to the sink node.

In the rest of this section we present the problem formulation of our proposed model in section 3.1. Then, in section 3.2 we introduce the routing process based on the above mentioned network model. Finally, we demonstrate the region failure analysis under single and dual region-failures in section 3.3.

3.1. Routing process

Nodes in a routing table are classified into three categories: 1) parent node, 2) sibling node and 3) child node. A parent node is a node in the transmission range of another sending node and having a hop count one less than the sending node. A sibling node is a node in the transmission range of another sending node and having the same hop count as the sending node. A child node is a node in the transmission range of another sending node and having a hop count one more than the sending node. After deploying sensor nodes into the network field, the routing tables of the underlying sensor nodes are established. To illustrate the process, a network is constructed in Fig. 3. The network has eight sensors: a sink node, two mission-critical sensors (A and F), and five regular sensors ($B, C, D, E,$ and G).

The routing process is started when the sink node broadcasts a setup packet to all nodes within its transmission range. The setup packet contains several parameters including the number

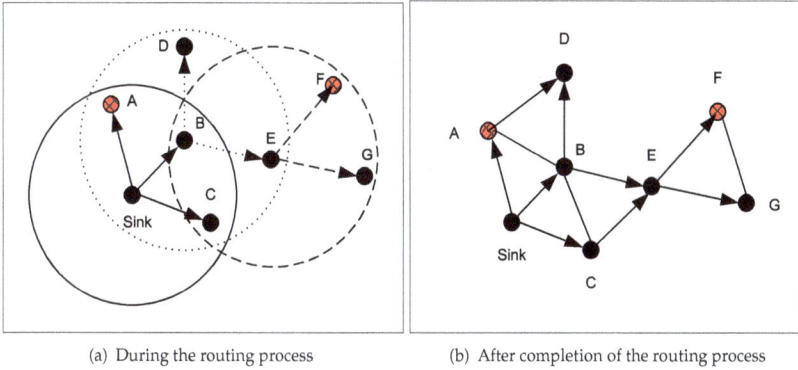

(a) During the routing process   (b) After completion of the routing process

Figure 3. Routing process configuration

of hops h between the sending node and the sink node. Each sensor node receiving the setup packet from the sink node sets its value of h to the incremented value of the received h and marks the sink node as a parent node. For the network shown in Fig. 3a, each of the nodes A, B, and C in the solid circle marks the sink node as its parent. Then, every node with hop count of 1 broadcasts a newly constructed setup packet. For example, suppose that the node B sends its setup packet. In this case, the sink node (belongs to the third category), the sensor nodes A and C (belong to the second category), and the sensor nodes D and E (belong to the first category) mark the node B as a child, sibling, and parent, respectively. Each node in the first category (in this case, D and E) sets its value of h to 2 (i.e., the value of received $h + 1$). Each of the nodes D and E constructs its setup packets and broadcasts it to its neighbors. This broadcasting process is repeated hop-by-hop by the nodes of the first category until all the deployed sensor nodes establish their routing tables. This will end up with the Fig. 3b, where a headed-arrow represents a link between a parent node and a child and it starts from the parent and heads into its child. A line with no arrow-head refers to a link between two sibling nodes.

3.2. Problem formulation

When considering circular cuts, we assume that a disaster results in a cut of radius r, which is centered at [x,y]. We define the worst-case region-cut of a network topology nt as $WCC(nt)$ which can be evaluated as follows.

$$WCC(nt) = \max_{\forall x,y}(rw_{x,y}) \tag{1}$$

Where $rw_{x,y}$ is the weight function of the region-cut centered at (x,y) and can be evaluated as follows.

$$rw_{x,y} = c \cdot \underbrace{\sum_{s_i \in ds_{x,y}} nw(s_i)}_{\text{Disconnected nodes}} + \underbrace{\sum_{s_j \in fs_{x,y}} pw(s_j)}_{\text{Failed nodes}} + \underbrace{\sum_{b_i \in br_{x,y}} bw(b_i)}_{\text{Failed links}} \tag{2}$$

Where c is a constant that gives the disconnected nodes higher weight than link-cuts, $ds_{x,y}$ is the set of disconnected sensor nodes due to the region-cut centered at (x,y), and $nw(s_i)$ is the weight of the sensor node s_i that differentiates between a regular node and a mission-critical node.

$fs_{x,y}$ is the set of failed sensor nodes within the region-failure centered at (x,y), and $pw(s_j)$ is the weight of the sensor node s_j on the path from a source node S to the sink node. The $pw(s_j)$ is used to distinguish between a node on a regular path or a mission-critical path. $br_{x,y}$ is the set of paths (bridges) connecting a disconnected network partition (due to the region-cut centered at (x,y)) with a connected network section. $bw(b_i)$ is the weight of the bridge b_i.

Equation 1 indicates that, the worst-case cut can be located by evaluating a weight function $rw_{x,y}$ for each region cut and the region-cut with maximum weight is considered the worst-case region-cut.

Equation 2 evaluates the weight of a region cut centered at $[x,y]$ by estimating the importance of three main factors: disconnected nodes, failed nodes, and link-cuts as described in Eq. 2.

The first term $\sum_{s_i \in ds_{x,y}} nw(s_i)$ is the summation of the weights of all disconnected nodes due to the region-failure. Through the node weight $nw(s_i)$, we provide more weight to mission-critical nodes than regular nodes as indicated by Equation 3.

$$nw(s_i) = \begin{cases} 1 & \text{if } s_i \in RS, \\ 2^c & \text{if } s_i \in MS. \end{cases} \tag{3}$$

Where RS is the set of all regular sensor nodes in the network field and MS is the set of all mission-critical sensor nodes.

The second term $\sum_{s_j \in rs_{x,y}} pw(s_j)$ represents the summation of all path weights of the failed nodes (all nodes within the region cut).

This weight function $pw(s_j)$ is used to grant higher weights to the failed nodes within the region failure that are used in forwarding mission-critical information as described in Eq. 4.

$$pw(s_j) = \begin{cases} 0 & \text{if } s_j \in RS \ \& \ s_j \notin MS, \\ 2^c & \text{if } s_j \in MS, \\ \sum_{s_k \in SC(s_j)} \frac{pw(s_k)}{pn_k} & \text{if } s_j \in RS \ \& \ s_j \in MP. \end{cases} \tag{4}$$

Where MP is the set of mission-paths which we refer to as paths carrying mission-critical information, SC is set of children of the sensor s_j and pn_k is the number of parents of the sensor s_k .

In Eq.4, the path weight $pw(s_j)$ is assigned a value of zero if a failed node is a regular node that is not located on a mission-path (i.e., it is not used to forward mission-critical information). on the other hand, if the failed node is mission-critical node it receives higher path weight. Finally, if the failed node s_j is a regular node that is located on a mission-critical path, then the path weight is evaluated recursively by the summation of the ratio of path weights $pw(s_k)$ of the child node k to the number of parent nodes pn_k of the child node k.

The last term $bw(b_i)$ is the weight of the bridge b_i. This weight function reflects the importance of a failed link due to a region-cut. Here, a failed link can participate in more than one bridge. Hence, we are interested in calculating the the For example, if all links on the failed bridge are not located on mission-path it is given a weight of 1.

on the other hand,if all links on the failed bridge are located on mission-path it is given the highest weight.

Finally, if not all of the links on the failed bridge are located on a mission-path, its weight is a percentage of the highest weight. This percentage can be evaluated by the ratio of the number of links carrying mission-critical information of the bridge ML_{b_i} to its total number of links TL_{b_i} as described in Equation 5

$$bw(b_i) = \begin{cases} 1 & \text{if all sensors in } b_i \notin MP, \\ c & \text{if all sensors in } b_i \in MP, \\ \frac{ML_{b_i}}{TL_{b_i}} \cdot c & \text{otherwise.} \end{cases} \tag{5}$$

where b_i is a linking path (bridge) connecting a disconnected node with a connected nodes and passing through failed nodes. Each of the connected and disconnected nodes should have a direct link with a failed node. The bridge direction should be from a child toward a parent (i.e. any bridge linking two sibling nodes is excluded). ML_{b_i} is the number of mission-critical links located on the bridge b_i, and TL_{b_i} is the total number of links of the bridge b_i.

To further illustrate the evaluation of the worst-case region-failure, we provide the following example shown in Fig. 4. In this example compare two different region failures as shown in Fig. 4c and Fig. 4d. Herefater, we also provide the corresponding analytical calculations.

For the network shown in Fig. 4a, to find the worst-case region-failure we have to apply the routing process discussed earlier to configure the routing tables by the category (parent, sibling, child) of each neighbor. The result of routing phase for the given network topology is depicted in Fig. 4b. As described earlier in this section, the headed-arrow represents a child node toward the head and a parent node toward the tail while a line without heads represents a sibling relationship. Hereafter in the following example, we assume that the value of the constant c is equal to 5.

For the failure *Region*1 shown in Fig. 4c, to find the worst-case region-cut, we apply Eq. 2 of our proposed model to get the weight of this region-failure as follows. For the calculation of the node weight $nw(s_i)$ based on Eq. 3, each node in the disconnected partition of the topology is assigned a value of 1. Hence, $\sum_{s_i \in ds_{Region1}} nw(s_i) = 6$. Note that, there is no mission-critical nodes in the disconnected partition. For the calculation of the path weight $pw(s_j)$ we use Eq. 4. As all the failed nodes within the region-failure are not carrying mission critical information, then $\sum_{s_j \in fs_{Region1}} pw(s_j) = 0$. Finally, for the bridge weight $bw(b_i)$ we use Eq. 5. Here we have 6 different paths (bridges) crossing the region-failure and none of these bridges is carrying mission-critical informations. Hence, each bridge will be given a weight of 1. Consequently, $\sum_{b_i \in br_{x,y}} bw(b_i) = 6$. Finally, since the constant c is assumed to have the value of 5, we have $rw_{Region1} = 5 * 6 + 0 + 6 = 36$. Note that, any bridge that has a link connecting two sibling nodes will be execluded.

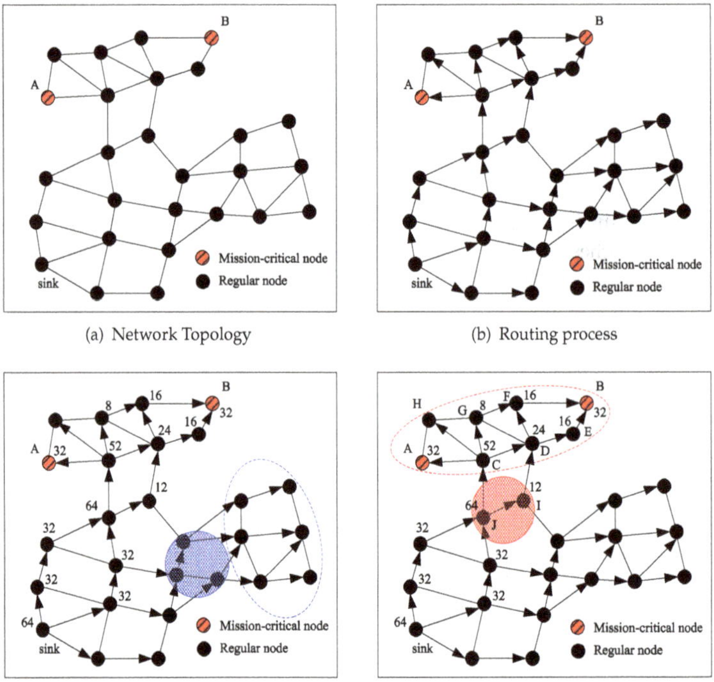

(a) Network Topology

(b) Routing process

(c) Region 1

(d) Region 2

Figure 4. Calculations of the worst-case region-cut

Similarly, For the failure $Region2$ shown in Fig. 4d, we apply Eq. 2 to get the weight of this region-failure as follows. For the calculation of the node weight $nw(s_i)$ based on Eq. 3,each node in the disconnected partition of the topology is assigned a value of 1 except each of the mission-critical nodes A and B will have the value of 32. Hence, $\sum_{s_i \in ds_{x,y}} nw(s_i) = 64 + 6 = 70$.

For the calculation of the path weight $pw(s_j)$ we use Eq. 4. As the failed nodes within the region-failure are carrying mission critical information of node A and B, we have to find the weight of each node on a mission-path to the sink node. The estimation of the $pw(s_j)$ is performed as follows. first, each mission-critical node such as A and B receives a weight of 32 based on equation 4. Since the node B has only two parent nodes (node F and node E), each of them will receive half of the weight of the node B (i.e 16). Node D will receive its $pw(D)$ weight of 24 which is the sum of 8 (half of the $pw(F)$ as node F has two parent nodes) and 16 (the $pw(E)$ as node E has only one parent node) received from its child nodes F and E, respectively. In the same way, Node G will have a $pw(G)$ weight of 8 (half of the $pw(F)$ of its child node F), Node C will have a $pw(C)$ weight of 52 (the sum of $pw(G)$, $pw(A)$, and half of $pw(D)$), Node H will receive a $pw(H)$ of zero since the node H is not on a mission-path, Node I will receive a $pw(I)$ weight of 12 which is half of the $pw(D)$ weight of its child node D, Node J will receive a $pw(J)$ of 64 that is the sum of $pw(I)$ and $pw(C)$. This process continues in the same way up to the sink node as shown in Fig. 4d. Then,

$\sum_{s_j \in f s_{Region2}} pw(s_j) = pw(I) + pw(J) = 76$. Note that, In Fig. 4c and Fig. 4d, a node z without a $pw(z)$ weight indicates that this node is not on a mission-path.

Finally, we calculate the bridge weight $bw(b_i)$ for each bridge using Eq. 5.In Fig. 4d, we have 4 different paths (bridges) crossing the region-failure and all of them carrying mission-critical informations. Hence, each bridge will be given a weight of 5 according to Equation 4. Consequently, $\sum_{b_i \in br_{Region2}} bw(b_i) = 20$. Finally, we have $rw_{Region2} = 5 * 70 + 76 + 20 = 446$.

Based on Eq. 1 of the proposed model, the region-failure with maximum weight will be chosen as the worst-case region-failure ($Region2$ in this example).

3.3. Region failure algorithms

In this section we introduce the main algorithms used to find out the worst-case region-failure under the single and dual region-failure scenarios in Section 3.3.1 and Section 3.3.2, respectively.

Hereafter, we present the main algorithms used in our model. The algorithm shown in Fig.5 finds all sensors within a region-failure. The algorithm depicted in Fig. 6 removes the failed sensors from the network topology (nt). The algorithm shown in Fig. 7 finds all paths between a given source/destination pair.

```
findSensorsWithinRegionAlgorithm (i, j, r, nt)
1      swr ← { } // swr is the set of sensors within the region
                 // centered at (i, j) with the radius r.
2      foreach sensor s ∈ nt
                 // calculate the distance from the region center (i, j)
                 // to the position of the sensor s (xₛ, yₛ).
3          d ← SQRT ((xₛ − i)² + (yₛ − j)²)
4          if  d ≤ r
                 // s is within the region so its ID is added to the set swr.
5              swr ← swr U idₛ
6          end
7      end
8      return swr // return the set of sensors found within the region.
end
```

Figure 5. Find sensors within a region-failure Algorithm

The algorithm shown in Fig. 5 demonstrates how to find sensor nodes located within the region-failure. The input parameters to this algorithm are the network topology nt, center (i,j) and radius r of the region-failure. In this algorithm, we first calculate the distance d from the center of region failure to the sensor node. If d is less than the radius of region-failure, then the sensor is located within the failed region.

The algorithm shown in Fig. 6 presents our strategy for removing a sensor node from the routing table (rt) of all sensors in the network field. The input parameters to this algorithm are the network topology nt_1 and the list rs of sensor nodes to be removed. In this algorithm we perform the following steps. First, for each sensor node in the removed sensor list (rs) we search for that sensor id. If the sensor ID. is found in any routing table of a sensor node that

---

removefailedSensorsAlgorithm (*nt_1*, *swr*)
1   **foreach** sensor *s* ϵ *swr*
2   **foreach** sensor *sn* in *nt*₁ & *sn* ϵ *rt* of *s*
                    // rt refers to a routing table
3   *nt_1* ← remove *s* from *rt* of *sn* in *nt_1*
4   **end**
5   *nt_1* ← *nt*₁ ∩ *s* *// remove s from the nt_1*
6   **end**
7   return *nt_1* *// return the updated network topology*
end

Figure 6. Remove sensors from the routing tables algorithm

belongs to the network topology, then it is removed from the routing table and the network topology is updated with the changes made.

---

findAllPathsAlgorithm (*paths, p, s, d*)
1   **if** *s* = *d* *// A path p is found*
2   *paths* ← *paths* U *p* *// the path p is added to the set of paths*
3   return *paths*
4   **else**
5   **foreach** sensor *sn* ϵ parent or sibling of *s*
6   **if** *sn* not ϵ *p*   *// to prevent cycle*
7   *findAllPathsAlgorithm* (*paths, p* U *sn, sn, d*)
8   **end**
9   **end**
10   **end**
end

Figure 7. Find All Paths from *s* to *d* Algorithm

The algorithm shown in Fig. 7 presents how to check if a sensor node *s* is still connected or not after a region-failure happens. In other words, we need to find all paths *p* from the given node *s* to the destination node *d* and add these paths to the empty list *paths* of paths. This is accomplished by checking first if the path already exists in the paths list *paths*. If it does not exist, a path search is initiated from each parent of the node *s*. If a path is found, the search is terminated, the found path is added to the paths list and the node *s* is considered a connected node. On the other hand, if no path exists from the parent node of node *s*, then a path search is initiated from the sibling node of node *s*. Finally, if no path is found from both the parent and the sibling nodes of node *s* the node is marked as disconnected node.

3.3.1. Single region-failure

The algorithm shown in Fig. 8 demonstrates how to estimate the number of disconnected nodes due to a single region-failure. The input parameters to this algorithm are the network topology *nt*, the radius of the failure region *r*, the increment value Δr for the radius *r*, the threshold of disconnected nodes ns_{th}, and the network field's length *nfl* and width *nfw*. In this algorithm, we first generate a failure region with radius *r*. then we find all sensors located within the region by the algorithm shown in Fig. 5. Sensors located within the failure region are then removed from the routing tables of all nodes in the network topology by executing the algorithm shown in Fig. 6. For the rest of the remaining sensors, we investigate the availability of a path from each node to the sink node by carrying out the algorithm depicted in Fig. 7. If a path is found the node is marked as connected node and disconnected otherwise. then we calculate the number of disconnected nodes due to the failure region. The above mentioned scenario is repeated by incrementing the coordinates of center of the region-failure until th whole topology is scanned by region-failures. Finally, Equation 1 is applied to estimate the worst case region-failure.

singleRegionFailureAlgorithm (r, Δr, nt, nfl, nfw, ns_{th})
// i and j are the center of a region with the radius r
1   *regions* ← { }
2   **for** $i \leftarrow r$ to nfl, i incremented by Δr
3   **for** $j \leftarrow r$ to nfw, j incremented by Δr
4                $id_r \leftarrow i \mid\mid : \mid\mid j$ // id_r is the ID of the region
5   *swr* ← { } // the set of sensors within the region id_r
6   *swr* ← *findSensorsWithinRegionAlgorithm* (i, j, r, nt)
7   nt_1 ← *removeFailedSensorsAlgorithm* (nt, *swr*)
8   *dis* ← { } // the disconnected sensors within the region id_r
9   **foreach** sensor $s \in nt_1$
10   *paths* ← *findAllPaths* ({ }, {s}, nt_1, s, *sink*)
11   **if** *paths* is empty // no path exists
12   *dis* ← *dis* ∪ s // the node s is disconnected
13   **end**
14   **end**
// number of disconnected nodes are above certain threshold
15   **if** *length* (*dis*) > ns_{th}
16   *regions* ← *regions* ∪ {id_r, *dis*}
17   **end**
18   **end**
19   **end**
// return the set of failed regions with their associated disconnected nodes ns_{th}
20   return *regions*
end

Figure 8. Algorithm for Finding the number of disconnected nodes under single region-failure

3.3.2. Dual region-failure

Under dual region-failure scenario, we propose the following algorithm shown in Fig. 9 to estimate the number of disconnected nodes within the given network topology.

The algorithm shown in Fig. 9 demonstrates our strategy to determine the number of disconnected nodes under dual region-failures scenario.The input parameters to this algorithm are the same as that of a single region failure shown in Fig. 8. In this algorithm, at the beginning, we follow similar steps to that used to find the number of disconnected nodes of a single region failure where we first generate a failure region with radius r. then we

```
twoRegionsFailureAlgorithm (r, Δr,  nt, nfl, nfw, ns_th)
              // i and j are the center of one region with the radius r
1         regions ← { }
2         for  i ← r to nfl, i incremented by Δr
3               for  j ← r to nfw, j incremented by Δr
4                     id_{r1} ← i || : || j    // id_{r1} is the ID of the first region
5                     swr_1 ← { } // the set of sensors within the region id_{r1}
6                     swr_1 ← findSensorsWithinRegionAlgorithm (i, j, r, nt)
7                     nt_1 ← removeFailedSensorsAlgorithm (nt, swr_1)
                      // k and l are the center of the second region with the radius r
8                     for  k ← i to nfl, k incremented by Δr
9                           for  l ← r to nfw, l incremented by Δr
10                                id_{r2} ← k || : || l  // id_{r2} is the ID of the second region
                                  // twoReg_{id} represents an ID for the underlying two regions
11                                twoReg_{id} ← id_{r1} || id_{r2}
                                  // the dis is the set of the disconnected nodes associated with
                                  // the two regions identified by the ID twoReg_{id}
12                                dis ← { }
13                                swr_2 ← { } // the set of sensors within the region id_{r2}
14                                swr_2 ← findSensorsWithinRegionAlgorithm (k, l, r, nt_1)
15                                nt_2 ← removeFailedSensorsAlgorithm (nt_1, swr_2)
16                                foreach sensor s ∈ nt_2
17                                      paths ← findAllPaths ({ }, {s}, nt_2, s, sink)
18                                      if  paths is empty     // no path exists
19                                            dis ← dis U s // the node s is disconnected
20                                      end
21                                end
                                  // number of disconnected nodes are above certain threshold ns_{th}
22                                if  length (dis) > ns_{th}
23                                      regions ← regions U {twoReg_{id}, dis}
24                          end
25                    end
26              end
27        end
          // return the set of dual-failed regions with their associated disconnected nodes
28        return regions
end
```

Figure 9. Algorithm for Finding the number of disconnected nodes under dual region-failures

find all sensors located within the region by the algorithm shown in Fig. 9. Sensors located within the failure region are then removed from the routing tables of all nodes in the network topology by executing the algorithm shown in Fig. 6. The above mentioned steps are repeated for the second region-failure. Now, we have come up with a network topology without the failed nodes due to the dual region-failure. Then, we examine the path availability from each node in the network topology to the sink node by executing the algorithm depicted in Fig. 7. If no path is available from a node to the sink node the node is marked as a disconnected node. On the other hand, if a path is found the node is marked as connected node. then we calculate the number of disconnected nodes due to the failure region. The above mentioned scenario is repeated by iterating all the possible combinations of the regions centers until the whole topology is visited by region-failures. As the number of disconnected nodes due to dual region-failures is usually large, we use a threshold such that we get the dual region-failures that lead to a number of disconnected nodes greater than the predefined threshold.

Finally, Equation 1 is applied to estimate the worst case region-failure.

4. Results

In this section, we present our simulation results. In our simulations we consider the network topology shown in Fig. 10 in which node 0 is the sink node. The failure information of different region failures generated during our simulations results are shown in Table 1 and Table 2, respectively.

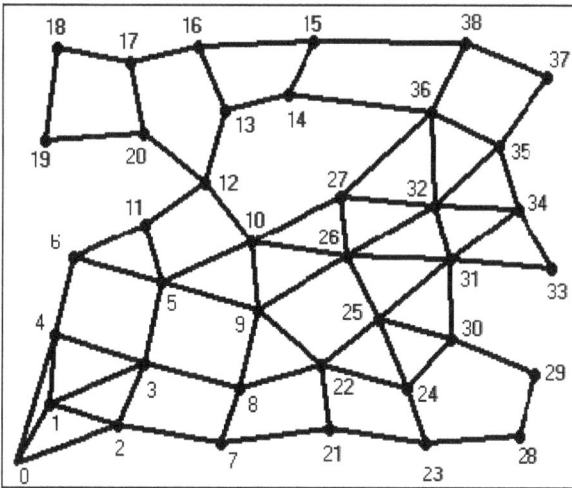

Figure 10. Network Topology

Note that, due to the large number of region-failures generated, we present only region failures that lead to disconnecting more than two and eight nodes for the single and dual failure scenarios, respectively.

Hereafter, we investigate the location of the worst-case region-failure under Link-based, Node-based (without a mission critical node) and Node-based with a mission critical node in Fig. 12, Fig. 14 and Fig. 16, respectively. In the Node-based with a mission critical node, the node 28 is chosen as a Mission-Critical(MC) node.

The results shown in Fig. 11 show that, based on the traditional definition of the worst-case region cut, the failure region with id 5 is considered the worst-region cut as its failure leads to having the maximum number of failed links. The location of this failure-region is depicted

Region id	Failed Nodes IDs	No. of Failed links	No. of Disconnected Nodes
1	12	4	8
2	11, 12	6	8
3	12, 13	6	7
4	2, 3	8	5
5	31, 32	11	3
6	13	3	3
7	7, 8	6	3
8	7	3	3
9	21	3	2

Table 1. Region-failure information

Region id	Failed Nodes' IDs at Region 1	Failed Nodes IDs at Region 2
1	6	2, 3
2	4	2, 3
3	1, 4	2
4	1, 4	2, 3
5	2	12
6	2	11, 12
7	2	12, 13
8	2, 3	5, 11
9	2, 3	12
10	2, 3	11, 12
11	2, 3	12, 13
12	5, 11	7, 8
13	12	7
14	12	7, 8
15	12	31, 32
16	11, 12	7
17	11, 12	7, 8
18	11, 12	31, 32
19	12, 13	31,32

Table 2. Failed Nodes due to dual region-failure

in Fig 12. It is also notable that, according to Table 1 the failure of this region results in disconnecting 3 sensor nodes namely, node 33, 34, and 35 as depicted in Fig. 12.

The results shown in Fig. 13 clearly indicate that, in absence of mission-critical nodes within the network topology the proposed model is very clever to find out the worst-case region-failure. The worst-case region-failure is the region-failure that has the maximum

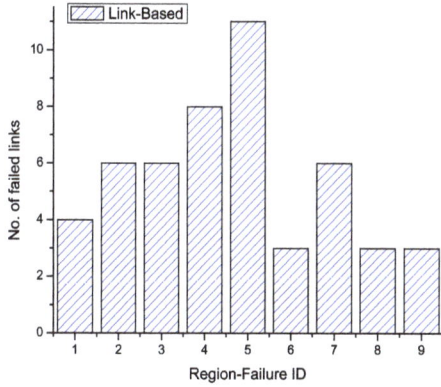

Figure 11. Single Failure-region weights under link-based approach.

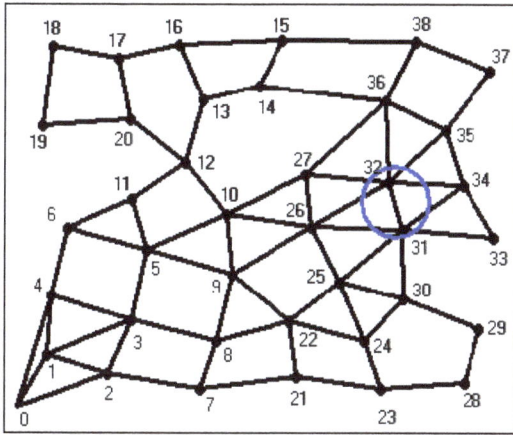

Figure 12. Worst-case region failure under Link-based approach.

impact on the network performance which is region 2 in this case as its failure results in disconnecting 8 sensor nodes, namely, 13, 14, 15, 16, 17, 18, 19 and 20. However, according to Fig. 11, the failure of this region leads to failure of 6 links only.

The results shown in Fig. 15 indicate clearly that, with the proposed model of a region-cut, introducing a mission-critical node into the network topology leads to a change in selecting the worst-case region cut which is region 4 as its failure results in disconnecting 5 sensor nodes, namely, 7, 8, 21, 23, and 28 as shown in Table 1. However, according to Fig. 11, the failure of this region leads to failure of 8 links.

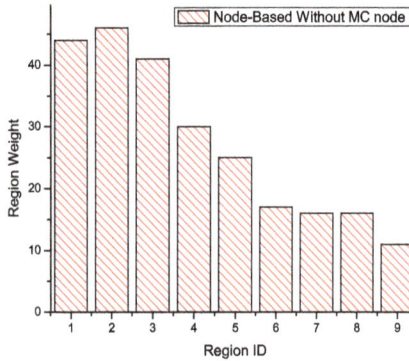

Figure 13. Single Failure-region weights under node-based approach without MC-node.

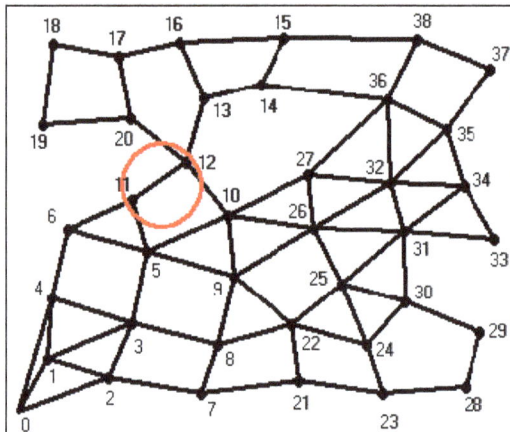

Figure 14. Worst-case region cut using Node-based approach.

Therefore, we claim that, using the failed links as the only criteria for defining the worst-case region cut is impractical as it ignores the case that the failure of some nodes may lead to failure of few links however its impact on the network is more severe due to disconnecting larger number of nodes. Moreover, it disregards the fact that some network nodes have higher priority than others.

Hereafter, we investigate the location of the worst-case dual region-failures without including a MC-node and with a MC-node in Fig. 17 and Fig. 18, respectively.

The locations of the worst-case dual region-failures shown in Fig. 17 indicate that, using the link-based approach (blue regions including nodes 11,12 and 31, 32) lead to cutting of 17 links and disconnecting 12 nodes. On the other hand, under the dual region failure scenario,

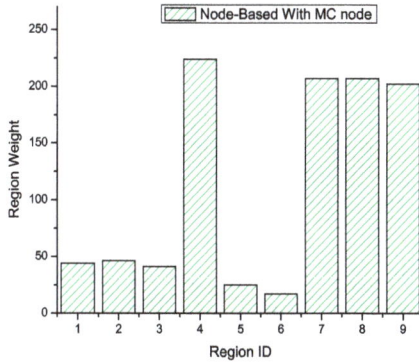

Figure 15. Worst-case single region failure under Node-based approach with MC node

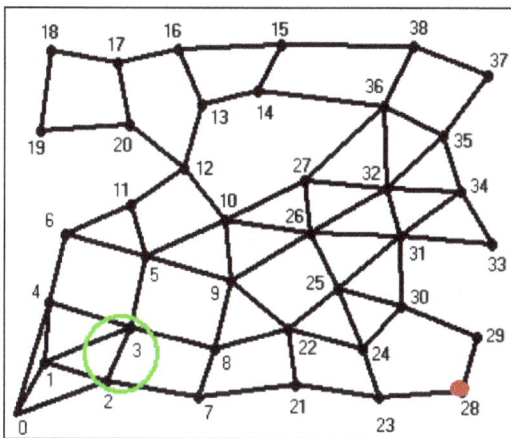

Figure 16. Worst-case region failure underNode-based approach with MC node

using the proposed Node-based approach, the worst-case region-failure depicted in Fig. 17 (red regions including nodes 1,4 and 2) have only 8 link-cuts however, it results in a full disconnection of all the network nodes. This is due to the fact that, based on our approach, the worst-case region cuts are located near by the sink node and the failure of these nodes lead to isolating the sink node from the whole network. The results shown in Fig. 18 show that, by introducing the MC-node number 28, the worst-case dual region failure remains unchanged as the earlier case without MC-node. This is due to fact that , by introducing the MC-node, no further influence can affect the network as the maximum impact has been already happened by having a complete disconnected sensor network.

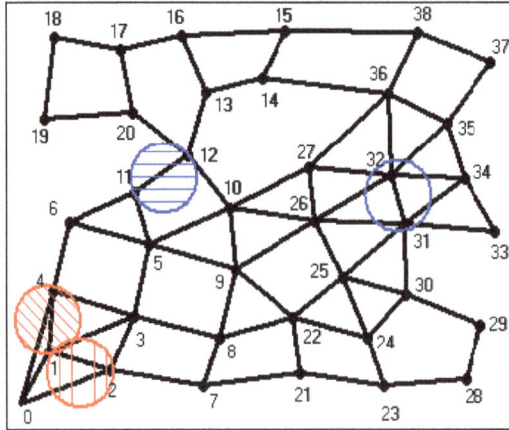

Figure 17. Worst-case Dual-Region Failures

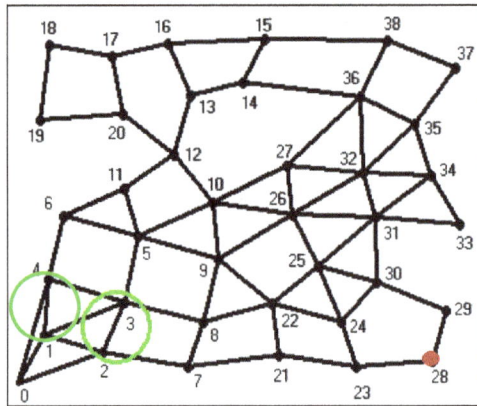

Figure 18. Worst-case Dual-Region Failures with a mission node

5. Conclusion

In this chapter, we introduced a new model for for a worst-case cut (partition) due to failure regions. The proposed model takes into consideration the physical correlation among the locations of the network nodes and the possible priority of some nodes over the others. Based on the proposed model, We identified the location of a disaster that would have the maximum impact on a sample test network under single and dual region-failure scenario. Extensive Simulation results indicate that, using the number of failed links as the main criteria for defining the worst-case region cut, underestimates the impact of a region failure on the overall network performance.

Author details

Mohamed M. A. Azim and Aly M. Al-Semary
Taibah University, Saudi Arabia

6. References

[1] Liotine, M. (2003). *Mission-Critical Network Planning*, Artech House Publishers, ISBN: 1-58053-516-X, USA.

[2] Markopoulou, A.; Iannaccone, G.; Bhattacharyya, S.; Chuah, C., & Diot, C. (2004). Characterization of failures in an IP backbone, *IEEE INFOCOM 2004* , pp. 2307-2317.

[3] A. Azim, Mohamed Mostafa and Kabir, M. N.,(2011), Availability Study of M:N Automatic Protection Switching Scheme in WDM Networks, International Journal of High Speed Networks, Vol. 18, No. 1, pp. 1-13, IOS Press, Dec. 2011.

[4] Lee, K.; Lee, H.-W.; Modiano, E., (2011), "Reliability in Layered Networks With Random Link Failures," IEEE/ACM Transactions on Networking, vol.19, no.6, pp.1835-1848, Dec. 2011.

[5] Kini, S.; Ramasubramanian, S.; Kvalbein, A.; Hansen, A.F.;(2010) , "Fast Recovery From Dual-Link or Single- Node Failures in IP Networks Using Tunneling," IEEE/ACM Transactions on Networking, vol.18, no.6, pp. 1988-1999, Dec 2010.

[6] A. Azim, Mohamed Mostafa and El-semary, M. Aly,(2012) Vulnerability Assessment for Mission Critical Networks Against Region Failures: A Case Study, International Conference on Computing and Information Technology (ICCIT2012), pp. 838-843, March 2012.

[7] Sen A., Shen B., Zhou L., & Hao B. (2006). Fault-tolerance in sensor networks: A new evaluation metric *INFOCOM, 2006*, pp. 1-12.

[8] Sen A., Murthy S., & Banerjee S. (2009). Region-based connectivity: A new paradigm for design of fault-tolerant networks *High Performance Switching and Routing 2009*, pp. 94-100, Paris, France.

[9] Sen A., Banerjee S., Ghosh P. & Shirazipourazad S. (2009). Impact of region-based faults on the connectivity of wireless networks . *Allerton conference on communication control and computing 2009*, Monticello, IL, 2009, pp. 1430-1437.

[10] Jiajia Liu; Xiaohong Jiang; Nishiyama, H.; Kato, N.; , "Reliability Assessment for Wireless Mesh Networks Under Probabilistic Region Failure Model," Vehicular Technology, IEEE Transactions on , vol.60, no.5, pp.2253-2264, Jun 2011

[11] S. Neumayer, G. Zussman, R. Cohen, and E. Modiano, Assessing the vulnerability of geographically correlated network failures, in Proc. MILCOM, 2008, pp. 1-6.

[12] S. Neumayer, G. Zussman, R. Cohen, and E. Modiano, Assessing the vulnerability of the fiber infrastructure to disasters, in Proc. INFOCOM, 2009, pp. 1566-1574.

[13] S. Neumayer and E. Modiano, Network reliability with geographically correlated failures, in Proc. INFOCOM, 2010, pp. 1658-1666.

[14] M. M. Hayat, J. E. Pezoa, D. Dietz, & S. Dhakal, Dynamic load balancing for robust distributed computing in the presence of topological impairments, in Handbook of Science and Technology for Homeland Security. Hoboken, NJ: Wiley, 2009.

[15] W. Wu, B. Moran, J. Manton & M. Zukerman, Topology design of undersea cables considering survivability under major disasters, in Proc. WAINA, 2009, pp. 1154-1159.

[16] K. Kim and N. Venkatasabramanian, Assessing the impact of geograph- ically correlated failures on overlay-based data dissemination, in Proc. GLOBECOM, 2010, pp. 1-5.

[17] P. Agarwal, A. Efrat, S. Ganjugunte, D. Hay, S. Sankararaman, and G. Zussman, Network vulnerability to single, multiple, and probabilistic physical attacks, in Proc. MILCOM, 2010, pp. 1824-1829.

[18] P. Agarwal, A. Efrat, S. Ganjugunte, D. Hay, S. Sankararaman, and G. Zussman, The resilience of WDM networks to probabilistic geographical failures, *INFOCOM, 2011*, to be published.

[19] Myounggyu Won & Stoleru, R.; , "Destination-Based Cut Detection in Wireless Sensor Networks," Embedded and Ubiquitous Computing (EUC), 2011 IFIP 9th International Conference on , vol., no., pp.55-62, 24-26 Oct. 2011.

[20] Barooah, P., Chenji, H., Stoleru, R. & Kalmar-Nagy, T., (2012) Cut Detection in Wireless Sensor Networks, *IEEE Transactions on Parallel and Distributed Systems*, vol.23, no.3, pp.483-490.

[21] Turner, D., Levchenko, K., Snoeren, A. C., & Savage, S.,   *SIGCOMM Computer Communcations*, California fault lines: understanding the causes and impact of network failures, Rev., vol. 40, pp. 315-326, August 2010.

[22] S. Kini, S. Ramasubramanian, A. Kvalbein, and A. F. Hansen, Fast recovery from dual link failures in ip networks, in Proceedings INFOCOM 2009, pp. 1368 - 1376, 2009.

[23] H. Choi, S. Subramaniam, and H. ah Choi, On double-link failure recovery in WDM optical networks,- in in Proceedings of IEEE INFOCOM, 2002, pp. 808-816.

MAC Protocols

Chapter 7

Preamble-Based Medium Access in Wireless Sensor Networks

Alexander Klein

Additional information is available at the end of the chapter

1. Introduction

Medium access protocols in the context of wireless sensor networks have to deal with a large number challenges resulting from hardware limitations, event-driven traffic characteristics, node density, unreliable radio links and requirements of the target application [3]. For these reasons, the design of MAC protocols is still a popular field of research [4] since protocol developers always try to optimize the communication as much as possible. A couple of years ago, the research focus was mainly laid on energy efficiency rather than Quality of Service (QoS). However, this has changed due to the technical progress which allows to employ more complex MAC protocols on the sensor nodes which suit the requirements of mission critical applications [5] and provide QoS [6].

In order to achieve energy efficient communication, the main goal of MAC protocols is to turn off the transceiver as often as possible since it is the part of the node which consumes most of its energy. Therefore, the protocols try to avoid overhearing due to the fact that overhearing is the main cause of energy consumption in duty-cycled networks. The term overhearing addresses the issue that a node receives data which is not dedicated for this node.

The medium access in duty-cycled networks can be achieved in various ways. A common approach is to make use of a Time Division Multiple Access (TDMA) based protocol which allows to efficiently use the radio resources by avoiding typical issues of energy consumption such as idle listening, overhearing, overemitting and collisions. The disadvantage of this approach is that it requires synchronization mechanisms due to the high clock drift of the low power hardware.

Another approach is represented by protocols which divide the time in common active and sleep periods. These approaches require less precise synchronization compared to their TDMA-based counterpart. However, the synchronization mechanisms still results in additional protocol overhead.

The last group is represented by random access protocols with duty-cycle support. These protocols make either use of packet retransmissions or preamble sampling to ensure that the receiver listening to radio channel and thus able to receive the transmission. Table 1 shows an overview of preamble sampling protocols whereas the protocols are categorized according to their medium access strategy.

Strategy	MAC Protocol
Long Preamble	Aloha with Preamble Sampling [1], B-MAC [7], BP-MAC [8], CSMA with Preamble Sampling [9], LPL [2]
Short Preambles	BPS-MAC [10], CSMA-MPS [11], MFP-MAC [12], PR-MAC [13], SEESAW [14], SpeckMAC-D [15], Ticer [16], X-MAC [17]
Short Preambles with Synchronization	CSMA-MPS [11], MixMAC [18],SyncWUF [19], WiseMAC [20]
Short Preambles with adaptive Duty Cycle	BEAM [21], LWT-MAC [22], MixMAC [18], MaxMAC [23], WiseMAC(more bit) [20]
Short Preambles for Contention Resolution	BPS-MAC [10], PR-MAC [13]

Table 1. Overview of MAC Protocols using Preamble Sampling

In recent years, preamble sampling techniques became more and more popular due to the fact that they do not require additional mechanisms for synchronization. This techniques can be applied in many ways as outlined by Cano et al. [24] in their survey of preamble sampling MAC protocols. The basic principle of preamble sampling is shown in Figure 1 which is adopted from [24]. The figure shows typical preamble sampling strategies and also points out the overhearing caused by these access procedures.

The first preamble sampling approach [1] followed the access procedure as described in Figure 1.1. Nodes wake up at periodic time intervals and listen to the radio channel for a short time. If a busy radio channel is detected, nodes continue listening to the channel. Otherwise, they switch off their transceiver and wait for the next active period. Thus, a node, that wants to communicate with another node, has to send a preamble which is longer than the maximum idle period in order to assure that the receiver is listening. This approach has a clear advantage of simplicity. However, the long preamble comes with several disadvantages such as high protocol overhead and overhearing costs. As a result of the long preamble, it is likely that a large number of nodes receive a transmitted preamble and stay awake even though they are not part of the receiver group. Moreover, collisions become very costly since the retransmission of packets involves the transmission of the long preamble which increases the overhearing. The transmission of a long preamble is not supported by every low-power transceiver. Most transceivers, like the CC2420 or CC2500, only support a maximum packet/preamble size of 128 Bytes due to hardware constraints. After the transmission of a packet/preamble, the transceiver switches automatically back to receive mode which results in a gap between consecutive packets/preambles.

Later approaches [11, 16, 17, 20] introduced the mechanism of short preambles to reduce overhearing and the utilization of the radio channel. In addition, preamble sampling access strategies, which use short preambles, can be deployed on any low-power transceiver as long as the gap between two consecutive short preambles is chosen with respect to the hardware characteristics in terms of Clear Channel Assessment (CCA) delay and Turnaround Time

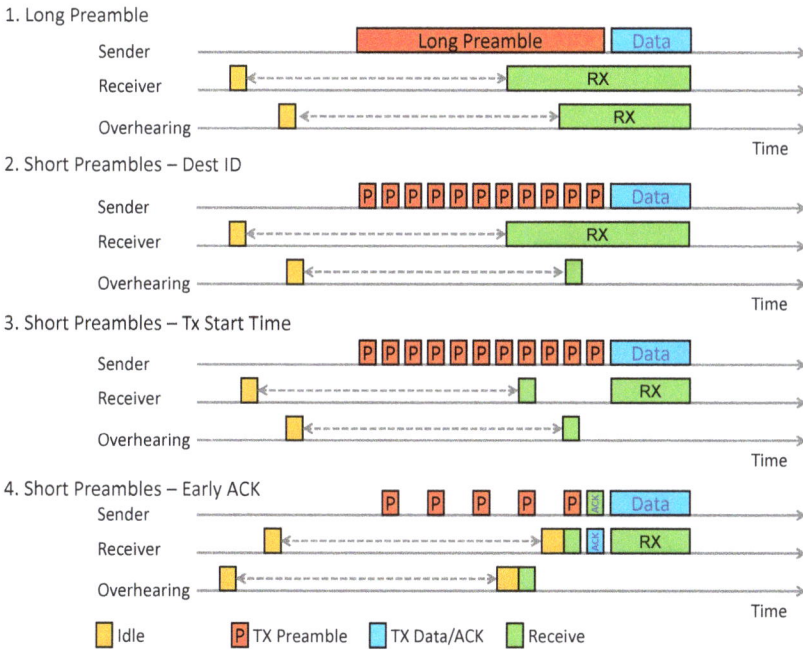

Figure 1. Overhearing in Wireless Networks depending on the Preamble Sampling

(TT). CCA delay specifies the time that a transceiver has to listen to the medium in order to determine whether the medium is busy or idle. The TT corresponds to the time interval that a transceiver requires to switch between receive and transmit mode and vice versa. Both issues and their impact on the performance of MAC protocols are discussed in Section 4.

Instead of using the preamble solely as reservation signal, it is possible to include useful information in the preamble to minimize overhearing as shown in Figure 1.2. Some protocols store the address of the destination in the short preamble which allows nodes that are not involved in the transmission to turn off their transceivers.Nevertheless, the destination node has to continue listening to the medium until the data transmission starts which represents overhead.

The protocol overhead can be further reduced if the start time of the data transmission is encoded in the preamble in addition to the destination address. In this case, it is sufficient for the destination to receive a single short preamble. The destination may than switch off its receiver until the transmission starts as outlined in Figure 1.3.

A new approach that is based on short preambles with destination information was introduced by Buettner et al. [17] in 2006. The idea of their approach is to apply a gap between consecutive preamble in order to allow the destination to respond with an early acknowledgment as shown in Figure 1.4. Upon reception of the early acknowledgment, the sender starts to transmit the data which further reduces the energy consumption of the sender and the protocol overhead.

The information in the preambles can also be used to enable synchronization [18–20], resolve contention on the radio channel [8, 13, 23] or to provide priority-based medium access for service differentiation [10]. These mechanisms are typically more complex and are therefore discussed in more detail in Section 2.

2. Survey of preamble-based MAC protocols

The number of preamble sampling MAC protocols is still increasing very fast since most of them are optimized for a certain scenario or application. In the previous section, the protocols were classified according to their used medium access scheme. Moreover, the problems of the different access schemes were highlighted and some solutions were introduced. The focus of this section lies on a more detailed description of a selection of popular preamble sampling protocols. The selected protocols either provide a basis for a large number of other preamble sampling protocols or introduce new mechanisms for contention resolution and priority-based medium access. In the following, the advantages and drawbacks of each protocol will be discussed.

2.1. CSMA-PS

The traffic load in WSNs is low compared to other wireless networks since nodes sleep most of the time to reduce their energy consumption. For this reason, nodes switch off their transceivers as often as possible since the transceiver usually is the most power-consuming part of a sensor node. Moreover, sensor nodes are often unsynchronized due to the high clock drift of the micro controllers. The CSMA-Preamble Sampling [1, 9] protocol was introduced by El-Hoiydi in 2002. The nodes in the network periodically activate their transceiver in order to listen to the medium for a short time interval. If a node senses a busy channel, it stays awake until the current data transmission has finished. Otherwise, the node switches off its transceiver and waits for the next wake-up interval. Therefore, a node transmits a preamble before its data transmission. The duration of the preamble has to be longer than the wake-up time interval to be sure that the destination node is listening to the medium. A medium access example of the CSMA-PS protocol with acknowledgments is shown in Figure 2.

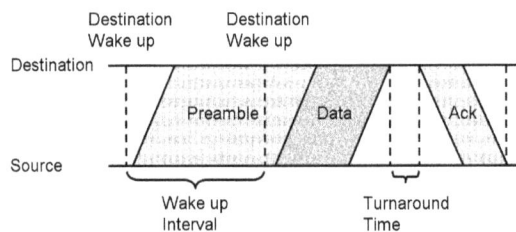

Figure 2. Medium Access Example - CSMA-PS with Acknowledgment

Acknowledgments are still required and strongly recommended for reliable data exchange due to the fact that hidden nodes may still interfere the communication. Furthermore, neighbor nodes could also disturb the current transmission if they start their own transmission during the gap between the reception of the last data packet and the transmission of the acknowledgment. The minimum gap duration is represented by the turnaround time of the transceiver. The idea of CSMA with preamble sampling is adopted by a large number of protocols to prolong the lifetime of WSNs. Nonetheless, the performance of CSMA-PS based protocols is strongly affected by the network characteristics, the hardware limitations, and the traffic pattern. Especially, the duty-cycle and the turnaround time have a large impact on the performance of the protocol. CSMA-PS can be further improved by using scheduled wake up after transmission as introduced by Cano et al. in [22].

2.2. X-MAC

The X-MAC [17] protocol is designed for asynchronous low-power duty-cycled WSNs. It uses strobed preambles to achieve a better performance than ordinary Low Power Listening (LPL) [2] based protocols. The short strobed preambles are used instead of a single large preamble. Moreover, the short preambles contain the address of the destination. Thus, a destination node may recognize its own address immediately and transmit an acknowledgment in the next gap after the preamble which reduces the medium access delay since the originator does not need to transmit all short preambles. Figure 3 shows the difference between the medium access of LPL and X-MAC.

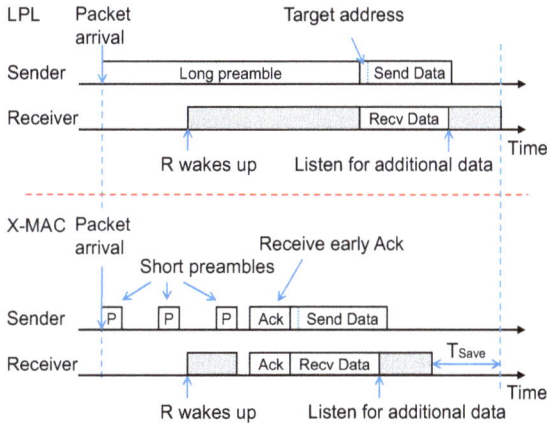

Figure 3. X-MAC - Medium Access

The advantage of X-MAC over LPL is that the destination node can respond immediately instead of listening to the whole preamble. The originating node stops the preamble transmission and starts its data transmission after receiving the early acknowledgment from the destination node during one of the gaps.

As a result, the medium access delay is reduced by approximately 50% even in the case that there is no contention on the radio channel. The difference may become larger depending on the preamble duration, the traffic load, and the packet size. The efficiency of the protocol depends on the CCA delay and the switching time of the transceiver between rx and tx mode since these hardware limitations are responsible for the length of the short preamble and the duration of the gaps. In addition, the medium access delay is strongly affected by the hardware limitations due to the fact that they also limit the length of the duty-cycle.

The protocol takes advantage from data sniffing. A destination node stays awake a short time after it has received a data transmission. Therefore, it can respond quickly with an early acknowledgment if another node wants to send packets to it. This feature may look unimportant at first glance. However, traffic patterns in WSNs are typically data-centric and event-driven. For this reason, data sniffing significantly affects the performance of the X-MAC protocol. Moreover, the acknowledgment covers the function of a CTS message if received by a node which is not the originator of the preamble. Thus, it reduces the collision probability in multi-hop networks caused by the hidden-node problem. The protocol is able to improve its energy efficiency depending on the traffic load since a node switches off its transceiver if it receives a preamble or an acknowledgment which is not dedicated for it. As a result, the corresponding node safes energy which prolongs its lifetime.

2.3. Wise-MAC

The Wise-MAC [20] protocol was developed by the Swiss Center for Electronics and Microtechnology as part of the WiseNET platform [25]. The protocol is optimized for energy efficiency in low traffic WSNs. The medium access is based on synchronized preamble sampling. In addition, the protocol is designed for infrastructure communication where more powerful and less energy-constraint nodes cover the task of base stations.

Nodes that are energy-constraint only communicate directly with the base station. In the following, these nodes are referred to as subscribers or subscriber nodes. If a subscriber node wants to transmit a packet to another node, it sends the packet to the base station. The base station transmits the packet to the destination node if the destination node is registered at this base station. Otherwise, the packet is forwarded to the corresponding base station where the destination node is registered.

In infrastructure networks, different MAC protocols and different radio channels can be used for the downlink and for the uplink since a base station will not switch off its transceiver in contrast to the subscriber nodes. Therefore, the downlink - from the base station to the subscriber nodes - represents the challenging part in low-power infrastructure WSNs due to the asynchronous sleep scheduling of the subscriber nodes. Wise-MAC is designed to optimize the downlink in terms of energy consumption and delay. It is based on preamble sampling like many other MAC protocols [1, 9]. However, the difference to other protocols lies in the fact that the base station learns the sampling schedule of its neighbor nodes. Thus, the idle listening time of the subscribers can be reduced if the base station starts to transmit the wake-up preamble in respect to the wake-up period of the corresponding subscriber. The medium access of the Wise-MAC protocol is shown in Figure 4.

Subscriber nodes sense the medium with a wake-up period of T_W. If a base station wants to transmit data to one of its subscriber nodes, it starts to transmit the wake-up preamble right before the wake-up period of the subscriber node. The transmission of a data frame is started as soon as the base station is assured that the subscriber is listening. Note that a frame may contain one or more data packets. The frame starts with the address of the subscriber. Thus, other subscribers can switch off their transceivers in order to avoid idle listening caused by overlapping wake-up intervals. The address field is followed by a data field which holds one data packet. Each frame ends with a frame pending bit to signalize to the subscriber station whether the base station has additional data frames pending for it. As a result, the energy efficiency of the protocol is increased since the subscriber is able to switch off its transceiver as

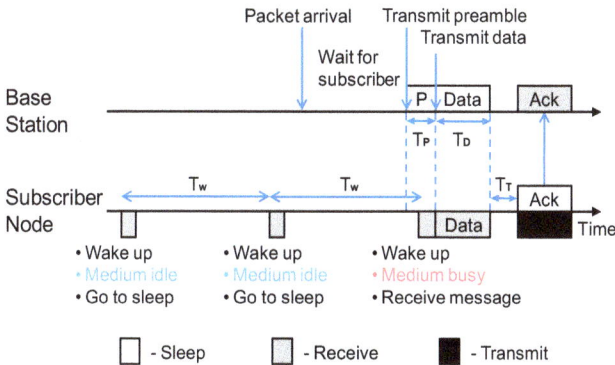

Figure 4. Wise-MAC - Medium Access

soon as possible. The subscriber node responds with an acknowledgment to the base station in the case that the base station has indicated that no additional frames are pending. The acknowledgment of the subscriber contains the information about the remaining time until the subscriber senses the medium again. This information is then used by the base station to keep its sampling scheduling information table up-to-date. The base station also stores the time when the acknowledgment was received in order to take the clock drift of the oscillator of the micro controller into account.

2.4. BPS-MAC protocol

Random access based MAC protocols are not able to reliably exchange data in dense WSNs with correlated event-driven traffic if they solely rely on the sensing capabilities of the low power transceiver due to the fact that the transceivers cannot detect a transmission that has been started within an interval that is shorter than the CCA delay and the turnaround time. The BPS-MAC protocol addresses this problem by using backoff preambles with variable length before transmitting data. The duration of the preamble is a multiple of the CCA delay or the turnaround time of the transceiver. Thus, a node is able to detect a synchronous preamble transmission of another node provided that they choose a backoff preamble with a different number of slots. Furthermore, the slot duration has to be larger or equal than the CCA delay and the turnaround time in order to leave the nodes enough time to switch the transceiver mode and/or to sense the medium. An example of the medium access procedure with two backoff sequences is introduced in Figure 5.

The example shows a scenario in which three nodes compete for the medium access. As mentioned in the previous paragraph, the BPS-MAC protocol divides the time during the medium access into time slots. A node that wants to transmit data senses the radio channel for duration of three slots. If the medium has been idle during the three slots, the node switches its transceiver from receive to transmit mode which requires an additional slot. Then, the node chooses a backoff duration and starts to transmit the backoff preamble. After the transmission of the preamble is completed, the node switches its transceiver back to receive mode and senses the medium. If a node senses a busy medium after the preamble transmission, it restarts the medium access procedure after a random number of slots. In the case that the medium is free after the preamble transmission, the node switches its transceiver back to tx

(a) Synchronous Access

(b) Collision

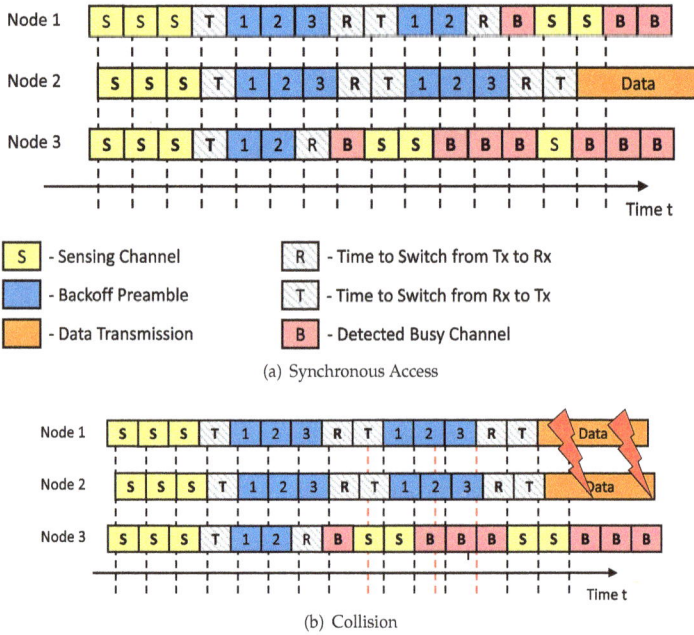

Figure 5. Sequential Contention Resolution

mode in order to proceed with the next sequence of the contention resolution. A node is only allowed to start its data transmission if it has sensed an idle medium after the transmission of the last backoff preamble. Note, the time between two consecutive preambles is two slots. For that reason, the nodes sense the medium for a duration of three slots at the beginning of the medium access process to assure that there is no ongoing data transmission.

The introduced procedure reduces the collision probability in case of synchronous medium access in a significant way. However, collisions may still occur if two or more nodes start their preamble transmission at the same time and chose the same number of preamble slots in every backoff sequence. Figure 5b shows a collision example for a contention resolution with two backoff sequences. The figure points out that the collision probability can be decreased by either increasing the maximum backoff duration of a single sequence or by increasing the number of backoff sequences.

Nonetheless, the backoff procedure represents protocol overhead which limits the maximum throughput of the protocol. Therefore, both parameters have to be chosen in respect to the node density and the traffic pattern. The sequential contention resolution represents an extension of the medium access procedure that is introduced in [8].

3. Implementing QoS strategies

Quality of Service requirements in the Internet lead to the development of several approaches for realizing QoS guarantees. The best known approaches are DiffServ [26] and IntServ [27].

While these protocols can be used in the global Internet, they can hardly be applied in the context of sensor networks due to multiple reasons. Sensor networks which consist of small and resource-constrained devices cannot run resource-intensive protocols that have high requirements concerning computational power, memory and bandwidth. Moreover, unreliable links and time varying channel conditions make QoS in WSNs a difficult task [28].

Instead, light-weight protocols, which require little synchronization between the nodes involved in the communication, are more suitable in this context. In order to build such light-weight protocols, simple QoS strategies need to be employed. These strategies, which can be implemented using preamble sampling protocols, should on the one hand be as simple as possible while fulfilling the requirements of a large number of applications on the other hand. In Subsection 3.1, we will discuss a number of strategies that should be implemented by a QoS approach. Afterwards, we will highlight how such an approach needs to be configured to provide the QoS strategies in Subsection 3.2. The impact of preamble transmission in multi-hop wireless networks is discussed in Subsection 3.3 by comparing the performance of the BPS-MAC protocol and Zigbee in a WSNs with high node density.

3.1. Qos strategies

3.1.1. Topology-aware

The topology in WSNs is often built from two types of nodes: few powerful nodes with little energy constraints that form a backbone and a large number of nodes with limited hardware and energy resources which use this backbone. While these backbone nodes have a distinguished special role in the network, they use the same shared medium for communication as the other nodes. This results in a situation where the backbone nodes compete for medium access with the constrained nodes. In order to avoid this, mechanisms that prioritize the medium access for backbone nodes should be implemented, which could improve the overall network performance: Since the number of backbone nodes with access priority is very small, the medium access delay for these nodes is decreased. This allows backbone nodes to forward messages faster in the WSN, thus decreasing the delay while increasing the delivery ratio. Furthermore, this strategy gives the backbone nodes control of the medium access which improves the support for data aggregation mechanisms.

3.1.2. Network-aware

WSNs have gained popularity due to their self-organizing capabilities, which allows them to be easily and randomly deployed in many scenarios. This includes scenarios where nodes can become hardly accessible, e.g. due to radioactive contamination. Such scenarios do not allow for careful sensor placement and topology architecture, but rely on the self-organization topology of the WSNs. Furthermore, some scenarios do not allow the replacement or relocation of individual nodes or the complete network. Asymmetric links or network partitioning might also make reprogramming or shutdown of nodes very difficult or even impossible.

If a user is forced, as a consequence of such conditions, to deploy a new sensor network on top of an older one, a number of problems can arise due to the shared characteristic of the wireless medium. This can limit the performance of the newly deployed network which typically operates in the same area and on the same radio channel. Frequently transmitting older nodes will compete with the new nodes for medium access which increases the power consumption of the nodes in the newly deployed network. A priority-based medium access strategy, which allows to assign a higher priority to the newly deployed network, can mitigate the problem of co-existing networks that operate on the same frequency.

3.1.3. Traffic-aware

An increasing number of sensor networks perform different tasks at the same time. The traffic streams related to these different activities might have different priorities for a user. Thus, traffic-awareness within the MAC protocol can provide QoS guarantees for the different streams. Assume a WSN in which nodes generate traffic with different priorities, e.g. the stress and strain measurements of a structural health monitoring application, which has high QoS requirements, and temperature measurements which can be transmitted as best effort traffic. Assigning a higher priority to the traffic of the structural health monitoring than the temperature application would lead to faster forwarding of this kind of critical information.

3.1.4. Service-aware

Network and service virtualization has become an interesting topic within the last few years, with first implementations for WSNs already available [29]. Their key improvement is that they allow several different users access to the nodes and sensors in a shared manner. Resource allocation for each user on a node, e.g. computational power, memory, sensors, must be done properly in such environments, and has been studied in a variety of research work in the past years. However, as soon as the medium access has to be taken into account, the consideration of user priorities becomes a challenging task. Scheduling of packets according to priority on a single node can be easily done by applying predefined user priorities. Synchronization between users on different nodes is however very complex. The best scheduling algorithm implemented in the operating system of a node is useless if that node does not get access to the medium in order to transmit the carefully scheduled and queued packets.

3.1.5. Distance-aware

A typical WSN topology is configured in a way that allows the transmission of measured sensor data to a small number of data sinks adjunct to the network. These data sinks can then evaluate and process the data themselves or work as a gateway to another network. The topology of these networks is often arranged in a tree structure [30], which allows to take advantage from data aggregation mechanisms. While such a topology provides a number of advantages, it can be often observed that traffic load increases towards the sinks. Medium access can therefore play a critical role in these networks: A priority based medium access procedure that takes the distance to the sink into account, can support the data aggregation mechanisms to decrease the energy consumption of sensing nodes on the one hand or minimize delay on the other hand.

If nodes that are closer to the sink have a higher priority, the delay in event-based WSNs can be reduced since the node which is triggered by the event and is closest to the sink has the highest priority. It can therefore immediately access the medium to transmit its data. In addition, lower link delays can be achieved because the priority of the transmitted packets further increases on the path towards the sink.

In scenarios where energy consumption is a major constraint, e.g. more important than the problem of high delay, a different prioritization can be beneficial. A medium access strategy, that gives nodes further away from the sink a higher priority than nodes closer to the sink, can reduce the energy consumption of the transmitting devices. The nodes that are furthest away from the sink can transmit their data immediately. Afterwards, they can turn off their transceivers at the end of the transmission, thus saving valuable energy. Furthermore, such prioritization improves the potential of data aggregation: All children of an aggregation node in the tree have a higher medium access priority than their parent. As a result, the children can transmit their data to the parent before the parent gains access to the medium in order to forward the data. Thus, the aggregation node can aggregate more messages from its children and operate more efficiently which reduces the number of medium access attempts.

3.1.6. Energy-aware

Wireless sensor nodes have very limited energy resources, which should be taken into account when prioritizing medium access. Designers of communication protocols therefore work very hard to minimize the power consumption while still meeting the given requirements. Energy-aware routing protocols, which include energy consumption into their protocol, typically aim at avoiding nodes that have little energy left. Such mechanisms have been proven to balance the traffic load and prolong the lifetime of WSNs. However, access to the medium can become a costly factor in the communication process if a node has to compete for the medium access multiple times before it can finally send its data. It can be therefore beneficial if nodes that run low on power have a higher medium access priority. These nodes can therefore save energy by the fact that their average number of medium access attempts is reduced by assigning them a higher access priority.

3.1.7. Buffer-aware

The small amount of memory represents a serious issue in WSNs. Especially, if Internet Protocol (IP) stacks are deployed on the devices since actions such as IP packet fragmentation and packet forwarding have high demands on memory. Most sensor nodes, like the TelosB, T-Mote or Mica nodes, only have as little as 8 or 10 KB of ram, which posses problems when multiple large IP packets need to be buffered before they can be forwarded. In conjunction with event-driven traffic patterns in WSNs, temporarily high traffic spikes can occur in the network. This can in turn lead to the demand for buffering several packets in some forwarding nodes. While load-balancing routing protocols can mitigate the impact of this issue in multi-hop networks, a MAC protocol which is aware of the problem can further improve the network performance. It can do this by taking the nodes' waiting queues into account: Nodes that have more packets stored in their buffers should have a higher priority, which enables them to get faster access to the medium. They can therefore reduce the amount of data in their buffers quickly, thus targeting the resource exhaustion problem already at an early stage. As a consequence, the maximum waiting queue length and share of dropped packets due to buffer overflows can be decreased.

3.1.8. Data-rate aware

The latest generation of routing protocols for WSNs, e.g. the Collection Tree Protocol (CTP) [30], apply adaptive mechanisms to cope with frequent topology changes. In general, these protocols increase their beacon transmission rate if they detect changes in their neighborhood. Topology changes usually result from interference or mobility of the nodes. The latter may lead to frequent topology changes which significantly increase the routing overhead. In dense networks, the routing overhead can even result in temporary congestion of the network. Temporary congestion can also be caused by applications which generate event-driven traffic, e.g. intruder detection. For these kinds of applications, it is important to receive information from all devices which have detected the event to gain more precise information and minimize false positives. The priority of the medium access should depend on the transmission rate of the nodes. A fair medium access can be achieved if a higher transmission rate results in a lower access priority and vice versa. Thus, nodes which rarely transmit traffic have a high probability of gaining access to the medium immediately. However, nodes that frequently transmit traffic can utilize the whole bandwidth as long as no other nodes need access to the medium.

3.1.9. Combined Strategy

Finally, it could be beneficial to have a strategy which combines the properties of the previously discussed ones. Depending on the target scenario and application, a combined strategy could further improve the performance. For example, a combined strategy could employ both the traffic-aware and buffer-aware strategy. Such a combination would represent a trade-off between the delay of high priority packets and packet loss of packets due to buffer overflows. A function which performs the trade-off calculation must be derived which calculates a priority value for each node, depending on the type of traffic that it has to forward and its current buffer fill-level.

3.2. Configuring QoS strategies

QoS strategies can be easily integrated in preamble sampling protocols since the preamble may hold additional information about the subsequent transmission. Another possibility is to encode the medium access priority in the preamble duration which is done by the BPS-MAC protocol. In the following, the an example of priority encoding is introduced which can be directly applied to the BPS-MAC protocol.

The protocol can be used in two different modes: Collision-free and prioritized contention resolution mode. Both modes result in different usages of the preamble sequences: In collision-free mode, node IDs are directly mapped onto preambles, resulting in unique preambles for each node which renders additional contention mechanisms unnecessary. In cases where preambles are not unique per node, but a priority is assigned per node group, some preamble sequences are assigned to prioritize the medium access while the others are used to resolve possible contention among nodes which have the same priority.

The decision parameter that defines the priority needs to be mapped to the length of the preamble sequence in order to configure the priority of a group of nodes. In the following we give recommendations on how the different QoS strategies that have been presented before can be implemented by choosing a certain preamble configuration. Table 2 summarizes the types of the different strategies along with the properties that should be mapped to the sequence length. Static strategies should be configured by the user in advance before the nodes are deployed to their final location, e.g. by defining static IDs to properties. Higher property IDs yield longer preamble sequences and therefore result in higher access priorities. Mappings for static priority strategies are very straight forward: A user needs to define how

Strategy	Type	Decision parameter	Characteristics
Topology	Static	Node Type (Backbone, Constrained, ...)	Flow / traffic optimization
Network	Static	Network ID	Co-existence of networks
Traffic	Static	Application ID	Service differentiation
Service	Static	User ID	User prioritization
Distance	Dynamic	Hop count to data sink	Data aggregation
Inverse Distance	Dynamic	Hop count to data sink	Delay minimization
Energy	Dynamic	Battery fill level	Load balancing and life-time extension
Buffer	Dynamic	Buffer fill level	Fairness and reliability
Data Rate	Dynamic	Transmission rate	Fairness

Table 2. Configuration of different QoS strategies

many priority classes have to be supported. These priority classes have to be encoded onto a number of preamble sequences and the lengths of the sequences. The user may choose between providing a mapping of the priority to a single sequence or to multiple sequences, which in total use up to N slots. Let s be the number of sequences, and n_i the length of sequence i, then the total number of used slots will be:

$$N = 4 + \sum_{i=1}^{s} n_i + 2s \qquad (1)$$

since the first four slots being used for initially sensing the medium and switching from rx to tx mode, and the two pause slots between each sequence. While, at a first glance, choosing multiple sequences seems to be a bad decision due to the pause slot overhead, choosing multiple preamble sequences increases the number of priorities that can be encoded. The number of supported medium access priorities is given by the product of the maximum length in slots of each preamble sequence as shown in Figure 6. Now, consider a configuration that employs three sequences, each having a length of four slots. This configuration results in a total maximum medium access delay of 22 backoff slots according to Equation 1. If a single preamble sequence would have been chosen instead, maximum preamble duration of 18 backoff slots could be chosen to guarantee a total maximum medium access delay of 22 backoff slots. Thus, the single sequence configuration can only encode up to 18 priority classes whereas the configuration with three sequences can support up to 64 priority classes. This ratio further increases the more sequences are chosen.

3.3. Performance evaluation

The performance of MAC protocols for WSNs strongly depends on the characteristics of the network, e.g. the number of nodes, the node density, and the traffic pattern. Moreover, the

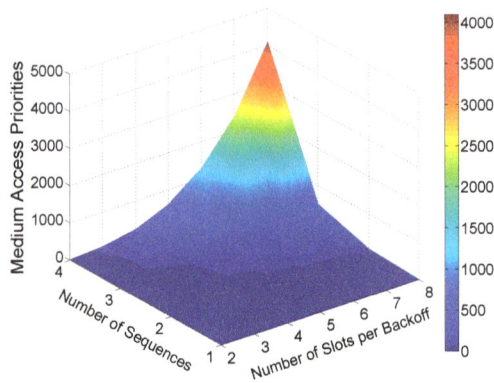

Figure 6. Medium Access Priorities depending on the Protocol Configuration

data rate and the sensing capabilities of the transceiver have a large impact on the network performance. In the following, it is assumed that the transceiver achieves a maximum data rate of 250 kb/s. Furthermore, a CCA delay and a turnaround time of 128 μs is assumed which represent typical values for low power transceivers. The OPNET Modeler [31] is used to simulate the performance of the protocols. Note that most simulation tools, like OPNET Modeler or ns-2 [32], simplify the physical layer in order to increase the simulation speed. Thus, their standard models simplify or even neglect important communication issues, e.g. the turnaround time of the transceiver and the CCA delay. For that reason, we modified the physical layer of the OPNET Modeler software such that it takes both communication issues into account. The transmission range is limited to 10 meters and the maximum interference range is set to 17 meters by modifying the so-called pipeline stages of OPNETs free space propagation model. These values reflect the average results from our first measurements with a small self-developed sensor board that uses a MSP430 micro controller and a CC2420 transceiver. The short range results from the fact that the nodes were placed inside the backrest of the seat. It is clear that these values may vary significantly depending on the position and orientation of the sensor node and the characteristics of the used antenna. Thus, the assumed values only fit to our particular example scenario.

The simulated scenario represents a typical middle-size airplane with six seats per row. A wireless sensor is placed in the backrest of each seat which monitors the state of the seat, e.g. whether the seat is occupied, the seatbelt is fastened, or the tray is secured. This information is reported periodically to a sink in the front of the plane. It has to be kept in mind that the simulated application is just an example application. There are currently a large number of applications under consideration to improve the existing flight cabin management system. A multi-hop network is required to enable connectivity between all nodes in the network due to the fact that large planes reach lengths of up to 60 meters. More powerful sensor nodes with routing capabilities are placed on the ceiling along alleyway approximately every 8 meters in order to connect the other sensors with the sink. An overview of the simulated scenario is shown in Figure 7.

The figure illustrates the high node density of up to 60 nodes.However, the large interference range has to be taken into consideration as well when specifying the application requirements.

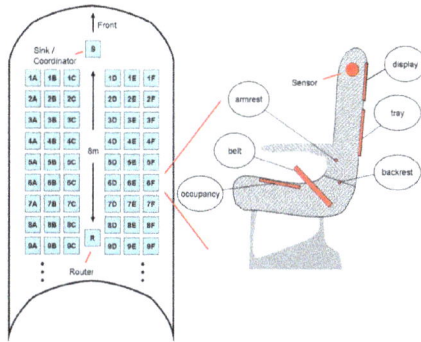

Figure 7. Overview of the Simulated Scenario

As a consequence of the high node density, the traffic pattern has a huge impact on the network performance in the simulated scenario. Data traffic is usually highly correlated in WSNs since it is often event-driven and data centric. Thus, we decided to simulate three different traffic patterns which are representative for a large number of popular intra-aircraft applications. The simulated traffic pattern are shown in Table 3. The number of (seat) rows is increased from 8 to 40 in order to find out how many nodes are supported by the protocols in the intra-aircraft scenario depending on the application. The results represent the 90 percent confidence intervals of the average end-to-end delay and packet loss that are collected from 20 simulation runs with a duration of 1000 seconds and different seeds.

The traffic pattern start after 80 seconds since the Zigbee model requires some time to build a tree topology. In addition, the traffic generation stops at 980 seconds to allow the nodes to empty their waiting queues. Thus, the packet loss is given by the fraction of generated packets and the number of packets that are successfully received by the sink. Zigbee is set to non-beacon mode. Zigbee implies network layer functionality. Thus, a directed-diffusion [33] based routing protocol is used in combination with the BPS-MAC protocol to support comparable routing functionality. The directed-diffusion based routing protocol is modified such that only routers retransmit the interest which minimizes the routing overhead. The BPS-MAC protocol uses three consecutive backoff preambles with a maximum number of four slots.

3.4. Scenario A

The introduced passenger monitoring application does not require a large amount bandwidth since some of the monitored characteristics, e.g. seatbelt fastened or unfastened, are logical. However, advanced monitoring features such as temperature or humidity can be considered. Furthermore, the sensed values are not time-critical. In scenario A, the nodes follow the low traffic pattern of application A which is introduced in Table 3. Figure 8a shows the average end-to-end delay between the nodes and the sink depending on the number of rows in the plane. The figure reveals that the end-to-end delay increases non-linearly which is the consequence of the multi-hop communication.

Moreover, the figure points out that the delay of the BPS-MAC protocol is higher compared to Zigbee if the number of rows is larger than 8. Nonetheless, the average end-to-end delay of the BPS-MAC protocol remains lower than 0.35 seconds even for the 40 row scenario which is

Pattern Name	Parameter	Distribution	Range / Values
Application A	Packet IAT	uniform	[9.99; 10.01] s
	Packet Size	constant	256 bit
	Start Time	uniform	[80;90] s
	Number of Rows	-	[8;16;24;32;40]
Application B	Packet IAT	uniform	[9.99; 10.01] s
	Packet Size	constant	256 bit
	Start Time	uniform	[80;81] s
	Number of Rows	-	[8;16;24;32;40]
Application C	Packet IAT	uniform	[3.95; 4.05] s
	Packet Size	constant	1024 bit
	Start Time	uniform	[80;84] s
	Number of Rows	-	[8;16;24;32;40]

Table 3. Traffic Pattern

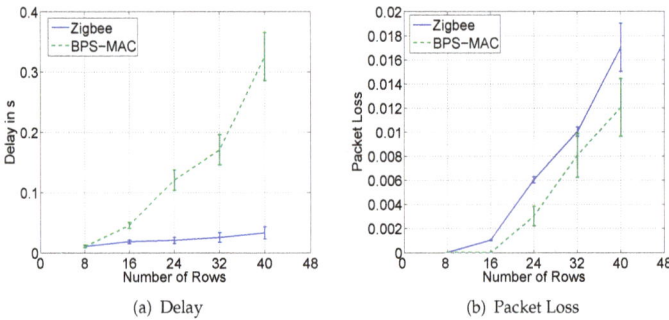

(a) Delay

(b) Packet Loss

Figure 8. Application A - Performance depending on the Number of Rows

quite acceptable for this kind of application. The BPS-MAC protocol achieves a lower packet loss than Zigbee in scenario A as shown in Figure 8b due to the fact that the medium access procedure is optimized for synchronous medium access. The probability increases that two or more nodes start their data transmission within an interval that is shorter than the CCA delay of the low power transceiver in- creases with the number of nodes in the networks. As a result, the packet loss increases almost linearly for both protocols but still remains below 2 percent. Therefore, both protocols represent an acceptable solution for application A.

3.5. Scenario B

Scenario B uses almost the same traffic pattern as scenario A. The only difference lies in the fact that the offset of the traffic pattern only varies uniformly distributed by 1 second. Thus, the probability that two nodes access the medium within an interval that is shorter than the CCA delay and the turnaround time is very high. The average end-to-end delay of the different protocols depending on the number of rows is shown in Figure 9a. Both protocols achieve a low delay for scenarios in which the number of rows remains below 24. The delay sharply increases if the number of rows exceeds 24 as a consequence of the multi-hop communication and the highly correlated traffic.

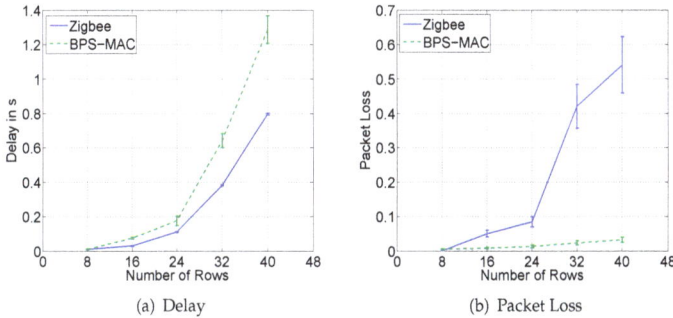

(a) Delay   (b) Packet Loss

Figure 9. Application B - Performance depending on the Number of Rows

Figure 9b shows a similar picture for scenarios with less than 24 rows. An extra ordinary high packet loss can be mentioned for the Zigbee protocol which results from the highly correlated traffic. Zigbee is not able to resolve the contention in this case due to the fact that the protocol is not addressing the problem caused by the CCA delay and the turnaround time. In contrast to Zigbee, the packet loss of the BPS-MAC remains on a low level such that it only increases to a maximum of 2 percent for the 40 row scenario.

3.6. Scenario C

In scenario C the performance of the protocols is simulated under a higher traffic load. The nodes in network generate traffic according to the traffic pattern of application C shown in Table 1. The traffic load is ten times higher than the load that is generated by application A or application B. Thus, the overall generated traffic load is 61.4 kB/s for the 40 row scenario. However, this calculation excludes the traffic that is required for forwarding data. It has to be kept in mind that some nodes require up to four hops to reach the sink in the 40 row scenario.

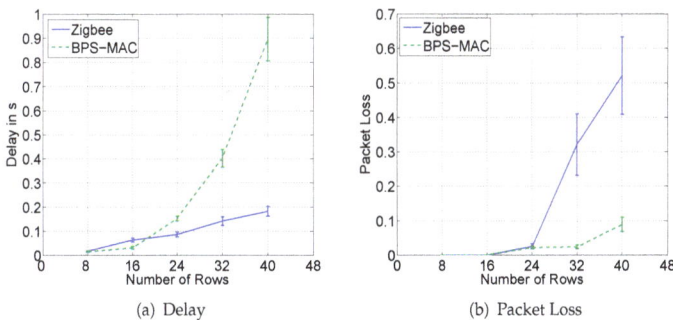

(a) Delay   (b) Packet Loss

Figure 10. Application C - Performance depending on the Number of Rows

Figure 10a shows the average end-to-end-delay in scenario C depending on the number of rows. The figure reveals that the BPS-MAC protocol achieves a slightly lower delay than Zigbee as long as the number of rows is smaller or equal than 16. The delay of Zigbee increases almost linearly while the slope of the delay graph of the BPS-MAC protocol shows exponential characteristic. This slope results from the high utilization of the medium and the large number of nodes in the network. Nonetheless, the average delay of the BPS-MAC protocol remains below one second.

The packet loss shown in Figure 10b points out that the BPS-MAC protocol in combination with a directed diffusion based routing protocol provides a better solution than Zigbee for scenario C. The figure indicates that Zigbee is not able to handle a network that is larger than 24 rows if the nodes generate traffic according to application C. In this case, the high traffic load in combination with the correlated traffic limit the performance of Zigbee since the MAC does not address the CCA delay and the turnaround time explicitly. The packet loss of the BPS-MAC protocol increases to approximately 2 percent in the 32 row scenario which is sufficient for non-mission critical data. If the number of rows exceeds 32 the packet loss of the BPS-MAC protocol increases to 9 percent as a consequence of the high utilization.

4. Implementation issues of preamble-based MAC protocols

Two communication issues are mainly responsible for the low performance of MAC protocols in WSNs. The first issue is represented by the interval that low-power transceivers require to switch between receiving and transmitting and vice versa. Thus, the switching time which is in the following referred to as turnaround time, specifies the time between the arrival of a packet and the beginning of the corresponding response [34]. During this time interval the transceiver is not able to detect the start of other transmissions.

The second issue is called Clear Channel Assessment (CCA) delay. The CCA delay specifies the interval that a transceiver requires to detect a busy medium provided that the transceiver is already in receive mode. A transceiver is not able to reliably detect the transmission of another node if the transmission has been started within an interval that is shorter than the CCA delay. A closer look is taken on the impact of the turnaround time and the CCA delay on the MAC performance in the following two subsections.

Another factor that limits the performance of MAC protocols in WSNs is represented by the limited hardware resources. Especially, the small receive buffer and the applied operating system have to be taken into account when designing preamble sampling protocols that rely on short preamble transmissions. As a consequence of frequent short preambles, the probability of buffer overflows increases which leads to loss of information.

4.1. Impact of the turnaround time

The turnaround time of transceivers has a direct impact on the efficiency of MAC protocols. However, the impact on the performance depends on the medium access procedure which is used by the MAC protocol. The importance of the turnaround time was first addressed in [35] by Pablo Brenner. In this work, he evaluated the wireless access method and physical specification of the IEEE 802.11 standard. The same topic is discussed in more detail by Johnson et al. [34] and Diepstraten [36] who describe the effect on the performance caused by several switching aspects. Diepstraten outlines the impact that the turnaround time has on the protocol efficiency. The efficiency decreases especially in the case that a quick mutual exchange of messages, e.g. RTS-CTS messages, data packets or short preambles with early acknowledgments, between the transmitter and the receiver is required. In addition, the time that a transceiver requires to switch from receive to transmit mode represents a vulnerable period for MAC protocols which rely on the CSMA functionality since transceivers cannot detect any transmissions that start during the switching period [37, 38]. Therefore, the developers of a preamble-based MAC protocols have to choose the gaps between consecutive preambles with respect to the turnaround time of the wireless transceiver.

4.2. The problem of clear channel assessment delay

CCA is a logical function which returns the current state of the wireless medium. It is provided by almost any low-power transceiver for WSNs in order to support CSMA functionality to the MAC layer. However, the transceivers require a certain period of time depending on their current state to reliably determine the state of the medium.

The CCA delay becomes the dominating performance limitation factor [39] for low-power transceivers which have a relatively high CCA delay compared to IEEE 802.11 transceivers. Typical low-power transceivers, like the CC2400 [40] and the CC2520 [41] (Texas Instruments) or the AT86RF231 [42] (ATMEL), have to listen to the medium for duration of 8 symbol periods to reliably detect an ongoing transmission. The chips average the Received Signal Strength Indication (RSSI) over the last 8 symbols in order to decide whether the channel is assumed to be busy or idle.

Technical aspects, like the CCA delay of low-power transceivers which have a large influence on the performance of wireless communication in sensor networks, are usually neglected. The impact of CCA delay on IEEE 802.15.4 networks is described by Kiryushin et al. [39]. The focus of their work lies on real world performance of WSNs and describes the impact of different kinds of communication aspects. Bertocco et al. [43] have shown that the performance of a wireless network can be improved by minimizing the CCA threshold. Nevertheless, the minimization of the threshold requires great knowledge of the radio channel, e.g interference and background noise, since a too small threshold will result in false positives which will significantly decrease the throughput. Thus, nodes will not transmit any data due to the fact that they falsely assume the channel to be busy. The latest generation of low-power transceivers supports different kinds of CCA methods. An intelligent cross-layer approach which takes advantage from different CCA methods is introduced by Ramachandran and Roy [44]. Their idea is to dynamically adapt the CCA method and parameters depending on the current channel conditions and the upper layer parameters.

4.3. Architecture of low-power transceivers

Wireless sensor nodes are very limited in terms of computational power and memory. The small receive buffer of low-power transceivers has to be considered when developing a preamble-based MAC protocol. Typical low-power transceivers, like the TI's CC2400 [40] and ATMEL's AT86RF231 [42], are only equipped with a 128 byte RXFIFO. The RXFIFO of the CC2500 [41] transceiver family is even limited to 64 bytes. Therefore, a received frame has to be immediately read from the FIFO in order to avoid buffer overflows caused by consecutive transmissions which would lead to loss of information. This issue can become a major problem for non real-time operating systems such as TinyOS which is only able to handle up to ≈ 170 packets per second. However, this packet reception rate only applies for small packets with a size of less than approximately 30 bytes. Moreover, there should no other time-consuming tasks be running on the sensor node. Otherwise, the packet reception rate drops down significantly as a result of the non-preemptive task scheduler.

4.3.1. Performance limitation factor - RXFIFO

In the following a brief introduction of the packet reception procedure for the CC2420 is given to provide MAC protocol designers useful information regarding the implementation. Upon detection of the Start Frame Delimiter (SFD) field, the chip begins to buffer the received data in its 128 byte RXFIFO [40]. In case the default Auto Cyclic Redundancy Check (AUTOCRC) settings are not changed, the chip replaces the first byte of the Frame Check Sequence (FCS) with a RSSI value estimated over the first 8 symbols after the SFD field. The second FCS byte is replaced by a 7 bit correlation value used for Link Quality Indication (LQI) computation and by a 1 bit field which indicates whether the frame was correctly received. The resulting data frame is shown in Figure 11.

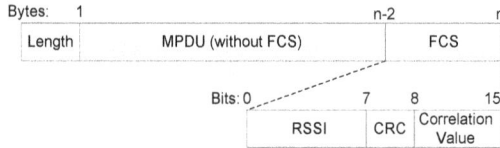

Figure 11. Data Frame in RXFIFO

The main limitation of this strategy is that damaged frames remain in the RXFIFO until they are read. In case of an overflow, the chip is not able to receive data until all correctly received frames are read and the buffer is flushed. Furthermore, the CC2420 does not protect the length field of the physical header. If the field is damaged and indicates, for example, that the stored frame is larger than the default frame size, TinyOS will *immediately* flush the buffer regardless of whether it contains correctly received frames since it does not know when the next correctly received packet starts. This means that if a MAC protocol fails to recognize a busy channel, a possible frame collision may cause correctly received frames to be flushed at the receiver before they are forwarded to the application. Listing 1 describes the reception routine in detail.

```
async event void RXFIFO.readDone (...){
    ...
    if (rxFrameLength + 1 > m_bytes_left) {
      flush ();
    }
    else {
      if ( !call FIFO.get () && !call FIFOP.get () ) {
        m_bytes_left -= rxFrameLength + 1;
      }
      if (rxFrameLength <= MAC_FRAME_SIZE) {
        // further frame processing
      }
      else {
        flush ();
      }
    }
    ...
}
```

Listing 1. Frame Length Field Processing

In general, a frame stored in the RXFIFO is read in three steps. The first step involves the interpretation of the first byte, which represents the length field, to detect the end of the frame. The second step includes the processing of the FCF header. During this phase the receiver may generate an acknowledgment when the acknowledgment request bit is set. The last step is needed to determine whether the received frame is correctly received and to signal reception to the upper layer components.

After evaluation of the length field, TinyOS tries to determine whether the received frame has a size larger than the RXFIFO. If this condition is true, the flush() function, which immediately erases the receive buffer, is called. Note that TinyOS uses the variable m_bytes_left to store the size of the buffer. This variable is initialized with a value of 128, but is decremented when an overflow occurs. With each frame which is still in the RXFIFO m_bytes_left gets decremented until the last incomplete frame. For this frame, the condition rxFrameLength + 1 > m_bytes_left is true and the queue is flushed. This action is required to enable reception of new data and reset m_bytes_left.
TinyOS checks whether the frame size is smaller than MAC_FRAME_SIZE. If the frame size is larger, the received frame is assumed to be corrupted and the RXFIFO is flushed. Thus, frames that were received correctly but have arrived after the corrupted frame will be lost.

4.3.2. Preamble transmission

Buffer overflows represent a serious problem in dense wireless sensor networks with event-driven traffic due to the fact that nodes often try to retransmit lost data. Moreover, frequent transmission of preambles increases the utilization of the medium and the probability of buffer overflows in the receiver. Before writing a packet in the receive buffer, a transceiver reads the SFD flag of a received frame. The packet is only stored in the receive buffer if the SFD flag indicates a valid frame. Otherwise, the frame is silently discarded. However, an invalid frame is still detected by transceiver. Thus, a preamble that is transmitted with an invalid SFD flag can be used as reservation signal. On the other hand, useful information cannot be stored in the preamble since receiving nodes immediately discard the frame without further evaluation. Therefore, this mechanism cannot be applied for protocols, like XMAC or WiseMAC, which store information in the preamble.

5. Conclusion

Preamble sampling protocol have many interesting characteristics which are of special interest for WSNs with their low traffic load and very limited low-power devices. In this chapter, we have discussed several preamble-based medium access strategies which either relied on long or short strobed preambles. The strategies were categorized with respect to the type of preamble (short or long), the information stored in the preamble and additional functionality such as synchronization and contention resolution. Furthermore, we outlined the impact of the different preamble sampling strategies on the energy consumption by focusing on the resulting protocol overhead. After describing the basic principles of preamble sampling, a closer look on a selection of preamble-based protocols was given. Due to the fact that preamble sampling protocols are ideal candidates for energy-constraint WSNs that require QoS support, a brief introduction on QoS in WSNs was given. Preamble sampling represents additional protocol overhead which has to be taken into account when deploying preamble-based protocols in dense wireless networks. For this reason, we compared the performance of a typical preamble sampling protocol with a CSMA-based in a large wireless network with high node density and event-driven data traffic. The results have shown that preamble sampling protocols usually have a higher delay compared to their CSMA-based counterparts. However, the preamble sampling protocols can be optimized and configured such that they achieve a higher reliability. Sensor nodes are usually very restricted in terms of memory which increases the probability of buffer overflows that lead to loss of information. This issue can be mitigated by setting the SFD flag of preambles to invalid if no information has to be stored in it.

Acknowledgements

The author would like to thank Christian Sternecker, Lothar Braun and Tsvetko Tsvetkov for many fruitful discussions regarding the implementation and Quality of Service extension of the BPS-MAC protocol. Furthermore, the author is very thankful for the support of Prof. Carle, the head of the Chair for Network Architectures and Services at the Technical University Munich.

Author details

Alexander Klein
Network Architectures and Services - Institute for Informatics, Technical University Munich, 85748 Garching, Germany

6. References

[1] A. El-Hoiydi. Aloha with preamble sampling for sporadic traffic in ad hoc wireless sensor networks. In *Proc. IEEE International Conference on Communications ICC 2002*, volume 5, pages 3418–3423, May 2002.

[2] Jason L. Hill and David E. Culler. Mica: A wireless platform for deeply embedded networks. *IEEE Micro*, 22(6):12–24, November 2002.

[3] I. F. Akyildiz, Weilian Su, Y. Sankarasubramaniam, and E. Cayirci. A survey on sensor networks. *IEEE Communications Magazine*, 40(8):102–114, August 2002.

[4] I. Demirkol, C. Ersoy, and F. Alagoz. Mac protocols for wireless sensor networks: a survey. *Communications Magazine, IEEE*, 44(4):115 – 121, April 2006.

[5] P. Suriyachai, U. Roedig, and A. Scott. A survey of mac protocols for mission-critical applications in wireless sensor networks. *Communications Surveys Tutorials, IEEE*, PP(99):1 –25, 2011.

[6] M. Aykut Yigitel, Ozlem Durmaz Incel, and Cem Ersoy. Qos-aware mac protocols for wireless sensor networks: A survey. *Comput. Netw.*, 55:1982–2004, June 2011.

[7] Joseph Polastre, Jason Hill, and David Culler. Versatile low power media access for wireless sensor networks. In *Proceedings of the 2nd International ACM Conference on Embedded Networked Sensor Systems*, SenSys '04, pages 95–107, New York, NY, USA, 2004. ACM.

[8] Alexander Klein, Jirka Klaue, and Josef Schalk. BP-MAC: a high reliable backoff preamble MAC protocol for wireless sensor networks. *Electronic Journal of Structural Engineering (EJSE): Special Issue on Sensor Network for Building Monitoring: From Theory to Real Application*, -:35–45, December 2009.

[9] A. El-Hoiydi. Spatial TDMA and CSMA with preamble sampling for low power ad hoc wireless sensor networks. In *Proc. Seventh International Symposium on Computers and Communications ISCC 2002*, pages 685–692, July 2002.

[10] Alexander Klein. Bps-mac: Backoff preamble based mac protocol with sequential contention resolution. In *Proceedings of the 4th International Conference on Multiple Access Communications*, MACOM'11, pages 39–50, Berlin, Heidelberg, 2011. Springer-Verlag.

[11] S. Mahlknecht and M. Bock. Csma-mps: a minimum preamble sampling mac protocol for low power wireless sensor networks. In *Factory Communication Systems, 2004. Proceedings. 2004 IEEE International Workshop on*, pages 73 – 80, September 2004.

[12] A. Bachir, D. Barthel, M. Heusse, and A. Duda. Micro-frame preamble mac for multihop wireless sensor networks. In *Communications, 2006. ICC '06. IEEE International Conference on*, volume 7, pages 3365–3370, June 2006.

[13] A. M. Firoze, L. Y. Ju, and L. M. Kwong. Pr-mac a priority reservation mac protocol for wireless sensor networks. In *Proc. Int. Conf. Electrical Engineering ICEE '07*, pages 1–6, 2007.

[14] Rebecca Braynard, Adam Silberstein, and Carla Ellis. Extending network lifetime using an automatically tuned energy-aware mac protocol. In *Proceedings of the Third European conference on Wireless Sensor Networks*, EWSN'06, pages 244–259, Berlin, Heidelberg, 2006. Springer-Verlag.

[15] Kai-Juan Wong and D. K. Arvind. Speckmac: low-power decentralised mac protocols for low data rate transmissions in specknets. In *Proceedings of the 2nd international workshop on Multi-hop ad hoc networks: from theory to reality*, REALMAN '06, pages 71–78, New York, NY, USA, 2006. ACM.

[16] E.-Y.A. Lin, J.M. Rabaey, and A. Wolisz. Power-efficient rendez-vous schemes for dense wireless sensor networks. In *Communications, 2004 IEEE International Conference on*, volume 7, pages 3769 – 3776, June 2004.

[17] Michael Buettner, Gary V. Yee, Eric Anderson, and Richard Han. X-MAC: A short preamble MAC protocol for duty-cycled wireless sensor networks. In *SenSys '06: Proceedings of the 4th International Conference on Embedded Networked Sensor Systems*, pages 307–320, New York, NY, USA, 2006. ACM.

[18] C.J. Merlin and W.B. Heinzelman. Schedule adaptation of low-power-listening protocols for wireless sensor networks. *Mobile Computing, IEEE Transactions on*, 9(5):672 –685, May 2010.

[19] Xiaolei Shi and G. Stromberg. Syncwuf: An ultra low-power mac protocol for wireless sensor networks. *Mobile Computing, IEEE Transactions on*, 6(1):115 –125, Januar 2007.

[20] A. El-Hoiydi and J.-D. Decotignie. WiseMAC: An ultra low power MAC protocol for the downlink of infrastructure wireless sensor networks. In *Proc. Ninth International Symposium on Computers and Communications ISCC 2004*, volume 1, pages 244–251, June 2004.

[21] M. Anwander, G. Wagenknecht, T. Braun, and K. Dolfus. Beam: A burst-aware energy-efficient adaptive mac protocol for wireless sensor networks. In *Networked Sensing Systems (INSS), 2010 Seventh International Conference on*, pages 195 –202, june 2010.

[22] C. Cano, B. Bellalta, A. Sfairopoulou, and J. Barcelo. A low power listening mac with scheduled wake up after transmissions for wsns. *Communications Letters, IEEE*, 13(4):221 –223, april 2009.

[23] Philipp Hurni and Torsten Braun. Maxmac: a maximally traffic-adaptive mac protocol for wireless sensor networks. In *Proceedings of the 7th European conference on Wireless Sensor Networks*, EWSN'10, pages 289–305, Berlin, Heidelberg, 2010. Springer-Verlag.

[24] C. Cano, B. Bellalta, A. Sfairopoulou, and M. Oliver. Low energy operation in wsns: A survey of preamble sampling mac protocols. *Computer Networks*, 55(15):3351 – 3363, 2011.

[25] C. C. Enz, A. El-Hoiydi, J.-D. Decotignie, and V. Peiris. WiseNET: An ultralow-power wireless sensor network solution. *Computer*, 37(8):62–70, August 2004.

[26] S. Blake, N. Carlson, Z. Wang, and W. Weiss. An architecture for differentiated services. RFC 2475, 1998.

[27] R. Braden, D. Clark, and S. Shenker. Integrated Services in the Internet Architecture - An Overview. RFC 1663, June 1994.

[28] A. Muneb, Umar Saif, A. Dunkels, T. Voigt, K. Römer, K. Langendoen, J. Polastre, and Z. A. Uzmi. Medium access control issues in sensor networks. *SIGCOMM Comput. Commun. Rev.*, 36(2):33–36, 2006.

[29] B. C. Donovan, D. J. Mclaughlin, M. Zink, and J. Kurose. Western massachusetts off-the-grid radar technology testbed. In *Proc. IEEE Int. Geoscience and Remote Sensing Symp. IGARSS 2008*, volume 5, 2008.

[30] Omprakash Gnawali, Rodrigo Fonseca, Kyle Jamieson, David Moss, and Philip Levis. Collection tree protocol. In *SenSys '09: Proceedings of the 7th ACM Conference on Embedded Networked Sensor Systems*, pages 1–14, New York, NY, USA, 2009. ACM.

[31] OPNET. Technologies, inc., OPNET modeler, university program, http://www.opnet.com/services/university/, May 2010.

[32] ns 2. Network simulator, http://nsnam.isi.edu/nsnam/, May 2010.

[33] Chalermek Intanagonwiwat, Ramesh Govindan, and Deborah Estrin. Directed diffusion: a scalable and robust communication paradigm for sensor networks. In *MobiCom '00: Proceedings of the 6th annual international conference on Mobile computing and networking*, pages 56–67, New York, NY, USA, 2000. ACM.

[34] E. E. Johnson, M. Balakrishnan, and Zibin Tang. Impact of turnaround time on wireless MAC protocols. In *Proc. IEEE Military Communications Conference MILCOM 2003*, volume 1, pages 375–381, October 2003.

[35] Pablo Brenner. The importance of the tx-rx switching time on the mac protocol, 1993.

[36] Wim Diepstraten. The importance of short rx-tx turnaround time, September 1993.

[37] G. Chalhoub, N. Hadid, A. Guitton, and M. Misson. Deference mechanisms significantly increase the MAC delay of slotted CSMA/CA. In *ICC*, pages 1–5, June 2009.

[38] A. Koubaa, M. Alves, and E. Tovar. A comprehensive simulation study of slotted CSMA/CA for IEEE 802.15.4 wireless sensor networks. In *IEEE International Workshop on Factory Communication Systems (WFCS)*, pages 1–10, July 2006.

[39] A. Kiryushin, A. Sadkov, and A. Mainwaring. Real-world performance of clear channel assessment in 802.15.4 wireless sensor networks. In *Proc. Second International Conference on Sensor Technologies and Applications SENSORCOMM '08*, pages 625–630, August 2008.

[40] Chipcon. Chipcon SmartRF CC2400 datasheet rev. 1.3, http://www.ti.com/, March 2006.

[41] Chipcon. Chipcon SmartRF CC2520 datasheet rev. 1.3, http://www.ti.com/, 2007.

[42] Atmel. AT86RF231 datasheet rev. b, http://www.atmel.com, February 2009.

[43] M. Bertocco, G. Gamba, and A. Sona. Experimental optimization of CCA thresholds in wireless sensor networks in the presence of interference. In *Proc. of IEEE EMC Europe 2007 Workshop on Electromagnetic Compatibility*, June 2007.

[44] I. Ramachandran and S. Roy. WLC46-2: On the impact of clear channel assessment on MAC performance. In *Proc. IEEE Global Telecommunications Conference GLOBECOM '06*, pages 1–5, November 2006.

Routing Protocols

Chapter 8

Cross-Layer Design for Smart Routing in Wireless Sensor Networks

.

Omar M. Sheikh and Samy A. Mahmoud

Additional information is available at the end of the chapter

1. Introduction

Wireless sensor networks (WSNs) are gaining market traction in numerous industrial segments because they are both cheaper and faster to deploy than their wired counterparts. They also provide incremental value as they are able to extend existing wireless mesh network (WMN) infrastructure to deploy commercial sensory applications on a per-need basis. This establishes a scalable network infrastructure that deploys WSNs in a distributed fashion; sensors gather sensory information from geographically distributed areas and feedback data to centralized controlling stations. These controllers either provide an automatic response based on internal logic or log data for manual response by an operator, if necessary.

There is, however, a need to revolutionize current routing methods to realize next-generation commercial applications for these networks. These innovative routing schemes, which we coin *smart routing*, are based on performance measure and energy optimization, as opposed to traditional routing schemes that typically only minimize energy consumption to prolong network lifetime. Smart routing - the selection of routing nodes that are best able to satisfy both performance and energy conservation requirements given current network conditions - is based on cross-layer considerations of the protocol stack. Cross-layer design streamlines communication between layers and provides response based on a more complete view of the stack. These cross-layer factors include the application's requirements, available network routes, transmission channel quality and energy distribution in the network.

The consideration of application performance is complicated by sensors that have critical power constraints. As mentioned, this has typically resulted in the optimization of these networks taking the form of the minimization of energy consumption, or the maximization of network lifetime, as the primary objective [3–6]. However, this typically occurs to the detriment of application performance. Certain studies do strive to reach a maximum delay requirement [7, 8, 10]; however, it is unknown if we can do better as performance is not optimized. Other studies perform rate control in WSNs but do not model the power cost of using a transmission link in terms of the achievable throughput level [5].

Next-generation sensor networks require performance optimization by considering both the potential performance that can be achieved and the corresponding impact on a node's energy capacity. This enables nodes to make more informed resource allocation decisions.

With that said, the dependencies of next-generation applications on various performance and energy factors vary. Many of these applications are critical and require immediate response such as those for physical security, industrial processes and infrastructure monitoring; however, those for temperature control and ambient light measurement, for example, are less critical and are able to conserve energy at the expense of less performance-heavy resource allocation. Hence, the aim is to create a flexible cross-layer platform for distributed WSNs that considers the *criticality* of the resource allocation for next-generation applications.

This chapter covers the main research areas that arise in designing smart routing protocols and require specific engineering attention:

- **Network Architecture** - determining the optimal configuration of the distributed architecture and the deployment of WSNs at areas of interest to extend the WMN;
- **Optimization Metrics** - identifying cross-layer performance and energy factors that impact resource allocation: application requirements, available routes, channel quality, battery life, physical (PHY) layer considerations (transmit power, operating channel and bandwidth), and the energy efficiency of the wireless communication protocol;
- **Criticality** - defining the dependency of commercial applications on performance and energy considerations;
- **Route Selection** - selecting the route with the optimal trade-off between performance and energy conservation for a given application criticality;
- **Coexistence** - providing connectivity between heterogeneous communication interfaces to bridge sensor and mesh technologies such as Bluetooth and WiMax, respectively; and,
- **Energy Harvesting** - quantifying the impact of replenishing energy reserves from kinetic, solar or heat energy on resource allocation.

Each of these topics will be covered in this chapter.

2. Network architecture

Wireless mesh networks (WMNs) are the architectural enabler for wireless sensor networks (WSNs). As mentioned, WMNs provide the opportunity to deploy WSNs in an incremental fashion to execute sensory applications at multiple locations of interest on a per-need basis; WMNs also provide an alternative to carrying Internet Protocol (IP) traffic in rural or hostile environments where access to fibre may not be available. This provides feedback of sensory data from a WSN to a centralized controlling station over a long haul through a mesh node that is assigned to govern a sensor cluster.

These specially-assigned mesh nodes, called *cluster-heads*, are selected based on proximity, or deployed to extend the network, to the sensory location(s) of interest. Cluster-heads provide a bridge to the mesh network and may assume supervisory control of their subordinate sensors, which are typically limited in their resources and computational capabilities. To perform these functions, cluster-heads are equipped with the additional resources to handle the traffic load, although they likely carry multiple types of traffic, only one of which may be sensory. A WMN enabling a WSN is presented in Figure 1.

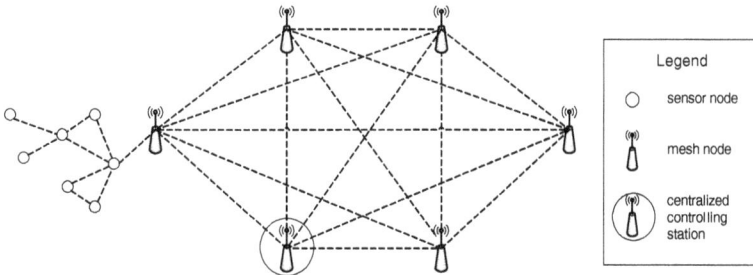

Figure 1. A Wireless Mesh Network Enabling a Wireless Sensor Network

We design a network architecture to analyze the impact of smart routing on resource allocation for WSN applications with varying requirements. This architecture, presented in Figure 2, consists of multiple sensor clusters and an overlay WMN that spans roughly a one kilometer area. In this network, one sensor cluster is formed of high-bandwidth Ultrawideband (UWB) sensors that are suited to data intensive applications such as video monitoring; UWB is a high-speed alternative to Zigbee for sensor networks with low power consumption but is inherently short range [2]. A second sensor cluster of Zigbee sensors is deployed to execute a low bandwidth application such as temperature monitoring. The WMN uses WiMax mesh technology to connect these geographically distributed clusters to the central controlling stations. This station is responsible for communicating with an outside controller or processing center, or is the processing center itself.

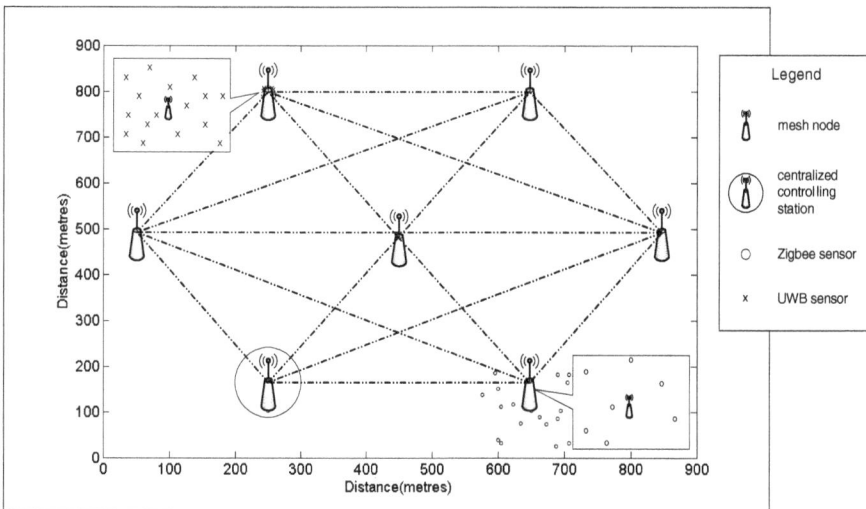

Figure 2. A Distributed Wireless Mesh Network of Zigbee and UWB Sensor Clusters

The communication technologies chosen are presented as a single scenario but typically depend on the range of communication required, node density and required bandwidth requirements of the applications of interest.

In terms of node interaction, cluster-heads perform supervisory control of sensors and optimize resource allocations for sensors that they govern. Sensors correspondingly inform their cluster-heads of the state of their resources periodically.

Ideally, mesh nodes, including cluster-heads, are organized in a hexagonal topology for maximum connectivity [1]. These mesh nodes are placed at the center of their clusters around which sensors are typically positioned randomly. However, while we would ideally like to maintain a hexagonal topology of mesh nodes, this may not always be practical because the organization of mesh nodes depends highly on the sensory locations of interest. For example, if a sensor cluster is deployed to monitor stresses on bridge infrastructure, or the military is interested in monitoring certain high security areas, it will likely not be possible to deploy a mesh node at an ideal location. It is important to note that, in networks for which a hexagonal mesh topology is not possible, network planners must be aware of potential single points of failure. In these cases, load balancing or redundancy should be explored to ensure that mesh nodes are not overburdened.

In this analysis, neither sensor nodes nor mesh nodes are wired to power sources. This allows us to explore a general architecture and expands the number of environments in which, and applications for which, the system can be deployed. In reality, certain mesh nodes may be connected to power sources if the locations in which the mesh nodes are deployed have power sources readily available. Another option is energy harvesting to replenish energy reserves over time, which we will cover later in this chapter.

2.1. Software radio

In wireless sensor networks (WSNs) that are based on multiple technologies, software radio is required to convert operating parameters between otherwise incompatible communication protocols. The conversion must consider the varying dependencies of these technologies on a number of characteristics that affect communication and performance. For example, in our network architecture, the cluster-head must convert transmission parameters between sensor and mesh communication technologies. These parameters include:

- Operating bandwidth,
- Transmit power,
- Transmission frequency, and
- Modulation scheme.

WSNs that are based on software radio enable for the deployment of large-scale and distributed systems that are designed with technologies that are most suitable to their applications. Various technologies may be selected based on throughput requirements, cost of deployment and energy efficiency. Software radio enables these systems to dynamically tune operating parameters around current networking conditions to improve capacity.

3. Cross-layer design

Layering systems are the norm in the design of communication protocol stacks. However, wireless systems are not always suited to the common layered protocol stack architecture. For example, in a layered architecture using the Transmission Control Protocol (TCP), a failed packet is considered a sign of congestion, as opposed to simply a lost or corrupted packet which is the case in wireless systems. For sensor networks, and smart routing specifically, given the need to conserve sensor energy and maximize application performance, cooperation between several layers in the protocol stack is crucial. This can only be achieved in a cross-layer architecture. Cross-layer design ensures that the route that best meets both performance and energy requirements can be determined.

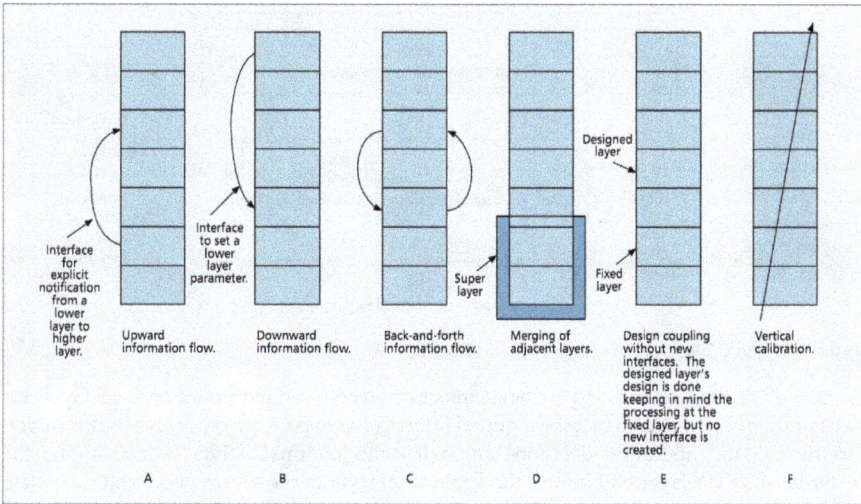

Figure 3. Various Cross-Layer Design Protocols [11]

Figure 3 presents a number of general ways in which a typical layered architecture can be modified by cross-layer design:

- Creation of new interfaces for information flow between non-adjacent layers (Figures 3a-c);
- Merging of adjacent layers for joint functionality and reduced overhead (Figure 3d);
- Design coupling between layers, i.e. one layer assumes information arriving from another (Figure 3e); and,
- Vertical calibration between layers (Figure 3f).

In next-generation wireless sensor networks (WSNs), a number of these protocols may be used. For example, upward information flow (Figure 3a) may be used to provide the application layer with available routes from the network layer, channel availability from the link layer and remaining energy information from the physical (PHY) layer. Furthermore, downward information flow (Figure 3b) or back-and-forth information flow (Figure 3c) may be used between the application layer and the PHY layer. For example, the application layer may inform the PHY layer of transmission parameters such as transmit power and operating frequency to use during transmission.

The design of a cross-layer optimization algorithm for WSNs that consider both performance and energy factors requires efficient communication between protocol stack layers such as the PHY, link, network and application layers. Direct signaling between application layers reduces latency in the communication between multiple layers and is crucial in the design of cross-layer optimization algorithms [9]. The direct signaling scheme for our protocol stack model is illustrated in Figure 4.

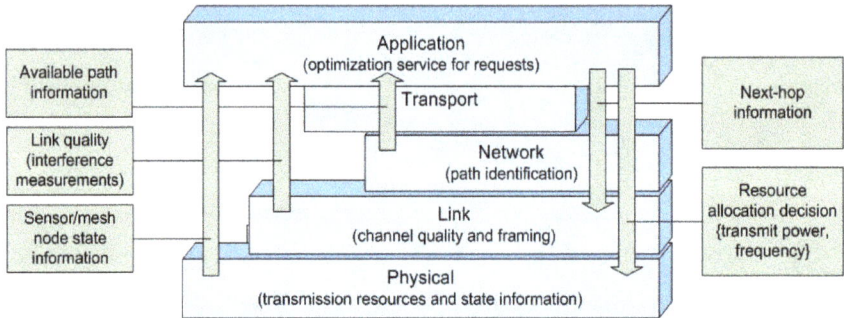

Figure 4. Protocol Stack Model to Enable Smart Routing

The goal of direct signaling is to exchange information between important protocol layers for smart routing. This ensures that the required information to perform cross-layer optimization is retrieved, and allocation decisions are sent, with minimal delay. For example, the cluster-head's PHY layer will inform the application layer of the sensor and mesh node state information which includes energy rating information, surrounding interference and more. State information and coordination protocols to provide feedback to the cluster-head are covered in Section 9. The link layer and network layer will also inform the application layer of the channel conditions and available path information, respectively. Furthermore, upon executing the cross-layer optimization policy, the application layer will inform the PHY layer of the necessary resource allocations and the link layer of the next-hop information. Direct signaling enables these interactions with minimal delay for optimized and timely responses in our distributed network.

For any two non-adjacent layers, l_x and l_y, the propagation latency T_{l_x,l_y}^{DS} for the direct signaling method is calculated as,

$$T_{l_x,l_y}^{DS} = \frac{T_{l_x,l_y}^{L}}{(n-1)} \qquad (1)$$

where T_{l_x,l_y}^{L} is the propagation latency between layers l_x and l_y in a traditional layered protocol stack with $(n-1)$ layers between them. Hence, the direct signaling method provides a speed-up factor of $(n-1)$ [9].

4. Problem statement

In solving the smart routing problem via cross-layer design, the goal is to solve the 4-tuple defined by,

- Path (or next hop);
- Operating channel;
- Transmission power; and,
- Throughput,

to identify the most suitable path that satisfies both performance needs of an application and energy conservation considerations; this path routes data from a source sensor to a centralized controlling station. In doing so, the cross-layer optimization policy focuses on functions at three layers: path selection at the network layer; channel selection at the link layer; and, transmit power allocation at the physical (PHY) layer. Hence, the 4-tuple defines the operating point for the network in solving the resource allocation for a single transmission.

The 4-tuple forms the basis of utility functions that model the preferences of nodes in allocating various PHY layer resources to throughput; the optimization problem exists since these PHY layer resources also impact network lifetime. Utility functions quantify the performance benefits and power costs associated with the allocation prior to selecting the optimal operating point. In doing so, the criticality of the application plays a major role in determining the necessary tradeoff for the given sensor.

5. Network definition

Define our distributed wireless sensor network (WSN) $N = (G(C_l(V_i, M_i), E_l), S)$ that is composed of a set of sensor nodes V_i and mesh nodes M_i to solve the smart routing problem. These sensor and mesh nodes are organized into clusters $C_l(V_i, M_i)$ and connected in a two-tier network via directed link set E_l. The nodes are connected according to a topology $G(C_l(V_i, M_i), E_l)$ that is presented in space S in Figure 2.

We make the following additional definitions:

- the set of all nodes in the network, $Z = V_i \cup M_i$;
- the set of technologies available to each node in the network, $T = \{t_{v_x} | \forall v_x \in Z\}$;
- the number of sub-networks (sensor and mesh networks) in the distributed WSN, S_N;
- the set of residual energies of all nodes in network, $E_{avail} = \{e_{v_x} | \forall v_x \in Z\}$;
- the set of initial energies of all nodes in network, $E_{init} = \{E_{v_x} | \forall v_x \in Z\}$;
- the set of channels available to all nodes of a given sub-network, $F_{S_N} = \{f_{t_{v_x}} | \forall t_{v_x} \in T\}$;
- the energy efficiencies of various communication protocols, $Q = \{q_{t_{v_x}} | \forall t_{v_x} \in T\}$;
- the central controlling station, $v_j \in M_i$;
- a cluster-head node, $v_m \in M_i$;
- the maximum number of hops in a path, K';
- the number of hops in a single path, $K \leq K'$;

- the average battery replenishment rate from energy harvesting, r_h;
- the path between any sensor v_i and central controlling station v_j through cluster-head v_m, $path_{v_i, v_j} = \{v_i, ..., v_m, ..., v_j\}$, where the number of nodes along a simple path between v_i and v_j can be computed as, $K = n(path_{v_i, v_j}) \leq (K' + 1)$; and,
- the $(K' + 1)$ x W x $|V_i|$ matrix, X, of all the paths from all sensor nodes in the network to the central controlling station, where W is the maximum number of paths over all nodes. $|V_i|$ is the cardinality of the sensor node set.

The selection of K' limits the maximum number of hops along a path and, hence, the number of nodes used to route traffic. Communication protocols have varying dependencies on multi-hop routing given their ranges of acceptable transmit powers and operating ranges, and as such they will likely have different path lengths. However, energy consumption characteristics between a transmitter and receiver suggest that it is better for the network to reduce the number of hops in a path at the expense of using a higher transmit power [3]. Hence, based on our network topology, we set a self-imposed limit to reduce the path length. This is configurable given the network scenario in question.

5.1. Assumptions

To perform this study, we make the following assumptions:

1. All clusters perform periodic data delivery according to a Poisson process with exponentially distributed inter-arrival times of events with parameter λ;

2. Sensor node are frequency-agile and can tune their frequencies to select different operating channels;

3. Mesh nodes use software radio to communicate with both sensor nodes and mesh nodes over different communication technologies;

4. All nodes within a cluster use omni-directional antennas with equal gains;

5. Neither sensor nodes nor mesh nodes are wired to power source so that we may explore a general architecture;

6. Mesh nodes are placed in a hexagonal topology for topology optimality and maximum network coverage as discussed in Section 2;

7. Sensor nodes are aware of their positions and are equipped with the Global Positioning System (GPS). Sensor localization will be considered in future work;

8. The central controlling station contains a mesh node and is the sink of transmissions in the network;

9. Single channels are not used end-to-end for a path and each hop chooses a different operating channel;

10. Each sensor node follows a $M/M/3/3$ queueing model with three operating channels and no queue. This limits the competition between all nodes to only three channels to leave sufficient capacity available to carry relay traffic; and,

11. The cluster-head has real-time channel state information, which can be assumed because our sensor nodes are fixed, allowing us to assume a slow fading model.

6. Link utility function

In forming our optimization functions via cross-layer criteria, we define the communication resources r_f associated with a wireless channel $f \in F$; F is the channel set of a given communication protocol. The capacity of a channel c_f is dependent on a number of factors including r_f, but in this study we focus on the case where $c_f = \phi(r_f)$ only. The transmission rate $R_{v_x,v_y,f} \leq c_f$ is defined as a fraction of the frequency division multiple access (FDMA) Shannon capacity for $r_f = (P^t_{v_x,f})$ as,

$$R_{v_x,v_y,f} = \phi(P^t_{v_x,f}) = w_f log_2\left(1 + \frac{P^t_{v_x,f}|H_{v_x,v_y,f}|^2}{w_f N_{v_y,f} + I_{v_y,f}}\right) \qquad (2)$$

where our resource of interest $r_f = (P^t_{v_x,f})$ is the selected power at transmitter v_x in sending data to receiver v_y on wireless channel f, $|H_{v_x,v_y,f}|^2$ is the channel gain between v_x and v_y, $N_{v_y,f}$ is the Gaussian noise power on the channel from the perspective of the receiver v_y, and $I_{v_y,f}$ is the interference of v_y on channel f. As we form our link utility function, it evaluates the allocation of our resources $r_f = (P^t_{v_x,f})$ for link $v_x - v_y$ in terms of the achievable transmission rate $R_{v_x,v_y,f}$ and the power cost associated with transmitting at rate $R_{v_x,v_y,f}$ over the link.

The use of the FDMA Shannon capacity allows us to perform adaptive resource allocations based on the real-time state of the network by relating the data rate to physical (PHY) layer parameters. In our study, $r_f = (P^t_{v_x,f})$ only because we consider that all channels have equal bandwidths w_f. However, since the Shannon capacity is theoretical, we limit the Shannon capacity with additional regulatory limits.

$$L_{v_x,v_y,f}(R_{v_x,v_y,f},t) = \alpha \log_{10}\left(1 + \frac{R_{v_x,v_y,f}}{C_{v_y,f}(t)}\right) - \psi(E)\frac{q_{t_{v_x}}lR_{v_x,v_y,f}}{m} + \eta \qquad (3)$$

Our link utility function $L_{v_x,v_y,f}(R_{v_x,v_y,f},t)$ is the basis of the resource negotiation between a prospective transmitter and receiver in network and is shown in (3). It is formed from the receiver's perspective as the difference between a benefit function and a cost function. Figure 5 shows a sample plot of the link utility function $L_{v_x,v_y,f}(R_{v_x,v_y,f},t)$, which is concave as a function of the $R_{v_x,v_y,f}$.

The first term of $L_{v_x,v_y,f}(R_{v_x,v_y,f},t)$ represents the benefit gained as a function of $R_{v_x,v_y,f}$. This is modeled by a logarithmic function which is monotonically increasing and follows the law of diminishing returns. As a result, an initial increase in $R_{v_x,v_y,f}$ is more important to a node than further increases in $R_{v_x,v_y,f}$ as the node approaches the incoming channel capacity, $C_{v_y,f}(t)$.

The second term is the cost function that models the power cost of utilizing a link in a path. The cost is a function of the energy efficiency coefficient of the technology used for the communication link, $q_{t_{v_x}} \in Q$. By multiplying $q_{t_{v_x}}$ by the ratio of l/m, we retrieve the energy efficiency for the full packet size including overhead. The power consumption in sending a packet over the prospective link (in *watts*) is found by multiplying $q_{t_{v_x}}$ (in *joules/bit*) by the transmission rate $R_{v_x,v_y,f}$ (in *bits/sec*). The result is the amount of energy over time, or power, used in transmitting a packet over the link. By considering the Shannon rate in (2), we observe that the link utility is a function of physical resources for both throughput and power cost.

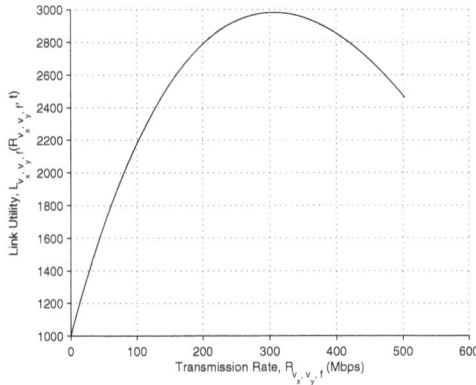

Figure 5. Example of a Link Utility Function $L_{v_x,v_y,f}(R_{v_x,v_y,f},t)$

$\psi(E)$ is a function of the transmitter and receiver's residual energy E, which represents the maximum of the inverse residual energy of the transmitter v_x and receiver v_y as,

$$E = max\left(\frac{E_{v_x}(t)}{e_{v_x}(t)}, \frac{E_{v_y}(t)}{e_{v_y}(t)}\right)$$

Hence, if either the transmitter or receiver of the link have a low residual energy, the cost of using the link increases. In this manner, we encourage the optimization algorithm to select nodes with higher residual energy in the network.

Since the range of available transmit power is small (on the order of milliwatts or microwatts), the algorithm is sensitive to small changes in link cost. Consider a situation where a node has half of its battery power remaining. While it should be able to operate normally, a doubling of the link cost, in the absence of a function like $\psi(x)$, forces the selection of zero transmit power. The effect is worse at larger R due to a higher link cost. As a result, $\psi(E)$ scales the power cost to operate within the limits of the optimality range.

The function $\psi(x)$ is of the form,

$$\psi(x) = \beta \ln(x) + \gamma$$

for a Ultrawideband (UWB) and WiMax transmission, and of the form,

$$\psi(x) = \beta x^2 + \delta x + \gamma$$

for a Zigbee transmission. The two forms of $\psi(x)$ are due to the achievable rates of the communication technologies of interest and the scale required to fit the power cost within the necessary operating range.

The α, β, γ, and δ parameters are coefficients of the empirical benefit and power cost that define the *criticality* of the application, which will be covered in Section 8. The parameter α scales the empirical benefit to provide a greater weight to the utility gained by achieving a higher transmission rate. Meanwhile, the three scaling factors, β, γ and δ are parameters in

$\psi(x)$ that may be obtained via curve-fitting after determining the desired operating points for a particular networking configuration as a function of E.

The optimized transmission rate $r^*_{v_x,v_y,f}$ is calculated by maximizing $L_{v_x,v_y,f}(R_{v_x,v_y,f},t)$ where $R_{v_x,v_y,f} = \phi(P^t_{v_x,f})$ as,

$$r^*_{v_x,v_y,f} = argmax_{R_{v_x,v_y,f}}\left[\alpha log_{10}\left(1 + \frac{R_{v_x,v_y,f}}{C_{v_y}(t)}\right) - \psi(E)\,\frac{q_{t_{v_x}}lR_{v_x,v_y,f}}{m} + \eta\right] \qquad (4)$$

Given $r^*_{v_x,v_y,f'}$ the optimized link utility for a single hop $v_x - v_y$ is calculated as,

$$L^*_{v_x,v_y,f}(r^*_{v_x,v_y,f},t) = \alpha log_{10}\left(1 + \frac{r^*_{v_x,v_y,f}}{C_{v_y,f}(t)}\right) - \psi(E)\,\frac{q_{t_{v_x}}l\,r^*_{v_x,v_y,f}}{m} + \eta \qquad (5)$$

Based on this information, it is necessary for the cluster-head to receive feedback of the real-time residual energies of sensors. This is achieved through the feedback of hello messages that will both announce presence in the network and provide feedback of node state information to the cluster-head. If the cluster-head is required to perform the cross-layer optimization in between two received hello announcements, the cluster-head will extrapolate the residual energies of both v_x and v_y based on the elapsed time since the last update to ensure accurate optimization.

The necessary coordination protocols will be covered in Section 9.

7. Objective function

The profit function, $U^a_{v_i,v_j}(t)$, calculates the suitability of candidate path $path_{v_i,v_j} = \{v_i, ..., v_m, ..., v_j\}$ to route a message from a source sensor v_i to the central controlling station v_j through cluster-head v_m. The profit function shown in (6) is our objective function where the goal is to find the candidate path with the maximum profit, $U^{a*}_{v_i,v_j}(t)$. The profit gained by using a candidate path is calculated as the average of the optimized link utilities, L^*, across each hop along a path.

$$U^{a*}_{v_i,v_j}(t) = max\;\frac{1}{K-1}\left[\sum_{x=1}^{g}L^*_{v_x,v_{x+1},f_x}(r^*_{v_x,v_{x+1},f_x},t) + \sum_{z=g+1}^{K-1}L^*_{v_z,v_{z+1},f_z}(r^*_{v_z,v_{z+1},f_z},t)\right]$$

$$
\begin{aligned}
s.t.\ \ (1)&\ \ P_{v_x,f_x} \cdot P_{v_{x+1},f_x} &= 0, &\qquad \forall v_x,v_{x+1} \in path_{v_i,v_j}\\
(2)&\ \ P_{v_z,f_z} \cdot P_{v_{z+1},f_z} &= 0, &\qquad \forall v_z,v_{z+1} \in path_{v_i,v_j}\\
(3)&\ \ P_{v_x,f_x},P_{v_z,f_z} &\geq max(P_{MIRS},P_{SNR},P_{min}), &\qquad \forall v_x,v_z \in path_{v_i,v_j}\\
(4)&\ \ P_{v_x,f_x},P_{v_z,f_z} &\leq min(P_{tech},P_{cap},P_{max}), &\qquad \forall v_x,v_z \in path_{v_i,v_j}\\
(5)&\ \ K \leq K'.&&
\end{aligned}
$$
$$\qquad (6)$$

$U^{a*}_{v_i,v_j}(t)$ is formed of two summations that separate the optimization of the sensor cluster and mesh network portions of a path. The first summation represents the sum of link utilities as computed via the optimized link utility function L^* across all nodes along a candidate path inside the cluster only; g represents the number of hops along the path within the cluster before reaching cluster-head v_m and is computed as $g = find(path_{v_i,v_j} == v_m) - 1$. The second summation represents the sum of link utilities for a candidate path.

We divide the utility sum by the hop count of the path to calculate the average link utility in the path. By doing so, we are able to more closely analyze the difference between a k-hop path and $(k+)$-hop path in choosing the optimal route. If we were to use total rather than average link utility, the algorithm would favor the $(k+)$-hop path as the summation of more link utilities leads to a higher U_{v_i,v_j}^a. In a power-constrained network, this over-utilizes already limited resources. Hence, by evaluating the average link utility, we are in fact reducing latency and conserving power. It should be noted, however, that, if a path with more hops has the highest average link utility $U_{v_i,v_j}^{a^*}$, it will be chosen by the optimization policy as the path with the best tradeoff between performance and energy consumption across all candidate paths.

7.1. Constraints

The constraints on the optimization as presented in (6) provide the boundaries for the selection of optimal resource parameters for both sensor and mesh nodes. The first constraint, $P_{v_x,f_x} \cdot P_{v_{x+1},f_x} = 0, \forall v_x, v_{x+1} \in path_{v_i,v_j}$, provides a power allocation restriction that a cluster node v_{x+1} in an end-to-end path cannot receive and transmit on the same channel f_x, that is $f_x \neq f_{x+1}$. This restricts node v_x and v_{x+1}, which are successive nodes in a path, from allocating power on the same transmission channel. The second constraint, $P_{v_z,f_z} \cdot P_{v_{z+1},f_z} = 0, \forall v_z, v_{z+1} \in path_{v_i,v_j}$, is the corresponding power allocation restriction for the mesh network.

The third constraint, $P_{v_x,f_x}, P_{v_z,f_z} \geq max(P_{MIRS}, P_{SNR}, P_{min}), \forall v_x, v_z \in path_{v_i,v_j}$, restricts the minimum power that the receiver can receive. P_{MIRS} corresponds to the minimum input signal power at the receiver, or the minimum input receiver sensitivity (MIRS), defined in Table 1. The P_{SNR} constraint is the minimum power required to reach the signal-to-noise ratio (SNR) threshold at the receiver. The third term, P_{min}, represents the lower bound on the transmit power that keeps $L_{v_x,v_y,f}(R_{v_x,v_y,f}, t)$ positive. There is a fourth factor that is the lower bound on the available capacity at the transmitter, which is zero and is thus ignored.

Standard	MIRS	Minimum SNR	Maximum EIRP
Zigbee	-85 dBm	-1.2 dB	0 dBm
UWB	-85 dBm	-1.59 dB	-14.3 dBm
WiMax	-83.2 dBm	9.8 dB	24 dBm

Table 1. Operating Parameters for Zigbee, UWB and WiMax for Constraint Modeling [12–15]

The fourth constraint, $P_{v_x,f_x}, P_{v_z,f_z} \leq min(P_{tech}, P_{cap}, P_{max}), \forall v_x, v_z \in path_{v_i,v_j}$, restricts the maximum power of the optimality range. P_{tech} corresponds to the maximum allowable transmit power on a transmission channel for the technology being used. From Table 1, P_{tech} is calculated as P_{tech} (dBm) = Maximum EIRP (dBm) - G_t (dBi), where the Effective Isotropic Radiated Power (EIRP) is the maximum allowable power that can be put on the transmission channel and G_t is the transmit antenna gain in dBi. P_{cap} represents the power that corresponds to the available outgoing capacity on the channel at the transmitter and ensures that the channel capacity is not exceeded. While the lower bound on the outgoing capacity is zero in the previous constraint, the upper bound P_{cap} is not zero unless the full channel capacity is being used by the transmitter. Meanwhile, P_{max} restricts the upper bound on transmit power as that which keeps $L_{v_x,v_y,f}(R_{v_x,v_y,f}, t)$ positive.

The final constraint limits a path to at most K' hops to reduce power dissipation due to routing.

7.2. Steps in resource allocation

The cluster-head executes the following steps to perform resource allocation for a sensor v_i:

1. **Path Identification**: The cluster-head forms the $(K' + 1) \times W$ path sub-matrix, $A \subset X$, of the available paths from v_i to the central controlling station v_j. For the purposes of this chapter, we will assume that X is known.

2. **Power Optimization**: The cluster-head optimizes the transmit powers across all links for each candidate path $a \in A$ for v_i using (4) and (5). The cluster-head stores the optimized $L^*_{v_x,v_y,f}$, rates $r^*_{v_x,v_y,f}$ and transmit powers $P^{t^*}_{v_x,f}, \forall v_x, v_y \in a$.

3. **Channel Optimization**: The cluster-head ranks channels according to their optimized link utilities to find the most preferred channels of a node along a candidate path. The cluster-head will attempt to allocate the most preferred channel at each hop. However, since this may not always be possible given channel constraints, it may be necessary to iterate over possible bin combinations to find the valid combination with the largest link suitability over all links. The optimal valid channels for path a are denoted as $f^*_1, \ldots, f^*_K$.

4. **Path Optimization**: The cluster-head assigns an overall suitability coefficient or profit, $U^a_{v_i,v_j}$, to each candidate path. $U^a_{v_i,v_j}$ is formed in (6) as the average link utilities using the optimal channel allocation $f^*_1, \ldots, f^*_K$ for the path.

5. **Path Selection**: The cluster-head ranks the paths by the suitability coefficient and selects the candidate path with the highest $U^a_{v_i,v_j}$. The suitability of the selected path is denoted as $U^{a^*}_{v_i,v_j}$ and the cluster-head will retrieve the corresponding frequencies $f^*_1, \ldots, f^*_K$, transmit powers $P^{t^*}_{1,f^*_1}, \ldots, P^{t^*}_{K,f^*_K}$, and rates $r^*_1, \ldots, r^*_K$.

6. **Routing Decision Propagation**: The cluster-head will inform each node v_x (sensor or mesh node) along the selected path of their necessary operating frequency $f^*_{v_x}$, transmit power $P^{t^*}_{v_x,f^*_{v_x}}$, and next-hop v_y in a routing decision update (RDU).

For node v_x at hop k, the 4-tuple $(v_y, f^*_k, P^{t^*}_{v_x,f_k}, r^*_{v_x,v_y,f_k})$ solves the smart routing problem.

8. Application criticality

The criticality c_r of an application is defined as,

$$c_r(\alpha, E) = \frac{\alpha}{\psi(E)} \tag{7}$$

which represents the ratio of the weights placed on the empirical benefit and power cost, respectively. At one extreme, $c_r(\infty, E)$ allocates maximum resources towards throughput performance, while, at the other extreme, $c_r(0, E)$ emphasizes energy conservation for *minimum energy routing.* The associated optimized link utility $L^*_{v_x,v_y,f}$ for minimum energy routing is,

$$L^*_{v_x,v_y,f}(r^*_{v_x,v_y,f},t) = -\psi(E)\,\frac{q_{t_{v_x}}\,l\,r^*_{v_x,v_y,f}}{m} \tag{8}$$

where the goal is to minimize the link's power cost. The corresponding objective function is,

$$U^{a^*}_{v_i,v_j}(t) = max \quad \frac{1}{K-1}\left[\sum_{x=1}^{g} L^*_{v_x,v_{x+1},f_x}\left(r^*_{v_x,v_{x+1},f_x},t\right) + \sum_{z=g+1}^{K-1} L^*_{v_z,v_{z+1},f_z}\left(r^*_{v_z,v_{z+1},f_z},t\right)\right]$$

$$
\begin{aligned}
s.t. \quad (1) \quad & P_{v_x,f_x}\cdot P_{v_{x+1},f_x} && = 0, && \forall v_x, v_{x+1} \in path_{v_i,v_j} \\
(2) \quad & P_{v_z,f_z}\cdot P_{v_{z+1},f_z} && = 0, && \forall v_z, v_{z+1} \in path_{v_i,v_j} \\
(3) \quad & P_{v_x,f_x}, P_{v_z,f_z} && = max(P_{MIRS}, P_{SNR}), && \forall v_x, v_z \in path_{v_i,v_j} \\
(4) \quad & K \leq K'. &&&&
\end{aligned}
\tag{9}
$$

where the effect of minimum energy routing is seen in the third constraint, which enforces that the minimum required power to transmit between node pairs is chosen.

The impact of application criticality on resource allocation will be analyzed by comparing the throughput performance of smart routing versus minimum energy routing.

9. Packet formats

In gathering the required node information for the cross-layer policy, we define the state of a node v_x that includes:

- **Node identifier (NID)**: the ID of sensor or mesh node v_x;
- **Sub-network identifier (SNID)**: the ID of the sub-network (sensor or mesh network) in which the node v_x resides;
- **Energy rating information (ERI)**: the remaining energy of node v_x in the form of a percentage of the initial energy capacity, e_{v_x}/E_{v_x}; and,
- **Surrounding interference temperature (SIT)**: the measurement of the surrounding interference plus noise $(I + N)$ energy as measured by node v_x.

State information is broadcasted to announce presence in the network and is also used in the propagation of resource allocation to nodes selected during the optimization process using a coordination channel. The state information above and the position of a node are all that a cluster-head requires to optimize a request.

From this point onwards, we shall denote V and M as the cardinalities of the sensor and mesh nodes sets as $V = |V_i|$ and $M = |M_i|$, respectively, to simplify equations.

9.1. Hello messages for presence broadcast

Node state information is used in the formation of hello messages sent between mesh nodes, and also between sensor nodes and their cluster-heads to announce presence in the network. The exchange of hello messages maintains accurate connectivity tables at the cluster-head. These hello messages also update the cluster-head's knowledge of the state of a sensor or mesh node v_x in terms of remaining energy capacity.

The hello packet format is presented in Figure 6 where the LAT and LONG fields represent the latitude and longitude of the node, respectively, using GPS. The node identifier (NID) field is used to identify the source of the hello packet and, hence, is the number of bits necessary to represent the node identifier. Thus, the NID field has a length of $ceil(log_2(V))$ or $ceil(log_2(M))$ bits depending on if the sender is a sensor or mesh node. The sub-network identifier (SNID) is the number of bits needed to identify the sub-network in which the node resides. The energy rating information (ERI) field is 14 bits in fixed point number representation to represent the percentage of initial energy capacity remaining (using a scaling factor of $1/100$), i.e. sending 14 bits that represent 10,000 in decimal yields an ERI of 100.00%; we use fixed point number representation for the ERI field, as opposed to single precision floating point (32 bits long), because the ERI has a fixed number of digits - two - after the decimal point. Hence, we reduce the number of bits needed to represent the residual energy.

NID	SNID	ERI	SIT	LAT	LONG	Padding
$ceil(log_2(V))$ or $ceil(log_2(M))$ bits	$ceil(log_2(S_N))$ bits	14 bits	32 bits	32 bits	32 bits	P_D bits

Figure 6. Hello Packet Format

The surrounding interference temperature (SIT) field, on the other hand, given real-time variations in the level of interference, is represented in full 32-bit single precision floating point format. Full 32-bit single precision floating point is also used to represent both the LAT and LONG fields. As nodes are stationary, it may only be necessary to include the LAT and LONG fields in the initialization phase to inform cluster-heads of node positions, after which it may not be required. Nevertheless, we include the LAT and LONG fields in the hello message, while P_D bits of padding may be used to fill out the packet.

Data aggregation is also critical in these networks to preserve sensor energy and reduce the amount of routed information in the network. Figure 7 illustrates an example of data aggregation in a distributed wireless sensor network (WSN) in which presence information is exchanged and specific data aggregation nodes are used to merge information from one or more neighbors. In the example presented, the identifiers of the sensors in the sensor cluster are aggregated at node 4 and node 5 to give a single message to the cluster-head at node 6 of the NIDs [1 2 3 4 5]. Data aggregation for presence information occurs in both the sensor and mesh networks where all mesh nodes are data aggregators and exchange information until there is a consistent view among all mesh nodes.

For a distributed WSN of $S_N = 3$ sub-networks (with two sensor clusters and the mesh network presented in Figure 2), $V = 1,000$ sensor nodes per cluster and a mesh network composed of $M = 7$ mesh devices, the hello packet has a length of 16 bytes. This includes ten

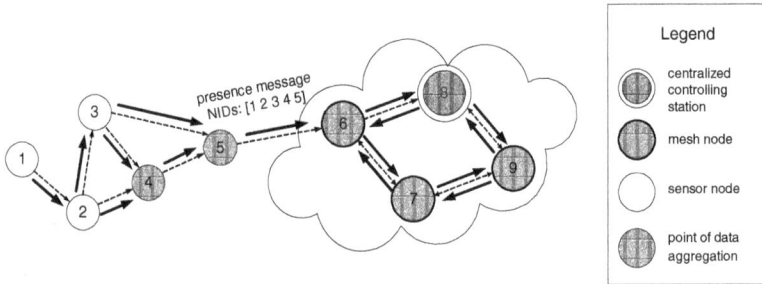

Figure 7. The Exchange of Presence Information in Distributed WSNs

or three bits for the NID field, and padding of six or thirteen bits, depending on if the source is a sensor or mesh node, respectively.

We design one hello packet format for both the sensor and mesh network to simplify the system design process and to reduce decoding complexity. A packet length of 15 bytes could have been used in the mesh network, but we choose to have a common packet format at the expense of a transmitting an extra byte of padding. With that said, separate packet formats may need to be considered depending on the number of sensor and mesh nodes in the network and the overhead associated with using a single format.

9.2. Routing decision updates

Upon determining the optimized resource allocation for a request from a source sensor v_i, a cluster-head v_m will propagate a routing decision update (RDU) to each node $v_x \in path_{v_i,v_j}$. The RDU is sent backwards along the path in the sensor network to the source v_i, and forwards along the path in the mesh network to the central controlling station v_j as shown in Figure 8. This ensures that all nodes along the path are aware of their necessary resource allocations.

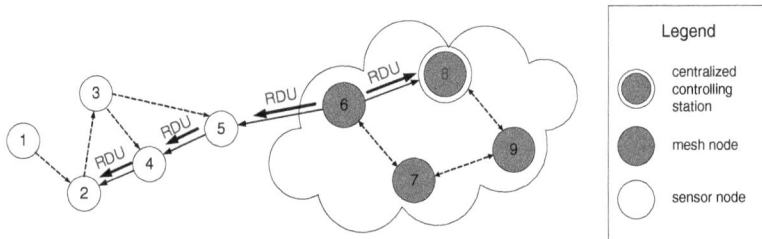

Figure 8. Propagation of RDU through Sensor and Mesh Networks

The cluster-head will include the following parameters such that all nodes can update their forwarding tables with up-to-date information:

- **Sending sensor identifier (SSID)**: the ID of source sensor v_i that is the origin of the request;

- **Request identifier (RID)**: the ID of the request being serviced at source sensor v_i given that sensors may support multiple requests;

- **Previous-hop identifier (PHID)**: the ID of node v_{x-1} from whom v_x will receive packets; and,

- **Next-hop identifier (NHID)**: the ID of node v_{x+1} to whom v_x will forward packets.

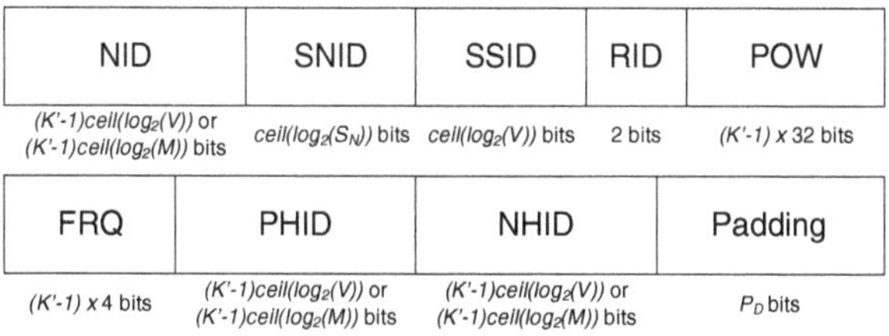

NID	SNID	SSID	RID	POW
$(K'-1)ceil(log_2(V))$ or $(K'-1)ceil(log_2(M))$ bits	$ceil(log_2(S_N))$ bits	$ceil(log_2(V))$ bits	2 bits	$(K'-1) \times 32$ bits

FRQ	PHID	NHID	Padding
$(K'-1) \times 4$ bits	$(K'-1)ceil(log_2(V))$ or $(K'-1)ceil(log_2(M))$ bits	$(K'-1)ceil(log_2(V))$ or $(K'-1)ceil(log_2(M))$ bits	P_D bits

Figure 9. RDU Packet Format

where the resulting format of the RDU is presented in Figure 9. Additionally, the cluster-head will include the node identifier (NID) of each node along the determined path such that they can retrieve their necessary operating parameters when they receive the routing decision. The message will also include the transmit power POW (each hop in 32-bit single precision floating point format), and operating channel FRQ (four bits to represent the channel number to be used per hop) determined during the optimization process, and the sub-network identifier (SNID) of the sub-network in which the node resides. P_D bits of padding may also be used.

The RDU packet is designed such that each RDU contains information for at most $(K' - 1)$ hops, with separate RDUs being sent backwards through the sensor network to the source and forwards through the mesh network to the central controlling station. Recall that, since the cluster-head acts as a bridge between the sensor cluster and the mesh network, at least one hop must reside in each sub-network. Hence, information for only $(K' - 1)$ hops is required.

For the RDU, we design our network with separate packet formats for the sensor and mesh networks, as the overhead associated with a single format is significant. In our case, the mesh network would need to transmit over eight bytes of overhead per RDU packet if we were to use a single format. As a result, given the same network conditions as for the hello packet, the RDU has a length of 27 bytes and 19 bytes for the sensor and mesh networks, respectively.

9.3. Data packets

In the data transmission process, upon receiving a data packet from the previous-hop identifier (PHID), an intermediate node $v_x \in path_{v_i, v_j}$ will set their next-hop based on the next-hop identifier (NHID) provided by the cluster-head in the routing decision update. Data packets will be formed of the transmitting node's node identifier (NID), originating sensor's identifier (SSID), request identifier (RID), and the following:

- **Payload information**: the gathered sensory information for feedback; and,

- **Transmission priority level (TPL)**: the priority of the transmission (high or low) for routing preference (optional).

While transmissions from different clusters may be of different priorities, typically transmissions from a given sensor cluster all have the same priority at any given time. Hence, it is optional and configurable for the network scenario in question.

10. Simulation

Our distributed network is formed of a V-sized Ultrawideband (UWB) sensor cluster and a V-sized Zigbee sensor cluster that are connected to a central controlling station via an M-sized overlay mesh network. The network spans a 1 km x 1 km campus area for which $V = 20$ sensor nodes (excluding the cluster-head) and $M = 7$ mesh nodes. We simulate a small network size without loss of generality. The UWB cluster is focused in a 10 m x 10 m area for a video feedback application, while the Zigbee cluster performs temperature sensing in a 75 m x 75 m area such as a computer server room.

Table 3 summarizes our selection of the α, β, γ, δ and η parameters that have been calibrated for our network scenario.

Sub-Network	α	β	γ	η	δ
UWB	47,000	5.22	409.2	1,000	-
ZigBee	10,000	-7×10^{-7}	1267.8	1,000	0.0517
WiMax	100,000	4.17	125.81	0	-

Table 3. Selection of Optimization Parameters in the Heterogeneous WSN

In terms of radio design, each Zigbee and UWB nodes uses a single transmit and receive antenna with gains of 0 dBi and 3 dBi, respectively. WiMax mesh nodes have transmit and receive antenna gains of 13 dBi and 16 dBi, respectively. For our operating parameters, we choose $F_1 = F_2 = F_c = 5$ as the number of sub-channels in the spectrum band with corresponding bandwidths of $w_1 = 75$ MHz, $w_2 = 12.5$ kHz and $w_c = 6.25$ MHz for UWB, Zigbee and WiMax, respectively. As can be seen in Table 4, UWB sensors also transmit four times the information per transmission than Zigbee sensors.

Sub-Network	Payload Length l (bytes)	Packet Length m (bytes)	Data Length (bytes)	Packets to Transmit, n	Number of Channels	Channel Bandwidth	Band of Operation
UWB	32	38	128	4	5	75 MHz	2.4 - 2.5 GHz
ZigBee	32	38	32	1	5	12.5 kHz	3.0 - 11 GHz
WiMax	48	256	128 or 32	4 or 1	5	6.25 MHz	5.25 - 5.725 GHz

Table 4. Communication Parameters for Three Sub-Networks

We also select the maximum number of hops $K' = 4$ for a single candidate path and the energy consumption coefficient $\zeta = 0.5$. According to Oppermann et al., "the amount of energy consumed while listening, receiving, and transitioning to receive mode is similar to that of transmitting, and cannot be ignored" and, as such, ζ is selected to divide the total energy consumption evenly between the transmitter and receiver [3]. The power consumption attributed to transmitting is higher than receiving in the communication between a transmitter and receiver; however, it should be noted that the energy consumed by listening for a transmission may be a dominant source of energy dissipation in these networks [16, 17].

10.1. Energy modeling

Initial energy capacities depend heavily on the energy efficiencies of the communication technology. This is intuitive since each wireless transmission depends on the amount of energy consumed during communication. As a result, we equip sensor and mesh nodes with different energy capacities. The energy efficiencies of Ultrawideband (UWB), Zigbee and WiMax are presented in Table 5, along with the expected data rates of the technologies. We also test our network with arrival rates of sensor node requests of $\lambda_1 = \lambda_2 = 0.1$ requests/minute.

Standard	Energy Efficiency q_{tv_x}	Standard Data Rate, R
ZigBee	0.5 μJ/bit	250 kbps
UWB	0.01 μJ/bit	500 Mbps
WiMax	7.0 μJ/bit	75 Mbps

Table 5. Energy Efficiencies and Data Rates for Various Standards [3] [18]

By considering each hop as a two-stage pipeline, we also implement holding times of $1/\mu_1 = 3.04$ μsec and $1/\mu_2 = 2.43$ msec per transmission. Hence, our network models traffic intensities of $a_1 = \lambda_1/\mu_1$ and $a_2 = \lambda_2/\mu_2$.

Table 5 presents the implemented energy capacities in the network for sensors, cluster-heads and mesh relay nodes given our system parameters. These energy capacities are determined as those to reach a one week network lifetime. Hence, we equip UWB and Zigbee nodes with $E_1 = 17.5$ J with $E_2 = 69.5$ J, respectively. The energy capacity of a mesh node is chosen as $E_m = 1$ kJ. We implement sensor energies slightly under the required levels to analyze the performance of these nodes in the final stages of the network lifetime.

Node Type	Required Energy Capacity	Applied Energy Capacity
UWB sensor	17.7 J	17.5 J
Zigbee sensor	69.9 J	69.5 J
Cluster-head (UWB)	750.3 J	1 kJ
Cluster-head (Zigbee)	684.8 J	1 kJ
Mesh relay	646.5 J	1 kJ

Table 5. Required and Applied Energy Capacities for Various Node Types

10.2. Energy harvesting

The most challenging factor facing the widespread deployment of wireless sensor networks (WSNs) is the power constraint faced by sensors that affects network lifetime and performance. Energy harvesting technologies are a significant enabler for smart routing because they relax the critical power constraint by replenishing energy reserves of sensors over time. This can be achieved by converting energy sources such as kinetic, solar or heat energy into usable battery energy.

The impact of energy harvesting can be observed by comparing the rate of energy replenishment to the rate of energy consumption in the sensor network. Define the rate of energy replenishment as r_h in joules per second, or watt. Given an energy consumption rate r_c of a sensor, the effect of energy replenishment on network lifetime fits into one of three categories:

- if $r_h = 0$, the network lifetime remains status quo;
- if $0 < r_h < r_c$, the network lifetime is prolonged but finite; and,
- if $r_h \geq r_c$, the network lifetime is theoretically infinite.

From the perspective of resource allocation, the effect of energy harvesting can also be observed by analyzing the network lifetime. The network is able to operate at peak performance as long as sufficient transmission resources are available; this occurs until nodes are unable to maintain a high level of performance because remaining energy capacities are insufficient. Since energy harvesting enables us to prolong the network, it also increases the period of time that the network operates at high levels of throughput performance. We will analyze the impact of the replenishment rate r_h on the network lifetime for smart routing.

11. Results

We conduct a performance evaluation of our policy based on the following metrics:

- Throughput performance;
- Spectral efficiency;
- Network lifetime based on finite energy capacity;
- Application criticality and performance improvement of smart routing vs. minimum energy routing;
- Blocking probability and its dependency on the number of operating channels in the network; and,
- Energy harvesting effects on energy capacity and network lifetime for various rates of energy replenishment.

11.1. Total throughput

Table 6 illustrates the ability of the smart routing protocol to meet performance requirements of a number of applications. The Ultrawideband (UWB) cluster, which executes a video monitoring application, achieves total network throughput that varies between 84.4 Mbps and 3.4 Gbps. Meanwhile, the Zigbee cluster achieves a maximum real-time throughput of 794.9 kbps in performing temperature monitoring; recall that Zigbee transmissions are only capable of achieving single throughputs of 250 kbps. The WiMax mesh network, in providing long-haul transmission to the centralized controlling station, achieves total network throughput of between 39.6 and 485.7 Mbps. This result illustrates the suitability of UWB for next-generation wireless sensor networks (WSNs) as UWB expands the range of applications that can be used for state-of-the-art resource management.

Network	Maximum Total Throughput	Minimum Total Throughput	Mean Total Throughput	Standard Deviation	Variance
UWB Cluster	3.4 Gbps	84.4 Mbps	1.8 Gbps	112.2 Mbps	1.3×10^{16}
ZigBee Cluster	794.9 kbps	240.1 kbps	770 kbps	2.2 kbps	5.0×10^6
WiMax Mesh	485.7 Mbps	39.6 Mbps	331.5 Mbps	10 Mbps	1.6×10^{13}

Table 6. Throughput Statistics

11.2. Spectral efficiency

Spectral efficiency provides an accurate metric to compare our three communication technologies in terms of the attainable transmission rate per Hz. This is presented in Figure 10 in units of bits/s/Hz. Zigbee provides an effective spectral efficiency of 250 kbps/2.5 MHz = 0.1 bits/s/Hz, Ultrawideband (UWB) attains an effective spectral efficiency of 480 Mbps/500 MHz = 0.96 bits/s/Hz, and WiMax provides improved spectral efficiency of roughly 75 Mbps/20 MHz = 3.75 bits/s/Hz over full channel bandwidths.

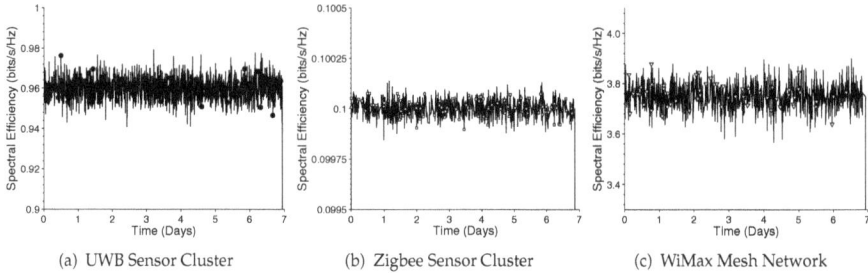

(a) UWB Sensor Cluster   (b) Zigbee Sensor Cluster   (c) WiMax Mesh Network

Figure 10. Spectral Efficiency (bits/s/Hz) over One Week Network Lifetime

11.3. Network lifetime

While the smart routing protocol indeed provides throughput performance benefits, we also analyze the ability of the policy to conserve energy and meet a desired network lifetime of one week. This enables us to evaluate whether the policy successfully meets both performance and energy conservation requirements.

Figure 11(a) presents the average remaining energy capacities of Ultrawideband (UWB) sensors, Zigbee sensors and the WiMax mesh nodes. The UWB and Zigbee sensors are able to survive for roughly the one week network lifetime, with the outages occurring just before the end of the simulation. At the end of the simulation, the WiMax mesh network has roughly 34% of its mean battery energy remaining.

These results meet our network lifetime expectations based on the initial energy capacities in Table 5. For example, we would expect that both the UWB and Zigbee clusters would lose connectivity in the last few hours of the network lifetime. We would also expect the mesh network to maintain roughly one-third of its energy capacity at the end of the simulation. This result is significant as it shows that we can indeed design wireless sensor networks (WSNs) to plan for predictable network lifetimes, while achieving significant throughput performance.

Figure 11(b) illustrates the remaining energy capacities in the final twelve hours of the simulation and the first nodes in each cluster to fully lose connectivity. Based on the initial energy capacities selected, the UWB cluster gave us almost two extra hours of connectivity over the Zigbee network. In terms of the first node outages, node 12 from the Zigbee cluster was the first node to lose connectivity; its remaining battery energy was just under that of the mean from the Zigbee cluster at 20:15. For the UWB cluster, node 6 experienced the first node outage and followed the mean battery energy of the UWB cluster quite strictly at 22:00.

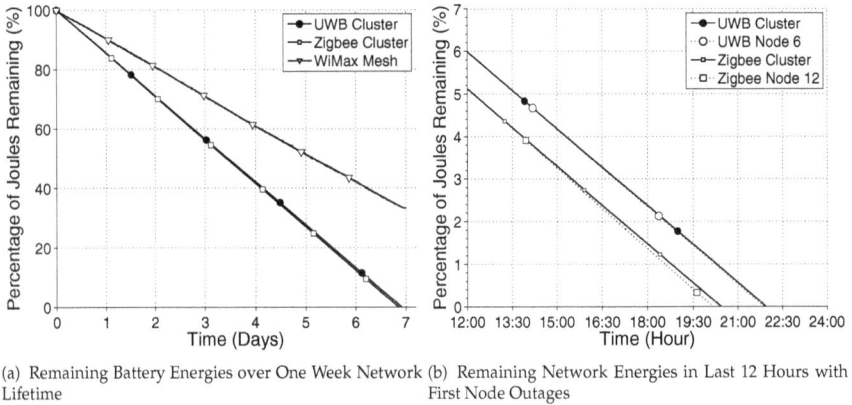

(a) Remaining Battery Energies over One Week Network (b) Remaining Network Energies in Last 12 Hours with
Lifetime   First Node Outages

Figure 11. Comparison of Remaining Battery Energies vs. Network Lifetime

11.4. Application criticality

The impact of application criticality on throughput performance is presented in Table 7 by comparing the performance of the smart routing protocol to minimum energy routing. Smart routing selects candidate nodes that are best able to satisfy both performance and energy conservation requirements given current network conditions. While smart routing is able to achieve total network throughput that varies between 84.4 Mbps and 3.4 Gbps, minimum energy routing only achieves throughputs of 49.2 Mbps to 501.2 Mbps. This is due to minimum energy routing basing its resource allocation decisions solely on ensuring minimum energy consumption; while lower resource consumption certainly has a positive effect on increasing network lifetime, minimum energy routing gives no consideration to the impact of resource allocation on application performance. As we observe in this performance evaluation, applications that have high performance demands require greater resources and, as a result, have shorter network lifetimes; energy-conserving systems, on the other hand, allocate resources to prolong the network lifetime at the expense of application performance.

Routing Policy	Maximum Total Throughput	Minimum Total Throughput	Mean Total Throughput	Standard Deviation	Variance
Smart Routing	3.4 Gbps	84.4 Mbps	1.8 Gbps	112.2 Mbps	1.3×10^{16}
Minimum Energy Routing	501.2 Mbps	49.2 Mbps	327.3 Mbps	20.8 Mbps	4.3×10^{14}

Table 7. Throughput Statistics of Smart Routing vs. Minimum Energy Routing

11.5. Blocking probability

Figure 12(a) illustrates the dependency between the network blocking probability and the number of operating channels for the smart routing protocol. This shows that, as the number of operating channels increases, the blocking probability decreases according to a logarithmic relationship. However, as the traffic intensity ρ and the number of channels increases, the blocking probability decreases at a slower rate. Figure 12(a) also illustrates that the blocking probability decreases as the traffic intensity decreases, which is expected. The sharpness of

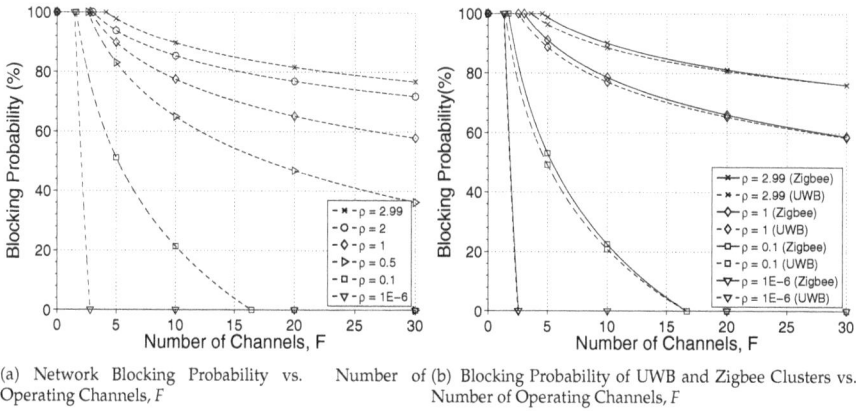

(a) Network Blocking Probability vs. Number of (b) Blocking Probability of UWB and Zigbee Clusters vs. Operating Channels, F Number of Operating Channels, F

Figure 12. Relationship Between Number of Operating Channels F and the Blocking Probability

the drop for a traffic intensity $\rho = 1E\text{-}6$ can be attributed to the near-zero blocking probability at extremely low traffic intensities.

Figure 12(b) illustrates the relationship between the blocking probability and traffic intensity separately for Ultrawideband (UWB) and Zigbee for $F = 5$, $F = 10$ and $F = 20$ channels. Given the same traffic intensity and number of operating channels, the UWB cluster has a blocking probability that is approximately 2% lower than Zigbee on average for $F = 5$. For $F = 10$ and $F = 20$, UWB also has a lower blocking probability than Zigbee but the improvement decreases as the number of channels is increased. This bodes well for next-generation commercial applications for wireless sensor networks (WSNs) that use UWB as the communication technology of choice.

11.6. Energy harvesting

The impact of energy harvesting on energy capacity is illustrated in Figure 13. Figure 13(a) presents the energy dissipation of a single Ultrawideband (UWB) node with no energy harvesting for the one week network lifetime. Two energy states are observed - sleep state and transmission state. In the sleep state, the impact on energy capacity is a regular dissipation of energy due to the sensor operating in a low power state. In the transmission state, we observe a sharp decrease in energy capacity for the duration of the transmission. The energy dissipation during the transmission state is positively correlated to the energy efficiency of the technology. For the given energy capacity $E_1 = 17.5\ J$ for UWB nodes presented in Table 5, we compute the rate of energy consumption as $r_c = 28.6\ \mu W$.

Figure 13(b) presents the impact of energy harvesting on the UWB node's energy capacity. We compare the energy capacity with $r_h = 0\ \mu W$ with replenishment rates $r_h = 22\ \mu W$, 25 μW and 30 μW. For the first two cases, we observe an increase in the energy capacity over time and, hence, a prolonged network lifetime. However, the network lifetime is finite. This is observed for all cases where $0 < r_h < r_c$. For $r_h = 30\ \mu W$, however, we seemingly have 100% energy capacity and hence an unlimited network lifetime. This is intuitive since the rate

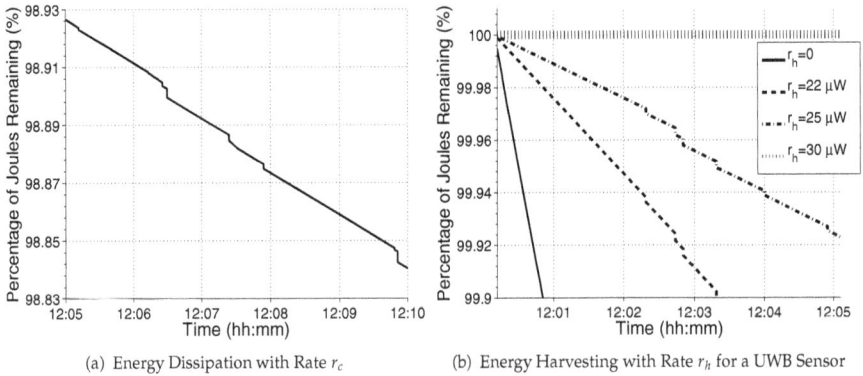

(a) Energy Dissipation with Rate r_c (b) Energy Harvesting with Rate r_h for a UWB Sensor

Figure 13. Energy Dissipation and Impact of Energy Harvesting

of energy replenishment is greater than the rate of consumption. In this case, the network is self-sustaining and can theoretically last forever.

12. Conclusions and future research

This chapter presented the smart routing protocol for large-scale networks that enables for the deployment of wireless sensor networks (WSNs) in geographically distributed locations of interest. Smart routing is based on performance measure and energy optimization using cross-layer considerations of the protocol stack. We presented the performance improvement of smart routing over minimum power routing in these distributed networks to illustrate the benefit of the smart routing protocol for enabling next-generation commercial applications. By doing so, we also presented the impact of application criticality on performance and network lifetime. Applications that have high performance demands require greater resources and, as a result, have shorter network lifetimes; energy-conserving systems, on the other hand, allocate resources to prolong network lifetime at the expense of performance.

We also covered energy harvesting and its impact on resource allocation. We determined that, since sensors are able to operate at peak performance as long as sufficient resources are available, energy harvesting enables us to maintain this level of performance for a longer period of time. If the replenishment rate r_h is greater than or equal to the consumption rate r_c, the network is self-sustaining and can theoretically survive on its own.

Future research shall explore a number of areas to further the smart routing protocol:

- **Software Radio Modeling**: Quantify the impact of packet conversion on energy reserves for multiple technologies including WiMax, WiFi, Ultrawideband (UWB) and Zigbee;

- **Distributed Source Coding (DSC)**: Design algorithms to reduce the amount of data that is routed from sensor networks based on the compression of multiple correlated sensor measurements. In this manner, energy may be conserved while maintaining performance;

- **Sensor Localization**:   Design sensor localization methods that are either triangulation-based or use a third-tier of nodes in the network, such as anchors, for which positions are known in the field. This reduces the dependency of the policy on GPS, until sensors equipped with GPS are made more readily available; and

- **Optimization Metrics**: Model additional performance metrics such as delay, and additional physical (PHY) layer parameters such as bandwidth and modulation.

Author details

Sheikh Omar M. and Mahmoud Samy A.
Faculty of Engineering and Design, Department of Systems and Computer Engineering, Carleton University, Ottawa, Canada

13. References

[1] Zhang H, Hou, J.C (2005) Maintaining Sensing Coverage and Connectivity in Large Sensor Networks, Ad Hoc & Sensor Wireless Networks. j. 1(1-2): 89-124.

[2] Cheekiralla S, Engels D.W (2005) A Functional Taxonomy of Wireless Sensor Network Devices, 2nd International Conference on Broadband Networks.j. 2: 949 - 956.

[3] Oppermann I, Stoica L, Rabbachin A, Shelby Z, Haapola J (2004) UWB Wireless Sensor Networks: UWEN-A Practical Example, IEEE Communications Magazine. j. 42(12): S27-S32.

[4] Chehri A and Fortier P, Tardif P.M (2009) Cross-Layer Link Adaptation Design for UWB-Based Sensor Networks, Elsevier. j. 32(13-14):1568–1575.

[5] Yin X, Zhou X, Li Z, Li S (2010) Cross-Layer Based Rate Control for Lifetime Maximization in Wireless Sensor Networks, Advances in Grid and Pervasive Computing. j. pp. 140-149.

[6] Zhou H Y, Luo D Y, Gao Y, Zuo D.C (2011) Modeling of Node Energy Consumption for Wireless Sensor Networks, Wireless Sensor Network. j. 3(1):18-23.

[7] Madan R, Cui S, Lal S, Goldsmith A (2006) Cross-Layer Design for Lifetime Maximization in Interference-Limited Wireless Sensor Networks, IEEE Transactions on Wireless Communications. j. 5(11):3142–3152.

[8] Krogsveen E, Wang C, Oien G E, Lindfors S (2007) Energy-Efficient Adaptive Route Configuration in Short-Range Wireless Ad-Hoc Networks. In: Patzold M, Jiang Y, Zhang Y, editors. Proceedings of 4th International Symposium on Wireless Communication Systems. Trondheim: IEEE. pp. 292-296.

[9] Wang Q, Abu-Rgheff M.A (2003) Cross-Layer Signalling for Next-Generation Wireless Systems. In: IEEE Wireless Communications and Networking Conference. New Orleans: IEEE. 2:1084-89.

[10] Wang C, Yin L, Oien, G.E (2009) Adaptive Route Configuration for Increased Energy Efficiency in Wireless Sensor Networks. In: Proceedings of 5th International Conference on Broadband Communications, Networks and Systems. London: IEEE. pp. 16-23.

[11] Srivastava V and Motani M (2005) Cross-Layer Design: A Survey and the Road Ahead, IEEE Communications Magazine. j. 43(12):112-119.

[12] IEEE 802.15.3b WPAN Task Group (2006) Part 15.3: Wireless Medium Access Control (MAC) and Physical Layer (PHY) Specifications for High Rate Wireless Personal Area Networks (WPANs) Amendment 1: MAC Sublayer.

[13] IEEE 802.15.4 WPAN Task Group (2006) IEEE Standard for Local and Metropolitan Area Networks - Part 15.4: Wireless Medium Access Control (MAC) and Physical Layer (PHY) Specifications for Low Rate Wireless Personal Area Networks (LR-WPANs).

[14] IEEE 802.16 Network Management Task Group (2004) IEEE Standard for Local and Metropolitan Area Networks - Part 16: Air Interface for Fixed Broadband Wireless Access Systems.

[15] Wan Q, Wang C (2011) Design of 3.1-10.6 GHz Ultra-Wideband CMOS Low Noise Amplifier with Current Reuse Technique, AEU-International Journal of Electronics and Communications. j. 65(12): 1006-1011.

[16] Sadler B.M (2005) Fundamentals of Energy-Constrained Sensor Network Systems, IEEE Aerospace and Electronic Systems Magazine. j. 20(8):17-35.

[17] Mergen G, Zhao Q, Tong L (2006) Sensor Networks with Mobile Access: Energy and Capacity Considerations, IEEE Transactions on Communications. j. 54(11):2033-2044.

[18] Baliga J, Ayre R, Sorin W.V, Hinton K, Tucker R.S (2008) Energy Consumption in Access Networks, Conference on Optical Fiber Communication/National Fiber Optic Engineers Conference, 2008. j. 2:1-3.

Chapter 9

Multihop Routing for Energy Efficiency in Wireless Sensor Networks

Elias Yaacoub and Adnan Abu-Dayya

Additional information is available at the end of the chapter

1. Introduction

Wireless sensor networks (WSNs) are attracting increasing research attention, due to their wide spectrum of applications, including military purposes for monitoring, tracking and surveillance of borders, intelligent transportation systems for monitoring traffic density and road conditions, and environmental applications to monitor, for example, atmospheric pollution, water quality, agriculture, etc. [26].

A WSN is composed of a number of sensor nodes (SN) transmitting wirelessly the information they capture. An SN is generally composed of a power unit, processing unit, sensing unit, and communication unit. Power consumption is the main limiting factor of an SN. In fact, SNs are in general required to operate autonomously and independently for a large period of time in areas where power infrastructure may not be available. Thus, battery-powered SNs should be able to operate with very low power consumption. Some SNs have batteries rechargeable by solar power, thus ensuring longer autonomous operation. The processing unit is responsible to collect and process signals captured from sensors before transmitting them to the network. The sensing unit is a device that produces a measurable response to a change in a physical condition like temperature or pressure. The wireless communication unit is responsible for transferring the senor measurements to the exterior world, e.g., to be stored on a server, where they can be distributed on the internet or accessed by specialized personnel. The wireless communication unit can also ensure a mechanism for ad-hoc communication between SNs forming a WSN [26]. In fact, in some scenarios, it might be more energy efficient to transmit a message via multihop communications over short distances instead of a single hop long distance transmission to the base station (BS).

In this Chapter, a protocol for energy efficient multihop communications in WSNs is presented and analyzed. In the presented approach, SNs form cooperative groups or clusters. Within each cluster, SNs communicate with each other over multihop links, and the SN at the last hop communicates with the BS by relaying the aggregated multihop data. Thus, cooperation between SNs is exploited for the benefit of energy efficiency. Hence, SNs use two wireless interfaces: one to communicate with the BS over a long-range (LR) wireless technology (e.g., UMTS/HSPA, WiMAX, or LTE), and one to communicate with other SNs over a short-range (SR) wireless technology (e.g., Bluetooth, ZigBee, or WLAN). In addition to freeing bandwidth at the BS and increasing network throughput [19, 20], SR collaboration between SNs leads to a reduced energy consumption [8, 31]. In fact, higher rates can be achieved over SR communications between SNs that are relatively close from each other in a single cooperating cluster. This leads to shorter transmission and reception times and hence less energy consumption from the batteries of the SNs.

In this Chapter, SNs are considered to be distributed throughout the cell area and can form several cooperating clusters. The energy minimization problem during cooperative content distribution in the multiple clusters case is formulated and the solution outline is presented. Multihop communications are studied, and remarkable energy savings are achieved even with the 2-hop scenario, corresponding to a clustering framework where a single SN, the cluster head (CH), is in charge of directly receiving the measurement data from each SN in the cluster on the SR, and for transmitting the aggregated data to the BS on the LR. A general formulation that incorporates both multihop and clustering is presented, and energy efficient suboptimal schemes are proposed.

The paper is organized as follows. Related work is presented and differences with the proposed approach are outlined in Section 2. The system model is presented in Section 3. The problem formulation and solution are discussed in Section 4. Suboptimal schemes leading to significant energy savings at reduced complexity are proposed in Section 5 for the multihop and clustering scenarios. The simulation results are presented in Section 6. Practical implementation aspects are discussed in Section 7. An application example of a WSN for air quality monitoring is presented in Section 8. Potential research directions for future investigation are described in Section 9. Finally, conclusions are drawn in Section 10.

2. Related work

This section presents an overview of related work in energy efficiency in multihop wireless communications. Differences with the approach investigated in this Chapter for energy efficient cooperative multihop data transmission are outlined.

Network topology design in order to achieve different requirements in a service-oriented framework is considered in [32]. Requirements include throughput maximization, delay constraints, security, and reliability. Energy minimization constraints are not considered. Topology control is also considered in [22], where energy constraints are taken into account via transmit power adjustments. Connectivity between nodes is determined based on distance considerations. In [23] and [16], energy efficiency is considered by having a minimum energy path between each pair of nodes in a wireless multihop network. Topology is controlled by varying the transmission power at each node, and the transmission power at the antenna is considered as the criterion for energy efficiency. In this Chapter, the energy drained from the sensors' batteries, not only the transmit power at the antenna, is used as the criterion for energy efficiency.

Processing capacity is studied in [25] for wireless sensor networks. A cross-layer collaborative in-network processing approach among sensors is adopted, where, in addition to processing information at the application layer, sensors synchronize their communication activities to exchange partially processed data for parallel processing. Sensor nodes are grouped into clusters, and operations are performed independently inside each cluster. Communications between clusters are performed using channels that are orthogonal to intra-cluster communications. Multihop communications are implemented inside each cluster to perform parallel computing of certain processing tasks. Thus, energy efficiency is considered in the sense of minimizing the processing power during task scheduling and implementation, not in the sense of transmissions and receptions for relaying measurement data of sensors, as is the case in this Chapter.

Small scale networks where sensor nodes are closely located are studied in [7]. TDMA is assumed as an access method. Both transmission and circuit-based energy consumption are considered. Perfect synchronization between nodes is assumed. The joint design of the physical, MAC, and routing layers to minimize network energy consumption is formulated into a convex optimization problem and the solution is provided. The approach presented in this Chapter does not make any assumptions concerning the channel accessing scheme or the scale of the sensor network.

In [13], energy efficiency is studied in wireless sensor networks. Sensors having data to transmit should relay this data to a single source using multihop. Nodes that do not have data to transmit or that are not relaying the data of other nodes can be put to sleep. Energy efficiency is achieved by reducing the number of active nodes. An energy efficient routing technique in multihop wireless sensor networks is presented in [28]. For each node, the energies consumed during reception, transmission, and sensing are considered in the analysis. In the model of [28], frame nodes relay the content of the source to the destination. If the communication fails between the source and a frame node, or between two frame nodes, assistant nodes come into play and relay the data to the next frame node. Hence the use of opportunistic transmissions depending on the fading conditions of the channel. The optimal number of nodes that should be included in a path is determined. The purpose is to reduce the energy consumption by reducing the number of nodes relaying the data from source to destination. In the scenario investigated in this Chapter, all nodes are assumed to have data to transmit, and hence cannot be put to sleep to achieve energy savings. This scenario corresponds, for example, to WSNs deployed for the purpose of air quality monitoring in a given area, where each sensor will periodically send measurement data to a central processing system.

In [3], multipath routing based on spatial relationships among nodes is considered. Stochastic geometric and queueing models are used for the evaluation of different types of scenarios. Energy aware routing with the possibility of energy replenishment of nodes in multihop wireless sensor networks is presented in [17]. An algorithm that only requires short term energy replenishment information is also presented. However, channel conditions are not taken into consideration in the approach of [17], conversely to the work in this Chapter where channel state information (CSI) is exploited in order to build the energy efficient routes from SNs to the BS.

Several papers in the literature consider implementation scenarios related to a particular standard. For short range multihop communications, IEEE 802.11s is receiving significant attention. In [6], a tutorial is presented for multihop communications and mesh capabilities in IEEE 802.11. Task group 802.11s is handling this issue. In the draft 802.11s proposal, the mesh network is implemented at the link layer and relies on MAC addresses instead of IP addresses, which provides layer-2 multihop communication. A survey of the unicast admission control schemes designed for IEEE 802.11-based multi-hop mobile ad-hoc networks (MANETs) is presented in [10], where different admission control protocols are discussed and analyzed. In [27], cooperative rate adaptation in multihop IEEE 802.11 is considered. The problem is formulated as an optimization problem and shown to be NP-hard. Thus, a suboptimal method is presented. Energy efficiency is considered in terms of reducing the transmission power at the SNs' antennas. Enhancements of the performance of IEEE 802.11-based multihop ad hoc wireless networks from the perspective of spatial reuse were surveyed in [2]. Techniques adopting transmit power control, tuning the carrier sensing threshold, performing data rate adaptation, and using directional antennas were discussed. In this Chapter, the presented approach is general and not confined to a particular standard, it does not only consider transmit energy at the antenna, but also the energy drained from the battery during transmission and reception. Compared to mesh networks, not every SN needs to communicate with all other SNs. Instead, each SN needs to transmit the measured data using an optimum energy minimizing path to the BS. This path remains the same as long as the channel conditions remain constant.

In addition to multihop, energy efficient clustering methods are also investigated in the literature. An algorithm is presented in [14] as an improvement on the methods in [12] and [15]. In [12, 14, 15], each node volunteers to be a cluster head in a probabilistic manner, and non-cluster nodes associate themselves with cluster heads based on the announcements received from these cluster heads. The actual energy drained from the battery of the device is considered. However, the problem is not formulated and solved as an optimization problem (as in this Chapter), but rather an efficient clustering algorithm that ensures fairness in energy consumption between nodes, due to the probabilistic selection, is presented. In [15], the use of a proxy node was added to the approach of [12], whereas in [14] the additional use of a main cluster head was implemented, with the main cluster head relaying the data from cluster heads to the BS. The work of [12] was extended in [4] to include multihop communications in addition to clustering. In addition, an approach to determine the optimal number of cluster heads is proposed. Clustering is performed on distance based criteria and a probabilistic random approach is adopted for the election of cluster heads. A cluster head selection based on proximity was adopted in [30], where the residual energy of the node is also considered in the selection process. A multihop time reservation using adaptive control for energy efficiency (MH-TRACE) is presented in [24]. Cluster formation is probabilistic and it is not based on connectivity information. In MH-TRACE, the interference level in the different time-frames is monitored continuously in order to minimize the interference between clusters. MH-TRACE clusters use the same spreading code or frequency and time division is adopted. In this Chapter, cluster head selection is not probabilistic or simply proximity based. Fading is considered in the selection approach since CSI affects the achievable rates and is thus incorporated in the optimization problem.

3. System model

The energy minimization problem in a WSN is considered. The data is to be delivered to the BS from K SNs distributed throughout the cell area of the BS. The SNs can communicate with the BS using a long range communication technology (e.g., UMTS/HSPA, WiMAX, or LTE), or with neighboring SNs using a short range technology (e.g., Bluetooth or WLAN). SNs form cooperating clusters for the purpose of energy minimization during cooperative

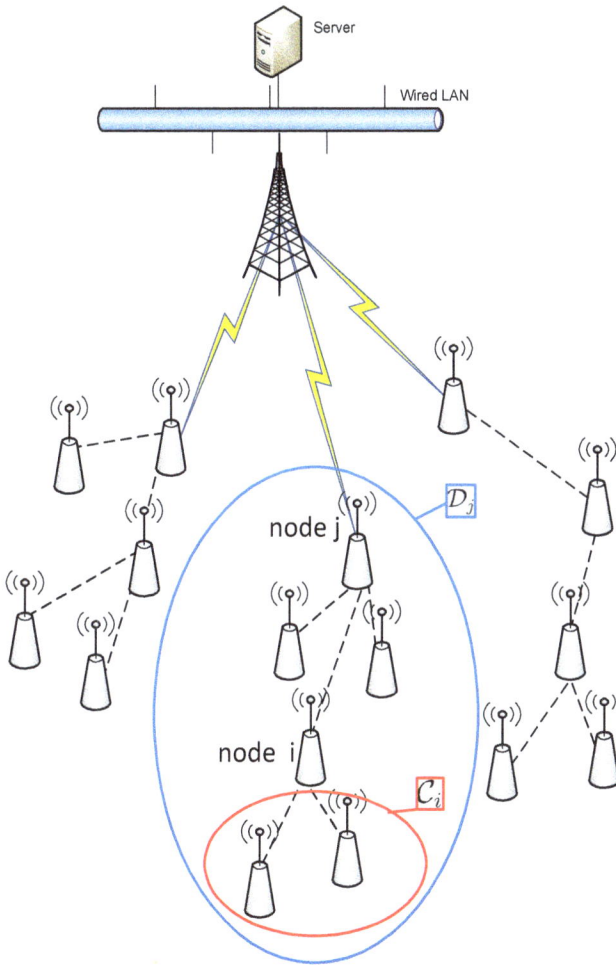

Figure 1. System model when multihop communications are allowed.

data transmission. Within each cooperating cluster, the data is delivered from the SNs in that cluster to the BS using multihop communications. Fig. 1 shows the scenario considered. The maximum number of hops allowed H can be specified as a parameter. With two-hop communications (case $H = 2$), the problem becomes a clustering problem that consists of finding the best grouping of SNs into cooperating clusters, as shown in Fig. 2.

Each SN transmits its measured data to a single destination, which could be either the BS or another SN. We consider the energy minimization problem with multihop/clustering. The BS and SNs are denoted as "nodes", with node $k = 0$ corresponding to the BS and nodes $k = 1, ..., K$ corresponding the SNs. As shown in Fig. 1, these nodes appear to form a direct

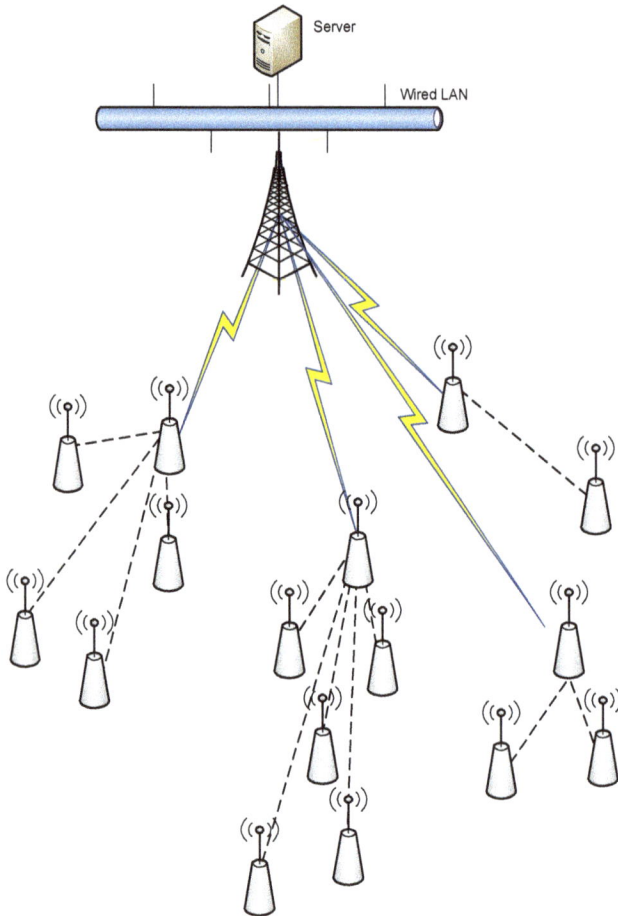

Figure 2. System model when 2-hops (clustering only) are allowed.

acyclic graph (DAG) starting from the node $k = 0$. If node j receives the data of node k on hop h, a parameter α_{kj}^h is set to one, marking the existence of an edge in the graph between k and j. Otherwise, α_{kj}^h is set to zero.

We define $\mathcal{C}_j$ as the set of children of j, i.e., the set of nodes sending their data directly to j:

$$\mathcal{C}_j = \left\{ k, \sum_{h=1}^{H} \alpha_{kj}^h = 1 \right\} \tag{1}$$

The set $\mathcal{D}_j$ is defined as the sub-DAG starting from j, i.e., having j as its root. It includes j, its children, the children of its children, etc. Thus, it can be expressed as:

$$\mathcal{D}_j = \{j\} \cup \bigcup_{k \in \mathcal{C}_j} \mathcal{D}_k \tag{2}$$

3.1. Data rates

Given for each node: the transmit power $P_{t,kj}$ that node k is using in order to transmit to node j, the channel gain H_{kj} of the channel between k and j, and the thermal noise power σ^2, the received signal-to-noise ratio (SNR) γ_{kj} on the link between k and j can be calculated following $\gamma_{kj} = \frac{P_{t,kj}H_{kj}}{\sigma^2}$. Given the target bit error rate P_e and the SNR, the bit rates on the link between any two nodes k and j can be calculated as follows:

$$R_{kj} = W_{kj} \cdot \log_2(1 + \beta\gamma_{kj}) \tag{3}$$

In (3), W_{kj} is the passband bandwidth of the channel between k and j, and β is called the SNR gap. It indicates the difference between the SNR needed to achieve a certain data transmission rate for a practical M-QAM system and the theoretical Shannon limit [9, 21]. It is given by: $\beta = \dfrac{-1.5}{\ln(5P_e)}$. The channel gain is expressed as:

$$H_{kj,\text{dB}} = (-\kappa - v\log_{10}d_{kj}) - \xi_{kj} + 10\log_{10}F_{kj} \tag{4}$$

In (4), the first factor captures propagation loss, with d_{kj} the distance between nodes k and j, and v the path loss exponent. The second factor, ξ_{kj}, captures log-normal shadowing with a standard deviation σ_ξ, whereas the last factor, F_{kj}, corresponds to Rayleigh fading (generally considered with a Rayleigh parameter a such that $E[a^2] = 1$).

4. Multihop problem formulations

With each SN transmitting the data in blocks of size S_T bits, the time needed to transmit this content on a link between nodes k and j having an achievable rate R_{kj} bps is given by S_T/R_{kj}. Denoting the power drained from the battery of node j to receive the data from node k by $P_{\text{Rx},kj}$, then the energy consumed by j to receive the data from k is given by $S_T \cdot P_{\text{Rx},kj}/R_{kj}$. Similarly, denoting by $P_{\text{Tx},kj}$ the power drained by the battery of node k to transmit the data to node j, then the energy consumed by k to transmit the content to j is given by $S_T \cdot P_{\text{Tx},kj}/R_{kj}$. It should be noted that $P_{\text{Tx},kj}$ can be expressed as:

$$P_{\text{Tx},kj} = P_{\text{Tx}_{\text{ref}},kj} + P_{t,kj} \tag{5}$$

where $P_{\text{Tx}_{\text{ref}},kj}$ corresponds to the power consumed by the circuitry of node k during transmission on the communication interface with node j, and $P_{t,kj}$ corresponds to the power transmitted over the air on the link from node k to node j.

In this section, a flexible formulation is presented that accommodates power adaptive or rate adaptive transmission. In the case of adaptive rate control, the node transmit power is constant, i.e., $P_{t,kj} = P_t$ and $P_{\mathrm{Tx},kj} = P_{\mathrm{Tx}}$. Consequently, the rate R_{kj} on the link between nodes k and j is the rate achievable with the transmit power P_t. It is varied adaptively depending on the channel conditions between nodes k and j. High data rates result in low energy per bit consumption, thus leading to a gain in total energy consumption. For example, the WLAN technologies apply rate control [11].

In the case of adaptive power control, the nodes communicate at a constant rate $R_{0j} = R_L$ on the LR or $R_{kj} = R_S$ (with $k > 0$) on the SR. The transmit power $P_{t,kj}$ is varied adaptively depending on the channel conditions between nodes k and j in order to achieve the target data rate R_L or R_S. Thus, nodes that are in proximity of each other will communicate with lower power than nodes that are further apart. This will result in a reduction of consumed energy. Some technologies such as Bluetooth apply power control [5].

Hence, the energy consumed during cooperative multihop content distribution can be expressed as follows:

$$
\begin{aligned}
E_{\mathrm{coop}} &= S_T \cdot \sum_{k=1}^{K} \sum_{j=0,j\neq k}^{K} \sum_{h=1}^{H} \frac{\alpha_{kj}^h \cdot |\mathcal{D}_k| \cdot P_{\mathrm{Tx},kj}}{R_{kj}} \\
&+ S_T \cdot \sum_{k=1}^{K} \sum_{j=1,j\neq k}^{K} \sum_{h=1}^{H-1} \frac{\alpha_{kj}^h \cdot |\mathcal{D}_k| \cdot P_{\mathrm{Rx},kj}}{R_{kj}} \\
&= S_T \cdot \sum_{k=1}^{K} \sum_{j=0,j\neq k}^{K} \sum_{h=1}^{H} \frac{\alpha_{kj}^h \cdot |\mathcal{D}_k| \cdot (P_{\mathrm{Tx},kj} + P_{\mathrm{Rx},kj})}{R_{kj}}
\end{aligned}
\tag{6}
$$

where the first term corresponds to the energy consumed by the nodes for transmission and the second term corresponds to the energy consumed by the nodes for reception. Hop $h = H$ corresponds to transmission on the LR and node $k = 0$ corresponds to the BS. The multiplication by $|\mathcal{D}_k|$, with $|\cdot|$ denoting set cardinality, is used to indicate that an SN aggregates the data of its sub-DAG before transmitting it on the next hop. To be able to write the last equality in (6), it is assumed that $P_{\mathrm{Rx},k0} = 0$ for all k. This corresponds to excluding the energy consumed at the BS to receive the data at hop H. In fact, power consumption of the BS is not considered in the energy minimization process since the interest is in the battery life of the SNs. This is justified by the fact that most BSs rely on power line cables and not on batteries and thus do not have as stringent power limitations as the SNs.

Consequently, the optimization problem can be formulated as follows:

$$
\min_{\alpha} \; E_{\mathrm{coop}} = S_T \cdot \sum_{k=1}^{K} \sum_{j=0,j\neq k}^{K} \sum_{h=1}^{H} \frac{\alpha_{kj}^h \cdot |\mathcal{D}_k| \cdot (P_{\mathrm{Tx},kj} + P_{\mathrm{Rx},kj})}{R_{kj}}
\tag{7}
$$

subject to

$$
\alpha_{k0}^h = 0 \text{ for } h < H \text{ and } k = 1, ..., K
\tag{8}
$$

$$
\sum_{j=0}^{K} \sum_{h=1}^{H} \alpha_{kj}^h = 1 \text{ for } k = 1, ..., K
\tag{9}
$$

$$
\alpha_{kj}^h \in \{0,1\} \forall k, j, h
\tag{10}
$$

	Power Adaptive	Rate Adaptive
SR	$R_{kj} = R_S \forall k, j \geq 1$	$P_{Tx,kj} = P_{Tx} \forall k, j \geq 1$
LR	$R_{0j} = R_L \forall j \geq 1$	$P_{t,0j} = P_t \forall j \geq 1$

Table 1. Parameter Values in Different Scenarios

The first constraint (8) indicates that transmissions to the BS take place at the last hop $h = H$ only. The second constraint (9) indicates that each SN should transmit its collected data exactly once to a single destination on one of the H hops (hop H on the LR and $H - 1$ hops on the SR). Finally, constraint (10) specifies that the optimization variable α_{kj}^h is a binary variable.

In the problem formulated in (7), the maximum number of hops can be specified as a parameter. Setting $H = K$ allows full multihop communications, although the actual hops might be less than K, and in this case the parameters α_{kj}^h corresponding to the unnecessary hops will be set to zero in the optimal solution. Setting $H = 2$ corresponds to reducing the problem into a clustering problem where SNs are grouped into clusters. In each cluster, an SN selected as cluster head (CH) in the optimal solution sends the data on the LR to the BS after aggregating the data it receives on the SR from the SNs in its cluster. Furthermore, setting $H = 1$ corresponds to the non-cooperative approach where all SNs send the data on the LR to the BS. In this case, the energy is denoted by $E_{No-coop}$. The normalized energy consumption η can be calculated as follows:

$$\eta = \frac{E_{coop}}{E_{No-coop}} \tag{11}$$

The value of η indicates whether the cooperation is beneficial in terms of energy consumption or not; if $\eta < 1$, then the cooperation results in a gain of energy consumption while $\eta > 1$ reflects a non-beneficial cooperation.

The formulation in (7) is applicable to any number of hops, allows communication using different wireless interfaces (different values of $P_{Rx,kj}$ and $P_{Tx,kj}$ can be set for each wireless link between any two nodes k and j), and permits any combination of power adaptive/rate adaptive transmissions. For example, a node may be transmitting to its parent in the DAG using rate adaptive transmission while another can be using power adaptive transmission. The values of the parameters in the different implementation scenarios are detailed in Table 1. Using, for each node in the network, the appropriate parameters from Table 1 according to its communication scheme adopted, then the formulation (7) can be customized to a huge variety of node combinations and hybrid wireless interfaces.

The problem formulated in (7) appears as a binary integer program that can be solved using known software solvers. However, this is not the case due to the dependence of $|\mathcal{D}_k|$ on the parameters α_{kj}^h, which makes the problem intractable. In addition, even when the problem can be considered as a binary integer program, the complexity of finding the optimal solution of the problem (7) using software solvers increases tremendously when the number of nodes increases and is not suitable for real time implementation. In fact, binary integer programming is known to be NP-hard. In the next section, low complexity suboptimal schemes are presented that are able to achieve efficient multihop routing of sensor data with significant energy savings compared to the non-cooperative approach.

5. Suboptimal energy-efficient WSN data routing methods

In this section, we present algorithms that perform energy efficient routing of sensor data. Section 5.1 presents a multihop approach whereas Section 5.2 presents a clustering-based approach. Section 5.3 presents a complexity analysis that applies to both methods.

5.1. Suboptimal multihop approach

In this section, we present an algorithm that performs energy efficient multihop routing of sensor data. Starting with the SNs having worst channel conditions on the LR (and hence worst achievable rates and highest energy consumption), we find for each SN k the parent p_k to which it can send the data with the minimum energy consumption. When the turn comes to SN p_k, a parent p_{p_k} is found to which p_k can send the data with the minimum energy consumption, thus leading to an additional hop if $p_{p_k} \neq 0$. The details of the proposed approach are presented below:

- **Step 1:** Sort the SNs in decreasing order of energy consumption without cooperation. After this step, SN $k = 1$ would be the one having the worst channel conditions on the LR and SN $j = K$ would be the one having the best channel conditions on the LR.

- **Step 2:** Start from SN $k = 1$.

- **Step 3:** For SN k, find the parent node (could be another SN or the BS) p_k to which k can forward the data with the least energy consumption. The search is done over the nodes j having better LR channel conditions than k, i.e., such that $j > k$. Energy consumption to distribute the content includes the energy of k to transmit and the energy of p_k to receive. i.e.:

$$p_k = \arg\min_{j;j>k} |\mathcal{D}_k| S_T \cdot \frac{(P_{\text{Tx},kj} + P_{\text{Rx},kj})}{R_{kj}} \tag{12}$$

- **Step 4:** break the connection of k with the BS and set p_k as the direct parent of k if $\frac{(P_{\text{Tx},kp_k} + P_{\text{Rx},kp_k})}{R_{kp_k}} < \frac{P_{\text{Tx},k0}}{R_{k0}}$, i.e., if it is more energy efficient for k to send the data to p_k rather than sending it directly to the BS. Then update $\mathcal{D}_{p_k}$ as: $\mathcal{D}_{p_k} = \mathcal{D}_{p_k} \cup \mathcal{D}_k$.

- **Step 5:** increment k and repeat Steps 3-5 on the SNs whose order is $> k$ in the sorted list.

- **Step 6:** After all the SNs have been assigned to their direct parent based on the most energy efficient path, we check if SN K can send the data with lower energy than sending it directly on the LR link, since it is still connected to the BS (due to sorting the SNs in decreasing order of LR energy consumption). Hence, if there exists an SN $x \neq K$ such that $p_x = 0$ (i.e. there is another path to the BS that does not go through SN K, which means that the link between the BS and SN K can be broken while still being able to send the data from the SNs to the BS), then for all SNs $j < K$ such that $p_j \neq K$, the parent of SN K is selected such that:

$$p_K = \arg\min_{j;p_j \neq K} |\mathcal{D}_K| S_T \cdot \frac{(P_{\text{Tx},Kj} + P_{\text{Rx},Kj})}{R_{Kj}} \tag{13}$$

- **Step 7:** We set p_K as the direct parent of K if $\frac{(P_{\text{Tx},Kp_K} + P_{\text{Rx},Kp_K})}{R_{Kp_K}} < \frac{P_{\text{Tx},K0}}{R_{K0}}$. Otherwise, we keep $p_K = 0$, i.e., the best destination for SN K to send the data to is the BS. If p_K is set as the parent of K, then update $\mathcal{D}_{p_K}$ as: $\mathcal{D}_{p_K} = \mathcal{D}_{p_K} \cup \mathcal{D}_K$.

The algorithm presented in this section does not impose a limit on the number of hops. The outcome could be any number H such that $1 \leq H \leq K$, where $H = 1$ indicates that all SNs send their data directly to the BS. This corresponds to a scenario where SNs are scattered to an extent such that collaboration is not energy efficient, and the best for each SN is to send the data directly to the BS. In the next section, we present a similar algorithm that performs node clustering ($H = 2$).

5.2. Suboptimal clustering approach

In this section, we present an algorithm that performs energy efficient clustering for sensor data transmission. The algorithm performs a grouping of SNs into cooperating clusters, with each cluster having an SN, the cluster head (CH), receiving the data from the SNs within its cluster and forwarding it to the BS, along with its own measurements. The algorithm could lead to situations where one or more clusters contain a single SN. In this case, that SN is the cluster head and sends its data on the LR without receiving from other SNs on the SR. This corresponds to a situation where other SNs are too far or the links with them are under severe fading, such that collaboration is not energy efficient, and the best solution for that SN is to send the data directly to the BS.

Starting with the SNs having worst channel conditions on the LR (and hence worst achievable rates and highest energy consumption), we find for each SN k the parent p_k to which it can send the data with the minimum energy consumption. If k is a cluster head, all members of $\mathcal{D}_k$ are moved to $\mathcal{D}_{p_k}$ if the data transmission form k and all the members of $\mathcal{D}_{p_k}$ to p_k is more energy efficient than having an independent cluster with k as cluster head. It should be noted that in the special case of clustering, we have $\mathcal{D}_k = k \cup \mathcal{C}_k$. The details of the proposed approach are presented below:

- **Step 1:** Sort the SNs in decreasing order of energy consumption without cooperation. After this step, SN $k = 1$ would be the one having the worst channel conditions on the LR and SN $k = K$ would be the one having the best channel conditions on the LR.

- **Step 2:** Start from SN $k = 1$.

- **Step 3:** For SN k, find the parent node (could be another SN or the BS) p_k to which k and all the members of $\mathcal{D}_k$ (if there are any SNs other than k) can send their data with the least energy consumption. Energy consumption to distribute the content includes the energy of p_k to receive and the transmission energy of the SNs in $\mathcal{D}_k$, i.e.:

$$p_k = \arg\min_{j;j>k} S_T \cdot \sum_{i \in \mathcal{D}_k} \frac{P_{\text{Tx},ij} + P_{\text{Rx},ij})}{R_{ij}} \tag{14}$$

- **Step 4:** break the connection of k with the BS, and the connection of all other members of $\mathcal{D}_k$ with k, and set p_k as the direct parent of k and all other SNs in $\mathcal{D}_k$ if

$$\sum_{i \in \mathcal{D}_k} \frac{(P_{\text{Tx},ip_k} + P_{\text{Rx},ip_k})}{R_{ip_k}} < \frac{P_{\text{Tx},k0}}{R_{k0}} + \sum_{i \in \mathcal{D}_k, i \neq k} \frac{(P_{\text{Tx},ik} + P_{\text{Rx},ik})}{R_{ik}}$$

i.e., move all members of $\mathcal{D}_k$ to $\mathcal{C}_{p_k}$ if this is more energy efficient than having an independent cluster with k as cluster head sending the data to the BS: $\mathcal{C}_{p_k} = \mathcal{C}_{p_k} \cup \mathcal{D}_k = \mathcal{C}_{p_k} \cup k \cup \mathcal{C}_k$.

- **Step 5:** increment k and repeat Steps 3-5 on the SNs whose order is $> k$ in the sorted list.
- **Step 6:** After all the SNs have been grouped into clusters based on the most energy efficient method, we check if SN K can send the data with lower energy than sending on the LR link, since it is still connected to the BS (due to sorting the SNs in decreasing order of LR energy consumption). Hence, if there exists an SN $x \neq K$ such that $p_x = 0$ (i.e. there is another cluster with cluster head other than SN K, which means that the link between the BS and SN K can be broken while still being able to send the data from the SNs to the BS), then for all SNs $j < K$ such that $p_j = 0$, the parent of SN K is selected such that:

$$p_K = \arg \min_{j; p_j = 0} S_T \cdot \sum_{i \in \mathcal{D}_K} \frac{(P_{\text{Tx},ij} + P_{\text{Rx},ij})}{R_{ij}} \tag{15}$$

- **Step 7:** We set p_K as the direct parent of K if

$$\sum_{i \in \mathcal{D}_K} \frac{(P_{\text{Tx},ip_K} + P_{\text{Rx},ip_K})}{R_{ip_K}} < \frac{P_{\text{Tx},K0}}{R_{K0}} + \sum_{i \in \mathcal{D}_K, i \neq K} \frac{(P_{\text{Tx},iK} + P_{\text{Rx},iK})}{R_{iK}}$$

Otherwise, we keep $p_K = 0$, i.e., SN K is a cluster head sending the data to the BS. If p_K is set as the parent of K, then we update $\mathcal{C}_{p_K}$ as: $\mathcal{C}_{p_K} = \mathcal{C}_{p_K} \cup \mathcal{D}_K = \mathcal{C}_{p_K} \cup K \cup \mathcal{C}_K$.

5.3. Complexity analysis

This section presents a complexity analysis that applies to both methods of Sections 5.1 and 5.2. Step 1 of the algorithms is a sorting step, and hence has a worst-case complexity $\mathcal{O}(K^2)$. In Step 3, the search involves K nodes when $j = 1$, it involves $(K - 1)$ nodes when $j = 2$, etc., and 2 nodes when $j = (K - 1)$. Hence, the complexity of Steps 2 to 5 is: $K + (K - 1) + \cdots + 2 = \frac{K(K+1)}{2} - 1$. In Steps 6-7, the search involves at most K nodes. Consequently, the worst-case complexity of the algorithms is: $K^2 + \frac{K(K+1)}{2} - 1 + K = \frac{3K^2}{2} + \frac{3K}{2} - 1$. This is a quadratic complexity of order $\mathcal{O}(K^2)$. Hence, the proposed suboptimal methods are significantly easier to implement than the optimal solution of the NP-hard problem of Section 4.

In the next section, we compare the methods of Sections 5.1 and 5.2 to each other and to the non-cooperative approach.

6. Results and discussion

In this section, simulation results are presented and analyzed. The simulation parameters are presented in Table 2. Channel parameters are obtained from [1], whereas energy consumption parameters are taken as in [18], where measurements are made with 3G communications on the LR, and 802.11 b on the SR using the rate adaptive approach.

In Sections 6.1 to 6.3, we investigate a scenario corresponding to multihop data transmission in a WSN. We consider that each sensor sends its measurement data in a file of size $S_T = 1$ Mbits, to be routed to the BS in an energy efficient manner. Two main SN deployment scenarios are investigated:

Parameter	Value
κ	-128.1 dB
υ	3.76
σ_ξ (dB)	8 dB
P_{Tx}	1.425 Joules/s
$P_{S,Rx}$	0.925 Joules/s
$P_{L,Rx}$	1.8 Joules/s

Table 2. Simulation Parameters

- In the first deployment scenario, SNs are assumed to be uniformly distributed in a rectangular area of size 200m × 200m, whose origin is at a distance d_{LR} m from the BS. Different values of d_{LR} are investigated in the simulations. This scenario corresponds, for example, to a WSN monitoring air pollution in a particular area of interest, e.g., near a power plant, or an area where a high density of lung disease was detected.

- In the second scenario, the SNs are assumed to be uniformly distributed throughout the whole cell. We consider a single BS placed at the center of a 1 × 1km cell. This scenario corresponds to a case where the whole cell needs to be monitored by the WSN, not a particular or specific area. This scenario will be referred to by "BS at center of 1 × 1km cell" in the figures.

Results are averaged over 50 iterations. In each iteration, new random SN locations are determined and 50 fading realizations are considered (thus results are averaged over $50 \times 50 = 2500$ fading realizations). We compare the methods of Sections 5.1 (denoted as "multihop" in the results) and 5.2 (denoted as "clustering" in the results) to the non-cooperative approach.

6.1. Example on the gap between the optimal and suboptimal methods

d_{LR} (m)	300	500	1000
Optimal	0.6761	1.2015	5.0010
Proposed Multihop	0.7342	1.2974	5.1023
Proposed Clustering	0.7423	1.3255	5.1455
No Cooperation	1.6185	5.3847	45.0133

Table 3. Energy (in Joules) Results for $K = 3$

In this section, the proposed methods of Section 5 are compared to the optimal multihop solution of Section 4 (with $H = K$) for a low number of SNs (in order for the optimal solution to be tractable). Selecting $K = 3$, all the possible cases are shown in Fig. 3. Hence, the optimal solution will be one of the 16 cases presented in Fig. 3, depending on the fading conditions. The results obtained after implementing the optimal solution and the proposed methods are listed in Table 3. It can be clearly seen that the gap between the suboptimal multihop and clustering results from the optimal solution is very small. In addition, Table 3 shows that the cooperative techniques lead to huge savings compared to the non-collaborative scenario.

Fig. 4 shows, for each of the 16 cases, the percentage of times that this case occurs as the optimal solution. When the distance to the BS is small, Case 1 (no collaboration) seems to be optimal for a significant percentage of the time. However, this percentage decreases as the

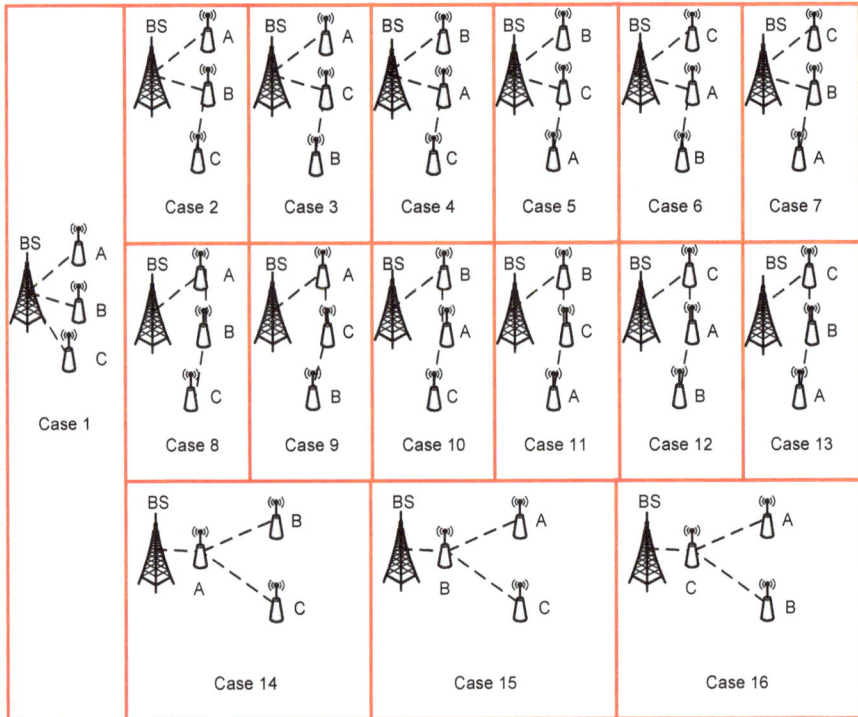

Figure 3. The 16 possible cases when $K = 3$.

distance to the BS increases. Cases 2-7 form a group of similar cases where the only variation is a permutation of the SNs involved in the connections. As expected, these cases have almost equal probability of being the optimal case for a given value of d_{LR}. The same reasoning applies for Cases 8-13 and Cases 14-16. Interestingly, Cases 8-13 were never optimal in the obtained results.

In fact, with Cases 2-7, and considering Case 2 as an example, SN A transmits S_T bits on the LR, SN C transmits S_T bits on the SR, and SN B transmits $2S_T$ bits (its own data in addition to the data of SN C) on the LR. With Cases 14-16, and considering Case 14 as an example, SN B transmits S_T bits on the SR, SN C transmits S_T bits on the SR, and SN A transmits $3S_T$ (its own data in addition to the data of SNs B and C) bits on the LR. In Cases 8-13, and considering Case 8 as an example, SN C transmits S_T bits on the SR, SN B transmits $2S_T$ bits (its own data in addition to the data of SN C) on the SR, and SN A transmits $3S_T$ (its own data in addition to the data of SNs B and C) bits on the LR. Since the SNs are deployed in a confined area of interest, and since SR transmissions in this case can occur at high rates due to the relative proximity of SNs, Cases 14-16 would generally lead to lower energy consumption than Cases 8-13, since both groups have the same LR energy consumption (due to transmitting $3S_T$ on the LR by one SN), but on the SR each of the other two SNs transmits S_T with Cases 14-16. However, with

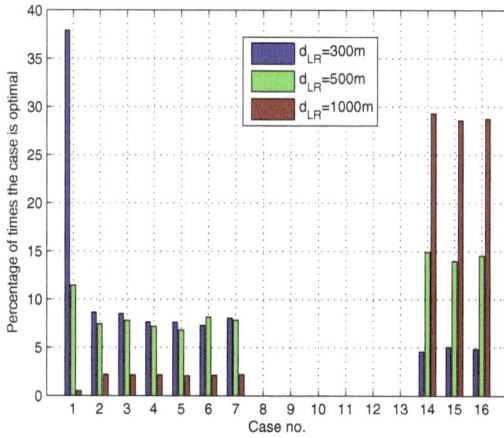

Figure 4. Percentage of having each of the 16 possible cases as the optimal solution when $K = 3$.

Cases 8-13, SR energy consumption is higher because one SN transmits S_T while the other transmits $2S_T$ on the SR.

Fig. 4 shows that as d_{LR} increases, Cases 14-16 become more favored than Cases 1-7. In fact, a large distance to the BS leads to spending most of the energy during LR transmission, since the achievable rates become significantly lower due to the increased distance. Thus, one LR transmission with an SN having favorable LR channel conditions in Cases 14-16 would be more energy efficient than two LR transmissions with Cases 2-7 (or three LR transmissions with Case 1).

6.2. Energy results

This section presents the energy savings achieved by using the proposed multihop and clustering methods, compared to the non-cooperative scenario. In the non-cooperative approach, each SN sends the data on the LR to the BS without any collaboration with other SNs on the SR. Fig. 5 shows the normalized energy results for the various investigated scenarios. Significant energy savings are achieved compared to the non-cooperative scenario, regardless of the number of hops allowed. In fact, the clustering approach corresponding to $H = 2$ and the multihop approach with $H = K$, thus representing the two extreme cases, have a very comparable performance in terms of normalized energy. Fig. 5 shows that the gains are reduced as the distance to the BS decreases. This is due to a reduction in the energy needed on the LR without cooperation and not to an increase in energy consumption with the proposed approach, since the LR distance was reduced. This leads to an increase in the ratio η.

In fact, the results of Fig. 6, presenting the energy consumption results without normalization, show that the energy is reduced when the distance to the BS is reduced, as expected. The results of the energy consumption in the non-cooperative scenario are shown in Fig. 6 for reference. Values for $d_{LR} = 1000$ m are not shown, since they are around an order of magnitude larger than the cooperative results, which makes all the plots of the various

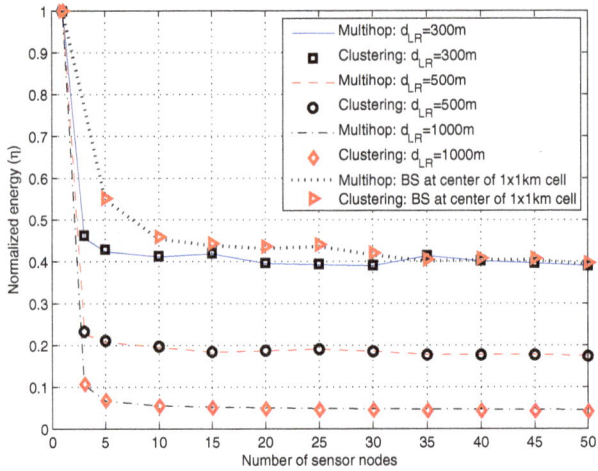

Figure 5. Normalized energy consumption vs. the number of SNs for different values of d_{LR}.

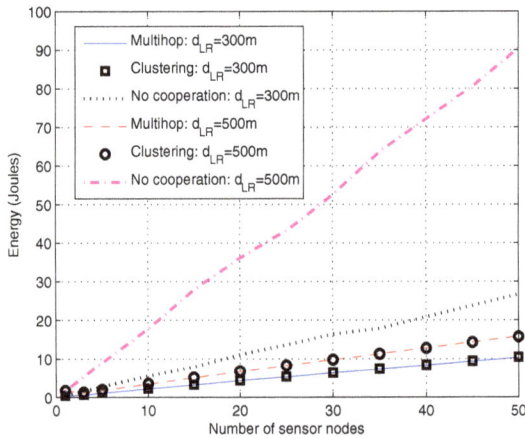

Figure 6. Energy consumption vs. the number of SNs for different values of d_{LR}.

cooperative scenarios appear to overlap. Thus, the combination of Figs. 5 and 6 allows to display both the gains of cooperation compared to the non-cooperative scenario and to understand the variation of the energy gains with the distance to the BS.

6.3. Delay results

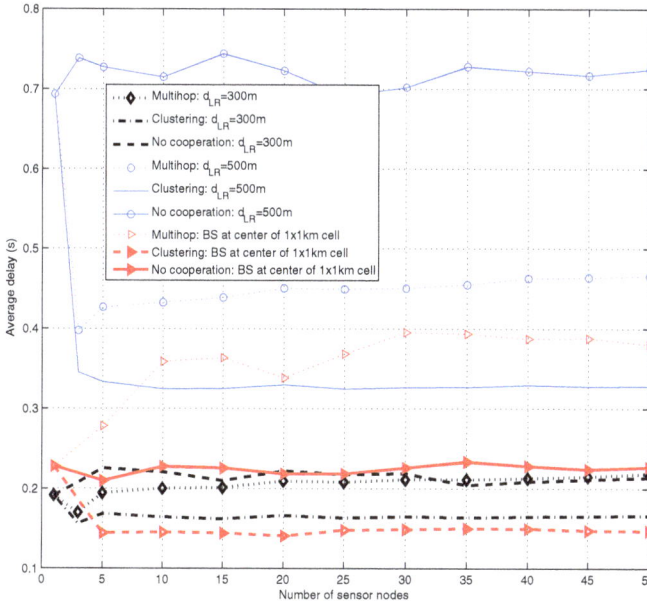

Figure 7. Average delay per SN vs. the number of SNs for different values of d_{LR}.

In this section, the impact of multihop-based energy minimization on delay performance is investigated. The transmitter at each hop is considered to wait until it receives all the data from the previous hop before starting transmission. In addition, at each hop, it is considered that transmission is done in parallel using orthogonal channels within the same cluster or within clusters at close proximity. The channels can be reused at clusters located further away. This corresponds, in practice, to the use of OFDMA with different subchannels allocated to each transmitter-receiver link, or to the use of CDMA with different orthogonal codes allocated to each transmitter-receiver link.

Fig. 7 shows the delay results averaged over the SNs. However, in delay sensitive applications, the interest is in the delay incurred by each SN. Therefore, Fig. 8 shows the maximum delay, i.e., the delay incurred by the last SN to send its data to the BS. In other words, this corresponds to the total delay needed to transmit the measurements of all SNs in the network, thus corresponding to the worst case result. Figs. 7 and 8 show that the delay increases with the distance to the BS, since a longer distance leads to lower achievable rates on the LR,

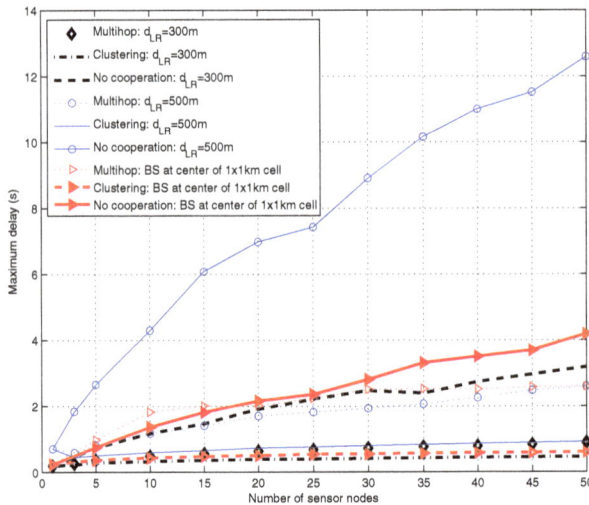

Figure 8. Total delay to distribute the content to all SN vs. the number of SNs for different values of d_{LR}.

which leads to an increase in data transmission time. In addition, the clustering approach outperforms the multihop approach by leading to shorter delays in all the investigated scenarios. Fig. 7 shows that when SNs are deployed in a confined area at a distance d_{LR} from the BS, the multihop approach leads to average delays comparable to the non-cooperative scenario when $d_{LR} = 300$m, and to better average delay performance when d_{LR} increase to 500m. The trend continues with larger distances. When the BS is placed at the cell center, with the SNs deployed throughout the cell area, the non-cooperative scenario leads to better average delay than the multihop approach, but not than the clustering approach.

Fig. 8 shows that the proposed cooperative methods significantly outperform the non cooperative case by leading to shorter maximum delay. Particularly, the clustering method leads to considerably shorter maximum delay compared to both the multihop approach and the non-collaborative scenario.

Thus, the suboptimal clustering approach leads to significant energy savings that are comparable to the multihop approach as shown in Figs. 5 and 6, and it leads to much shorter delays in transmitting the measurement data as shown in Figs. 7 and 8, and thus constitutes a suitable approach leading to both energy and delay efficiency in WSNs.

6.4. Bandwidth savings

d_{LR} (m)	300	500	1000	Centered BS
Number of clusters	27	16	8	35

Table 4. Number of Collaborative Clusters for $K = 50$

In this section, traffic offloading from the BS due to using the proposed approach is investigated. Table. 4 shows the number of SNs transmitting the data directly to the BS on the LR, in the case where a network of $K = 50$ SNs is deployed. This corresponds to the number of wireless channels needed on the LR. It should be noted that in the example of Table. 4, the multihop and clustering approach lead to the same number of clusters, although the transmission occurs on different routes inside each cluster. From Table 4 it can be seen that a significant portion of the LR bandwidth can be freed due to implementing the proposed approach. In fact, around 46%, 68%, and 84% of the bandwidth can be saved when $d_{\text{LR}} = 300$, 500, and 1000 meters, respectively. In addition, when the WSN is deployed throughout the cell area with the BS at the cell center, 30% of the LR bandwidth can be saved. When the proposed approach is implemented network wide, the significantly reduced loads of some BSs might be accommodated by other more loaded BSs. The initial BSs would be switched-off in this case. Hence, the proposed approach would contribute to green communications at the BS level, although its initial purpose was to save battery energy of SNs.

7. Practical implementation aspects

In this section, we discuss some practical limitations of the proposed techniques and propose methods to overcome these limitations.

7.1. CSI Exchange for algorithm implementation

In the proposed methods, the BS is assumed to be aware of the channel state information (CSI), and hence of the achievable rates R_{k0} on the LR links in addition to the CSI and rates R_{kj} ($j > 0$) on the SR links. Since the sensors considered are not assumed mobile, this can be achieved by a training phase that precedes the actual data transmission phase. The BS can know the CSI on the LR via feedback from the SNs, which is common in state-of-the-art wireless communication systems. On the SR, SNs can take turns in broadcasting pilot signals. Thus, each SN can estimate its CSI, and hence the rate R_{kj}, with every other SN within its transmission range, by measuring the received strength of the pilot signals. The SR pilot broadcasting process can be coordinated by the BS to avoid collisions. When each SN gets a CSI estimate on its SR links with the other SNs, it can feed-back this information to the BS on the LR link. After this training phase, the BS can then coordinate the data transmission process using the proposed methods. The same analysis applies in a limited mobility scenario, without necessarily having the sensors fixed. Hence, in the case of fixed SNs or in a low mobility scenario (portable SNs), the overhead due to the training phase can be considered low since a long time can elapse before the channel conditions change and the need arises to repeat the process.

In addition, it should be noted that SNs form cooperative clusters with other SNs when they can successfully hear their pilot transmission, i.e., when R_{kj} is high enough to allow efficient communication between SNs. When R_{kj} is too low between two SNs k and j, these will automatically be in different clusters. Thus, if no CSI feedback is received about the link between SNs k and j, then there will not be a possibility for direct communication between these SNs in the multihop approach of Section 5.1. Furthermore, in the clustering approach of Section 5.2, SN k cannot be a cluster head in a cluster of which j is a member and vice versa. This leads to eliminating several candidates in the search conducted in the schemes of Sections 5.1 and 5.2, and hence to a significant reduction in the complexity of the algorithms. Consequently, the results of Section 5.3 correspond to a worst-case scenario, and the complexity in practical scenarios is generally lower.

7.2. Fairness considerations

The multihop and clustering methods are based on selecting certain nodes that transmit the data of other SNs in addition to their own data. This could lead to an increase in energy consumption for some of these nodes compared to the non-cooperative scenario, although the overall energy consumption in the network is minimized. In [29], it was shown that, within a single cluster, fading variations lead to selecting a different cluster head for each fading realization, and this was shown to lead to fairness in energy consumption in the cluster on the long term. Thus, in the case of WSNs deployed for long term measurement and monitoring of certain parameters, different training phases (as explained in Section 7.1), will occur. Consequently, the techniques presented in this Chapter can be considered to be fair. In fact, different SNs will take turn to relay the SR data when the fading varies, which averages out the energy consumption levels among SNs.

8. Application example - air quality monitoring

The methods presented in this Chapter can be applied to several WSN deployment scenarios. An important application of WSNs is the monitoring of environmental parameters. With the advancements in the production of small, accurate, low power sensors, it is becoming more and more possible to deploy a WSN for continuous monitoring of air quality. The WSN would report the concentration of several pollutants in the atmosphere, and the reported measurements can be made available to the general public via dedicated websites, mobile applications, etc. In addition, the stored measurements can be made available to expert environmental scientists to analyze and assess pollution information in order to submit recommendations to the relevant authorities in order to take appropriate action.

In this section, we present a high level description of the system architecture for air pollution monitoring and describe the role of the SNs where the presented communication protocol will be applied. The system model for air pollution monitoring is displayed in Fig. 9. Each BS covers a cell of certain area, where several SNs are deployed to monitor environmental parameters. The architecture follows a three-tier approach:

1. The sensor nodes (SNs): these include the sensors, measuring pollutants to be monitored, e.g., CO, NOx, Ozone, and Particulate Matter (PM), in addition to other environmental parameters like relative humidity and temperature. An SN usually can accommodate one or more sensors, with each sensor measuring one of the mentioned parameters. The SNs transmit the measured data using the presented communication methods. Thus, the nodes can form cooperative clusters, and relay the data in a multihop fashion ensuring energy efficiency.

2. The database server: the data received at the BS is sent to a database server where it is stored using a common format in order to automate its extraction and analysis. The measured data might contain missing, noisy, or erroneous values. Appropriate data integrity checks should be performed before storing the data for subsequent use. Afterwards, the data becomes ready for analysis and display. Analysis techniques include statistics (for computation of daily, monthly, or yearly averages of a certain air pollutant),

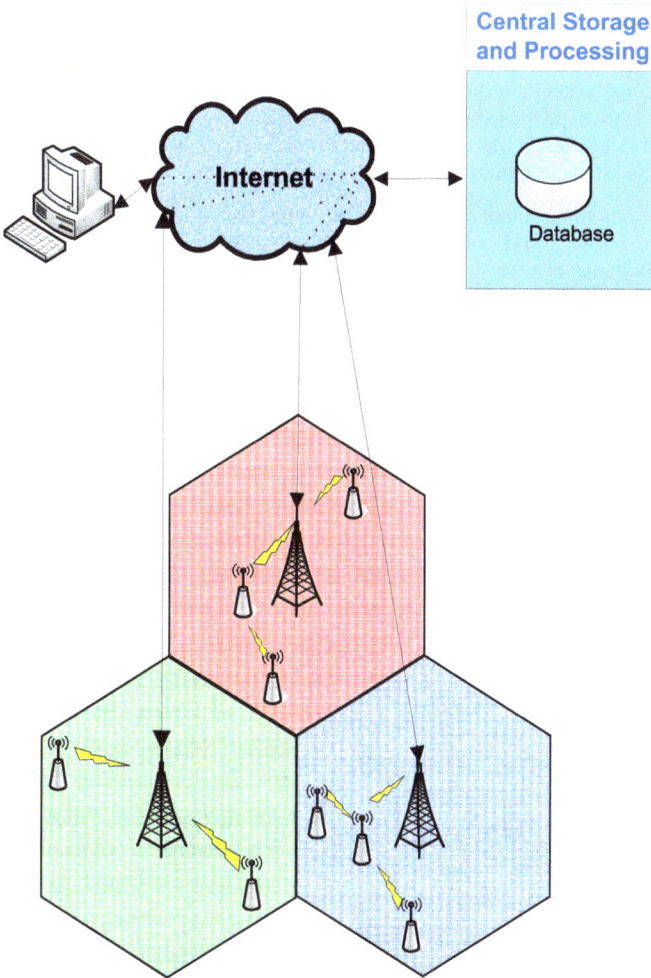

Figure 9. Implementation scenario for air pollution monitoring.

advanced interpolation, neural networks, principal component analysis, and data mining techniques.

3. The Client tier: it consists of client-side applications running on computers or mobile devices, e.g. smart phones. These applications access the network via the server, which forwards the stored data received from the sensors. Examples of applications include periodically updated web sites with data summaries and statistics, data visualization with display of sensor locations on a map (along with each SN's measurements), and data dissemination applications like SMS alerts relating to pollution levels in certain areas.

9. Future work

After describing the previous contributions in the literature and outlining the differences with the presented approach in Section 2, the problem was defined and formulated in Sections 3 and 4, then the novel proposed method to address the formulated problem was presented in Section 5, and its efficiency was demonstrated in Section 6. Hence, the role of this section is to introduce some interesting future research directions.

In addition to a more thorough and detailed investigation of the topics described in Sections 7 and 8, future work would consist of implementing the proposed methods in a sensor network testbed and of matching the simulation results with actual energy measurements. Another interesting research direction is to consider SNs with variable power sources, and to distinguish between battery powered SNs and SNs having access to renewable energy sources (e.g. solar powered) or mains powered. The problem can be reformulated by imposing a constraint that the latter SNs should be cluster heads since they can transmit large amounts of aggregated data on the LR without suffering from energy shortage.

10. Conclusions

Cooperative data transmission in wireless sensor networks was studied with the objective of energy minimization. The problem was formulated into an optimization problem, and efficient suboptimal methods were presented for the two scenarios: the multihop case where the maximum number of hops is allowed and the clustering case where sensors are grouped into cooperating clusters, each headed by a cluster head in charge of the communication with the base station. The two methods were shown to lead to significant energy savings compared to the non cooperative scenario, with the clustering approach leading to better delay results than the multihop approach. Practical implementation aspects were also discussed.

Acknowledgment

This work was made possible by an NPRP grant from the Qatar National Research Fund (a member of The Qatar Foundation). The statements made herein are solely the responsibility of the authors.

Author details

Elias Yaacoub and Adnan Abu-Dayya
QU Wireless Innovations Center (QUWIC), Doha, Qatar

11. References

[1] 3GPP [2006]. 3gpp tr 25.814 3gpp tsg ran physical layer aspects for evolved utra, v7.1.0.
[2] Alawieh, B., Zhang, Y., Assi, C. & Mouftah, H. [2009]. Improving spatial reuse in multihop wireless networks-a survey, *IEEE Communications Surveys and Tutorials* 11(3): 71–91.
[3] Baek, S. & de Veciana, G. [2007]. Spatial energy balancing through proactive multipath routing in wireless multihop networks, *IEEE/ACM Transactions on Networking* 15(1): 93–104.

[4] Bandyopadhyay, S. & Coyle, E. [2004]. Minimizing communication costs in hierarchically-clustered networks of wireless sensors, *Computer Networks* 44(1): 1–16.

[5] Bluetooth [2010]. Specification of the bluetooth system, bluetooth core specification version 4.0.

[6] Carrano, R., Magalhães, L., Muchaluat Saade, D. & Albuquerque, C. [2011]. Ieee 802.11s multihop mac: A tutorial, *IEEE Communications Surveys and Tutorials* 13(1): 52–67.

[7] Cui, S., Madan, R., Goldsmith, A. & Lall, S. [2007]. Cross-layer energy and delay optimization in small-scale sensor networks, *IEEE Transactions on Wireless Communications* 6(10): 3688–3699.

[8] Fitzek, F. & Katz, M. [2006]. *Cooperation in Wireless Networks: Principles and Applications*, Springer.

[9] Goldsmith, A. [2005]. *Wireless Communications*, Cambridge University Press.

[10] Hanzo, L. & Tafazolli, R. [2009]. Admission control schemes for 802.11-based multi-hop mobile ad hoc networks: A survey, *IEEE Communications Surveys and Tutorials* 11(4): 78–108.

[11] Haratcherev, I., Lagendijk, R., Langendoen, K. & Sips, H. [2004]. Hybrid rate control for ieee 802.11, *Proceedings of the International Symposium on Mobility Management and Wireless Access*.

[12] Heinzelman, W., Chandrakasan, A. & Balakrishnan, H. [2000]. Energy-efficient communication protocol for wireless micro-sensor networks, *Proceedings of the Hawaii International Conference on System Science, Maui, Hawaii*.

[13] Jin, Y., Wang, L., Jo, J., Kim, Y., Yang, M. & Jiang, Y. [2009]. Eeccr: An energy-efficient m-coverage and n-connectivity routing algorithm under border effects in heterogeneous sensor networks, *IEEE Transactions on Vehicular Technology* 58(3): 1429–1442.

[14] Kim, K., Lee, B., Choi, J., Jung, B. & Yong Youn, H. [2009]. An energy efficient routing protocol in wireless sensor networks, *Proceedings of the International Conference on Computational Science and Engineering (CSE)*.

[15] Kim, K. & Youn, H. [2005]. Peach: Proxy-enable adaptive clustering hierarchy for wireless sensor networks, *Proceedings of The 2005 International Conference On Wireless Networks (ICWN)*.

[16] Li, L. & Halpern, J. [2004]. A minimum-energy path-preserving topology-control algorithm, *IEEE Transactions on Wireless Communications* 3(3): 910–921.

[17] Lin, L., Shroff, N. & Srikant, R. [2007]. Asymptotically optimal energy-aware routing for multihop wireless networks with renewable energy sources, *IEEE/ACM Transactions on Networking* 15(5): 1021–1034.

[18] Mahmud, K., Inoue, M., Murakami, H., Hasegawa, M. & Morikawa, H. [2005]. Energy consumption measurement of wireless interfaces in multi-service user terminals for heterogeneous wireless networks, *IEICE Transactions on Communications* E88-B(3): 1097–1110.

[19] Peng, M. & Wang, W. [2008]. Investigation of cooperative relay node selection in heterogeneous wireless communication systems, *Proceedings of the IEEE International Conference on Communications (ICC 2008)*.

[20] Popova, L., Herpel, T. & Koch, W. [2007]. Improving downlink umts capacity by exploiting direct mobile-to-mobile data transfer, *Proceedings of the International Symposium on Modeling and Optimization in Mobile, Ad Hoc and Wireless Networks*.

[21] Qiu, X. & Chawla, K. [June 1999]. On the performance of adaptive modulation in cellular systems, *IEEE Transactions on Communications* 47(6): 884–895.

[22] Ramanathan, R. & Rosales-Hain, R. [2000]. Topology control of multihop wireless networks using transmit power adjustment, *Proceedings of IEEE INFOCOM*.

[23] Rodoplu, V. & Meng, T. [1999]. Minimum energy mobile wireless networks, *IEEE Journal on Selected Areas in Communications* 17(8): 1333–1344.

[24] Tavli, B. & Heinzelman, W. [2004]. Mh-trace: Multihop time reservation using adaptive control for energy efficiency, *IEEE Journal on Selected Areas in Communications* 22(5): 942–953.

[25] Tian, Y. & Ekici, E. [2007]. Cross-layer collaborative in-network processing in multihop wireless sensor networks, *IEEE Transactions on Mobile Computing* 6(3): 297–310.

[26] Vieira, M., Coelho, C., da Silva, D. & da Mata, J. [2003]. Survey on wireless sensor network devices, *Proceedings of the IEEE Conference on Emerging Technologies and Factory Automation*, pp. 537–544.

[27] Wang, K., Yang, F., Zhang, Q., Wu, D. & Xu, Y. [2007]. Distributed cooperative rate adaptation for energy efficiency in ieee 802.11-based multihop networks, *IEEE Transactions on Vehicular Technology* 56(2): 888–898.

[28] Wei, C., Zhi, C., Fan, P. & Ben Letaief, K. [2009]. Asor: An energy efficient multi-hop opportunistic routing protocol for wireless sensor networks over rayleigh fading channels, *IEEE Transactions on Wireless Communications* 8(5): 2452–2463.

[29] Yaacoub, E., Al-Kanj, L., Dawy, Z., Sharafeddine, S., Filali, F. & Abu-Dayya, A. [2011]. A nash bargaining solution for energy-efficient content distribution over wireless networks with mobile-to-mobile cooperation, *Proceedings of the Wireless and Mobile Networking Conference (WMNC 2011)*.

[30] Younis, O. & Fahmy, S. [2004]. Heed: A hybrid, energy-efficient, distributed clustering approach for ad hoc sensor networks, *IEEE Transactions on Mobile Computing* 3(4): 366–379.

[31] Zhang, Q., Fitzek, F. & Iversen, V. [2007]. Design and performance evaluation of cooperative retransmission scheme for reliable multicast services in cellular controlled p2p networks, *Proceedings of the IEEE International Symposium on Personal, Indoor and Mobile Radio Communications (PIMRC 2007)*.

[32] Zhang, T., Yang, K. & Chen, H. [2009]. Topology control for service-oriented wireless mesh networks, *IEEE Wireless Communications* 16(4): 64–71.

Security Mechanisms

Chapter 10

Reputation System Based Trust-Enabled Routing for Wireless Sensor Networks

A. R. Naseer

Additional information is available at the end of the chapter

1. Introduction

The issue of secure routing[1] in wireless and mobile computing is a major challenging design factor in different networking aspects. However, the problem gets more complicated when considering infrastructure-less networks that exhibit even more constraints and new types of attacks. Wireless sensor networks (WSN), which is an ad-hoc type of networks, is a clear representative case.

In the continuously and rapidly evolving area of wireless communication, the field of wireless sensor networks (WSN) comes into the picture as a very hot area of research in all its aspects. WSN is a multi-hop network that is actually one type of ad hoc networks. However, WSN draws the special attention of researchers due to the fact that it exhibits more constraints and critical conditions than normal ad hoc networks in terms of power sources, computing capabilities, memory capacity and other factors. This requires different approaches and protocol engineering directions from those applied to normal ad hoc networks.

One special aspect in WSN is the provision of secure routing. As mentioned previously, the nature of WSN complicates the security requirements and adds difficulties in solving security problems. In fact, secure routing in WSN is actually still not captured well in the research field. One main reason is that the design of a routing protocol is biased towards solving the problem of power limitations and reducing communication overhead, while keeping security concerns in a later phase to be integrated with the current routing solutions.

One specific class of security problems in routing aspects in WSN is the exposure to attacks that are related to nodes' activities and behavior in the network. Such attacks cannot be recognized by verifying nodes' identities because most of these attacks are launched by compromised nodes or insider attacks; i.e. nodes belong to the same network community. Among different approaches in solving this problem, reputation system based solution is one technique that has generated enough interest among WSN research community. Reputation systems attempt to provide security by allowing different nodes to rate each other based on their routing activities and behavior analysis. When a node has an experience profile about its neighbors, it may select the node that it trusts more, and, hence, achieve a secure routing operation.

In this chapter, a reputation system solution for behavioral based attacks at the network layer as a provision of secure routing in WSN is presented.

1.1. Motivation

In this work, we provide a reputation system based solution for routing security in WSN. We believe that such a solution approach is a feasible and applicable solution for the following reasons:

- Conventional security solution such as cryptography can successfully defend against outsiders' attack. The mechanics of such solutions fail when the attack is done by insiders or compromised nodes. Some of such attacks are intentionally performed like the misbehavior of selfish nodes and compromised nodes. Other attacks can be carried-out unintentionally by faulty nodes [2]. Thus, security systems like reputation based security solutions that have a mechanism to treat such attacks by behavior analysis are more suitable. This is especially true in networks where such misbehavior is very possible or even it is the dominant type of attacks, which is the case in WSN.
- In contrast to different secure routing mechanisms, reputation based systems provide a means for an adaptive and dynamic decision making and reaction at the individual node level behavior. Such features are needed in networks that exhibit dynamicity in nodes' behavior like that in WSN.
- Most WSN deployments and applications invite a very dynamic networking nature. The current conditions and statistics of the network will change from time to time. The security system, thus, must accept to tune itself to these changes at the network level.
- WSN life and operation depends on the cooperation of nodes like any other ad hoc network. This implies that the security interest of a node is not only about itself but also about the whole network. As a result, such networks will prefer to communicate security information in order to keep the network healthy. This is an important feature of reputation systems. Node rating is one type of information that contributes to node's decision making and can be communicated as second hand information. However, the node reaction is also important and affects other nodes' decisions. Thus, the security system should have the feature of a consulted and well-analyzed decision-making and behavior, which are core concepts in reputation systems.
- An interesting and important feature of any reputation system is that it follows a generalized and modular solution approach to fight against any attack in a general framework. The system then is customized to face a subset of these attacks. Thus, new

attacks will be tackled by modifications in the details of the module of interest that does not require a complete system revision. For example, a new attack might require adjusting the monitoring and detection phase without touching other parts of the system. WSN are deployed in very hostile conditions that expect new attempts and attacks. Thus, it is better to support reputation systems in that regard rather than other solutions that can be totally and entirely useless with the occurrence of new misbehavior strategies.

In literature, there are different, proposed reputation based solutions for secure routing in ad hoc networks. Very familiar examples include CONFIDANT (Cooperation Of Nodes – Fairness In Dynamic Ad-hoc Networks) [3], SORI (Secure and Objective Reputation-Based Incentive Scheme for Ad Hoc Networks) [4] and CORE (Collaborative Reputation Mechanism to Enforce Node Cooperation in Mobile Ad Hoc Networks)[5]. There are also other solutions that are close to the reputation systems but they do not follow the general mechanism. Examples are watchdog and pathrater as well as context-aware detection [6, 7].

As these solutions are applied to ad hoc networks, the conclusion of applying them to wireless sensor network as a type of ad hoc networks is not totally accurate. There are several reasons that show the need to have a special reputation system design and implementation that targets WSN. This differentiation comes from the following facts:

- *Resource Constraints*: An obvious difference between MANET and WSN is resource constraints. Resources include power, memory and processing capabilities. Although both networks suffer from resources deficiency, WSN are more constrained and limited by such resources, especially in power. Any protocol design and implementation targeting WSN from the physical to the application layer must consider resource usage optimization not as an additional feature in the system but as a main design goal. Therefore, an optimized approach must be considered when designing a reputation system for WSN.
- *Conditions and Applications:* Security conditions in WSN are different from general MANET networking type. As a result, the reputation based security system will be looking at providing solutions that satisfy these conditions that indeed implies different approaches for WSN and MANET as they differ in that aspect. Moreover, the risky environment that comes from the application types in WSN raises a remark of having security models that are different than MANET. What implies is a different view of reputation system for WSN.
- *Underlying Routing Protocol*: In contrast to MANET, DSR (Dynamic Source Routing) as a routing protocol is not the accurate or suitable choice for WSN for several reasons related to resource constraints and efficiency. Moreover, other routing protocols like GPSR (Greedy Perimeter Stateless Routing) [8] and GEAR (Geographic Energy Aware Routing protocol) [9] prove their outperformance compared to DSR. Thus, the implemented reputation systems in WSN should consider the operation of routing protocols that are more applicable than DSR.

1.2. Secure routing problem

Routing is a fundamental operation in almost all types of networks because of the introduction of inter-domain communication. Ensuring routing security is a necessary requirement to guarantee the success of routing operation. When we talk about secure routing, we are concerned with security problems that may occur due to improper actions from an assumed router. These undesired actions can be related either to the router identity or the router behavior. If the router has an undesirable identity or authorization, it is considered as an intruder who might perform serious attacks. Such attacks can be avoided by providing security services that validate the routers' identities. On the other hand, a router that misbehaves in the network by performing undesirable routing operations also contributes to the routing security problem. However, the attacks caused by misbehaving routers can be avoided by mechanisms that validate and evaluate the router behavior in the network.

In WSN, secure routing is more demanding due to the nature of the routing operation in WSN. Since WSN lacks an infrastructure, nodes depend on the cooperation among each other to route their packets. Thus, a router in WSN is simply any node that offers a routing service. This "any node" should be selected such that it will be the most secure choice to route the packet. To come up with a proper routing decision we need to understand first what security goals we are targeting.

1.3. Secure routing goals

Security problems in WSN at the network layer can be related to router identity or router behavior. These two issues highlight two main tasks when we would like to design a secure routing solution [1,10].

- Securing Packet Content: This task is concerned with identity related security problems. The goal of this task is to assure that the packet is not accessed by unauthorized nodes as it travels from the source to the destination. This task can be achieved if we can provide the following services:
  - Data Confidentiality: In this service, only the destination node should be able to access the packet content initiated from the source node. Any intermediate router must not have any access to such information. As we can see here, the access of the packet is restricted to the destination node. Thus, if a node other than the destination accesses the packet, it means that the destination identity has been compromised.
  - Data Integrity: When a destination node receives a message from a source, the destination should be able to detect any change that could occur in the message.

Securing packet content is obtained usually based on the idea of identity trust in which a routing decision is made after verifying that the selected node is authorized and has an acceptable identity according to certain criteria. This is achieved in literature by using crypto-based systems. However, any solution must obey WSN constraints of processing capacity, memory limits and energy consumption.

- Securing Packet Delivery: This task deals mainly with behavior related security problems. Its objective is to guarantee that any packet transmitted will be ultimately received at the target destination. Thus, a misbehaving router node should not be able to drop a packet, misroute a packet or deny the ability of routing of other nodes by denial of service attacks. This task can be interpreted in terms of a security service called data availability.
  - Data Availability: If a node A is authorized to get information from another node B, then node A should acquire this information at any time and without unreasonable delay.

There are different approaches to achieve this second task. However, as the first task, the designer should be aware of the suitability of the solution with WSN tight constraints such as energy scarcity. In this work, we are proposing a solution for securing packet delivery task with an account for energy efficiency. Our solution is based on the concept of behavior trust where nodes should trust the behavior of another node in order to select it as a router. This approach is well-known in literature as trust aware routing.

The rest of the chapter is organized as follows. Section 2 of the chapter provides the relevant background material covering an overview of WSN that includes WSN definition, sensor node structure, applications, etc. As WSN is a class of MANET, the main differences between WSN and MANET will be presented. These differences are explained in a way that emphasizes to the reader how they make WSN an independent research target as compared with MANET. Then we introduce the notion of trust and reputation in social networks, how these concepts can be applied smoothly to Wireless Sensor Networks to mitigate node misbehaviors, illustrate the issues in Modeling and Management of Trust & reputation, highlighting the importance of Trust-Aware Routing, and general concept of reputation systems. This will be followed by a detailed discussion on some of the important related work carried out in the area of Reputation system based trust-enabled routing for WSNs.

Section 3, being the Reputation System Overview section, will provide an overview of the proposed reputation system. The section will start by discussing the general reputation system framework clearly introducing the readers to various components of the Reputation system highlighting the functions to be performed by each component. This is followed by description of our customized reputation system- SNARE (Sensor Node Attached Reputation Evaluator)[82] that fits into the framework guidelines. Reputation-based solution will be discussed as a detection approach by presenting the general concept of reputation systems, followed by suggestions and approaches in reputation system solutions that can fit WSN secure routing requirements. In this section, we briefly describe our proposed monitoring component called Efficient Monitoring Procedure in Reputation Systems (EMPIRE)[84], a new rating approach for reputation systems in WSN called CRATER(Cautious RAting for Trust Enabled Routing)[85] and a simple but strong, independent and representative scale to evaluate reputation systems called *REputaion Systems-Independent Scale for Trust On Routing* (RESISTOR)[85].

In section 4, our enhanced routing protocol that aims to provide a secure packet delivery service guarantee by incorporating the trust awareness concept into the routing decision is presented. Our proposed protocol is called Geographic, Energy and Trust Aware Routing (GETAR) which is an enhanced version of the Geographic and Energy Aware Routing (GEAR) protocol[9]. GEAR is basically a geographic routing protocol in which the next hop is selected based on two metrics: the distance between the next hop and the destination and the remaining energy level the next hop owns. The new contribution in GETAR is to add a third metric in the next-hop selection process, i.e. the risk value of a node that is computed by the rating component, CRATER[85] in our case. In section 5, we present a comparison of our approaches with previous reported work and highlight our main contributions. The chapter finally concludes with a summary and future research directions in this field.

2. Background and literature survey

In this section, some background material is provided. It covers general aspects of WSN and then some specific discussions on routing protocols in WSN. This is followed by a general provision of the most familiar work related to the subject of secure routing, notion of Trust and Reputation, reputation systems and trust aware routing.

2.1. WSN: Definition and applications

Wireless Sensor Network (WSN) is one type of ad hoc networks that consists of a very large number of tiny devices equipped with signal processing circuits, microcontrollers, sensors and actuators and wireless transmitters/receivers. Nodes are deployed either randomly or in a grid-like structure according to the sensing and environmental conditions and requirements [11].

WSNs have different applications; most of them are critical mission applications, for example:

- Military Applications [11,12]
  - Monitoring friendly forces, equipment and ammunition
  - Battlefield surveillance
  - Reconnaissance of opposing forces and terrain
  - Targeting guidance
  - Battle damage assessment
  - Nuclear, biological and chemical (NBC) attack detection and reconnaissance.
- Environmental Applications [13,14,15,16]
  - Tracking the movements of birds, small animals, and insects
  - Monitoring environmental conditions that affect irrigation
  - Earth, and environmental monitoring in marine, soil, and atmospheric contexts
  - Forest fire detection
  - Meteorological or geophysical research
  - Flood detection

- Pollution study
- Health Applications[17,18,15]
  - Providing interfaces for the disabled
  - Integrated patient monitoring
  - Administration in hospitals
  - Tele-monitoring of human physiological data
  - Tracking and monitoring doctors and patients inside a hospital
- Commercial Applications[17,19,20,15]
  - Managing inventory
  - Monitoring product quality
  - Robot control and guidance in automatic manufacturing environments
  - Interactive museums
  - Monitoring disaster area
  - Smart structures with sensor nodes embedded inside
  - Vehicle tracking and detection

2.2. WSN node structure

The basic structure of WSN is that it is composed of sensor nodes and base stations. Sensor nodes, viewed as communicating parties in WSN, are more than simple sensing devices. In fact, every node holds an embedded system that performs three main functions:

- Sensing: Every node should have the ability to observe and/or control the physical environment.
- Computing: The collected data from physical environment through sensing function is processed to produce beneficial information.
- Communication: Every node should be able to communicate and exchange raw data or processed information among them.

Looking at the above functions, the requirements on the sensor hardware will be as follows [11,12,21,22,23,17]:

- *Sensors/Actuators*: Sensing and actuator units are usually composed of sensors, actuators, and analog to digital (for sensing) and digital to analog (for actuating) converters (ADC/DAC). The analog signals produced by the sensors based on the observed phenomenon are converted to digital signals by the ADC, and then fed into the processing unit or the controller. On the other hand, the digital signals produced by the controller are converted to analog signals by the DAC to feed the actuators.
- *Controller*: The controller consists of a processor and a memory system. The processor manages the procedures that make the sensor node collaborate with the other nodes to carry out the assigned sensing tasks. The memory system stores data, software and application programs required to run the node. Though the higher computational powers are being made available in smaller and smaller processors and controllers, processing and memory units of sensor nodes are still scarce resources. For instance, the

processing unit of a smart dust mote prototype is a 4 MHz Atmel AVR8535 micro-controller with 8 KB instruction flash memory, 512 bytes RAM and 512 bytes EEPROM [24]. TinyOS operating system is used on this processor, which has 3500 bytes OS code space and 4500 bytes available code space

- *Radio transceiver*: The radio transceiver unit is responsible for connecting the node to the network.

- *Power supply unit*: One of the most important components of a sensor node is the power unit. Since the sensor nodes are often inaccessible, power is considered a scarce resource and the lifetime of a sensor network depends on the lifetime of the power resources of the nodes. Power is also a scarce resource due to the size limitations. For instance, the total stored energy in a smart dust mote is of the order of 1 J [25]. It is possible to extend the lifetime of the sensor networks by energy scavenging [26], which means extracting energy from the environment. Solar cells are an example for the techniques used for energy scavenging.

- *Localization Systems; e.g. GPS(Global Positioning System):* Most of the sensor network routing techniques and sensing tasks require the knowledge of location with high accuracy. Thus, it is common that a sensor node has a location finding system like the global positioning system GPS.

2.3. Routing protocols in WSN

- *Data centric Routing*: Data-centric routing protocols have an architecture in which there is a sink that communicates with certain regions to collect data from the sensors located in the selected regions [27]. An example of such protocols is SPIN (Sensor Protocols for Information via Negotiation) [28] which is the first data-centric protocol that considers data negotiation between nodes in order to eliminate redundant data and save energy. Another famous example is Directed Diffusion [29]. In this protocol data is diffused through sensor nodes by using a naming scheme for the data. An enhanced version of Directed Diffusion is Rumor routing [30] that routes the queries to the nodes that have observed a particular event rather than flooding the entire network to retrieve information about the occurring events.

- *Hierarchical Routing*: Hierarchical routing attempts to efficiently maintain the energy consumption of sensor nodes by involving them in multi-hop communication within a particular cluster. Data is then aggregated and fused in order to decrease the number of transmitted messages to the sink[27]. LEACH (Low Energy Adaptive Clustering Hierarchy) [31] is one of the first hierarchical routing approaches for sensors networks in which clusters of the sensor nodes are formed based on the received signal strength. The cluster-heads are then used as routers to the sink. This will save energy since the transmissions will only be done by such cluster heads rather than all the sensor nodes. Other protocols are mainly inspired by this protocol, such as TEEN (Threshold sensitive Energy Efficient sensor Network protocol) [32] that is designed to be responsive to sudden changes in the sensed attributes such as temperature.

- *QoS-based Routing*: QoS-aware protocols consider end-tuned delay requirements while setting up the paths in the sensor network [27]. One famous example is SPEED protocol [33]. The main goal of SPEED is to provide soft real-time end-to-end guarantees. The protocol works by making each node maintain information about its neighbors and uses geographic forwarding to find the paths. In addition, SPEED strives to ensure a certain speed for each packet in the network so that each application can estimate the end-to-end delay for the packets by dividing the distance to the sink by the speed of the packet before making the admission decision.

- *Location-based Routing*: Most of the routing protocols for sensor networks require location information for sensor nodes. In most cases, location information is needed in order to calculate the distance between two particular nodes so that energy consumption can be estimated. Since, there is no addressing scheme for sensor networks like IP-addresses and they are spatially deployed on a region, location information can be utilized in routing data in an energy efficient way [27]. One example of such protocols is GPSR [8], which is a greedy protocol. In this protocol, every node selects the next hop as the closest neighbor to the destination. In case when the node of concern is farther to the destination than all its neighbors (such a case is called the void region case), it uses perimeter forwarding based on the planar graphs concept. This research work adopts another interesting protocol under this category, i.e. GEAR [9].

2.4. WSN vs. MANET

WSN is a kind of ad hoc network. From an abstract network view point, WSN is similar in most of the aspects to ad hoc networks. However, WSN is very special compared to other types of ad hoc network due to the following [12,34]:

- The number of sensor nodes in a sensor network can be several orders of magnitude higher than the nodes in an ad hoc network.
- Sensor nodes are densely deployed.
- Sensor nodes are prone to failures.
- The topology of a sensor network changes very frequently.
- Sensor nodes mainly use broadcast communication paradigm whereas most ad hoc networks are based on point-to-point communications.
- Sensor nodes are limited in power, computational capacities, and memory.
- Sensor nodes may not have global identification (ID) because of the large amount of overhead and large number of sensors.
- *Network terminal features*: In WSN, the nodes are tiny sensor nodes. The word tiny describes the node's size and functionality. Computing capabilities, memory capacity and scarce power resources are all limited and significantly smaller than normal terminals in ad hoc networks, which are mostly of a laptop class.
- *Network environment conditions*: WSN has the characteristic of interacting with the environment through sensing functions. As a result, WSN nodes are intentionally deployed to actively respond to physical conditions in the environment such as temperature, vibrations, acceleration, etc. This phenomenon adds a critical

consideration in WSN compared to other types of ad hoc network in the sense that WSN when deployed it is mainly focused on how to satisfy the environmental conditions.

- *Application specifications*: While normal ad hoc networks can be usually thought as general purpose networks, the whole WSN is built to serve a specific application. Therefore, WSN must satisfy the application requirements in addition to the environment conditions. This complicates the issue of finding general purpose solutions for many aspects in WSN.

2.5. Routing attacks

There are several attacks that target the network layer in WSN. For example, in the black-hole attack, adversary nodes do not forward packets completely, while it selectively forwards some packets in gray-hole attack. Another example is the sybil attack in which a node pretends multiple identities. Thus, such a node can virtually exist in different neighborhoods and drop more packets. Wormhole attack is a collusion based attack in which an agreement between two adversaries is made to perform other attacks like blackhole. In wormholes, one adversary misroutes a received packet and sends it to its partner by faking a good routing decision. A detailed explanation of these attacks and others can be found in [35].

2.6. Trust and reputation

In social networks, Trust and Reputation are generally the two important components which play a major role in establishing relationship between entities which have been studied mainly by social scientists for a long time. All kinds of daily transactions, interactions, and communications in human life are based on trust. In a human social community, trust between two individuals is developed based on their actions over time. When faced with uncertainty, individuals trust and rely on the actions and opinions of others who have behaved well in the past. When affairs are to be handled in social networks, people always consider trust and reputation of concerned parties as prime tools for decision making.

Trust in general is the level of confidence in a person or a thing. More precisely trust can be defined as: "the quantified belief by a trustor with respect to the competence, honesty, security and dependability of a trustee within a specified context" [36, 37]. Reputation is a notion sometimes confused with trust; it is defined as "the global perception about the entity's behavior norms based on the trust that other entities hold in the entity" [38]. Reputation is the opinion of one person about the other, of one internet buyer about an internet seller, and one WSN node about another. Trust is a derivation of the reputation of an entity. Based on a reputation, a level of trust is bestowed upon an entity. The reputation itself has been built over time based on that entity's history of behaviour, and may be reflecting a positive or negative assessment.

In Wireless Sensor Network routing approaches, Reputation system based trust models borrowed from human societies have been proposed to combat misbehaviors. Nodes establish trust relationships between each other and base their routing decisions not only on geographical or pure routing information, but also on their trust that their neighbors will sincerely cooperate. In the context of WSN, Trust is the confidence of one node on another node that it will perform the given task as expected with full cooperation without any deviation. To evaluate the trustworthiness of its neighbors, a node not only monitors their behavior, i.e., through direct observations also known as First Hand Information(FHI), but may also communicate with other nodes to exchange their opinions , i.e., through indirect observations also known as Second Hand Information(SHI). The methods for obtaining trust information and defining each node's trustworthiness are referred to as trust models. A trust model is mostly used not only for higher layer decisions such as routing [39,40], data aggregation [41], but also for cluster head election [42] and for key distribution [43]. Goal of the trust model is to improve security thereby increasing the throughput, the lifetime and the resilience of a wireless sensor network.

Trust in WSN plays an important role in constructing the network and making the addition or deletion of sensor nodes from a network very smooth and transparent. Trust in WSN has been studied lightly by current researchers and is still an open and challenging field. Trust is an old yet important issue in any networked environment that can solve some problems which lies beyond the power of traditional cryptographic security. A Trust Management System is required to support the decision making processes of the network. Trust management is fundamental to identify malicious, selfish and compromised nodes which have been authenticated. In wireless sensor network, trust management system aids the nodes termed as trustors to deal with uncertainty about the future actions of other participating nodes termed as trustees. By evaluating and storing the reputation of other members, it is possible to calculate how much those members can be trusted to perform a particular task. It has been widely studied in many network environments such as peer-to-peer networks, grid and pervasive computing and so on. However, in reality, sensor nodes have limited resources and other special characters, which make trust management for WSNs more significant and challenging. Various Trust models, Trust evaluation metrics and Trust Management schemes have been reported in the literature[36-59]. Current research on the trust management mechanisms of WSNs have mainly focused on nodes' trust evaluation to enhance the security and robustness. The practical applications of this method include the route, data integration and cluster head vote[44]. Although some existing approaches have played greater roles in improving security of other ad-hoc networks, trust management in WSNs still remains a challenging field.

The trust problem is a *decision problem under uncertainty*, and the only coherent way to deal with uncertainty is through *Probability*. There are several frameworks for reasoning under uncertainty, but it is well accepted that the probabilistic paradigm is the theoretically sound framework for solving decision problems with uncertainty. Some of the trust models introduced for sensor networks employ probabilistic solutions mixed with ad-hoc approaches. The problem of assessing a reputation based on observed data is a statistical problem. Some trust models make use of this observation and introduce probabilistic modeling that uses a Bayesian updating scheme known as the Beta Reputation System [65] for assessing and updating the nodes reputations. The use of the Beta distribution is due to the binary form of the events considered. For example, RFSN[2] uses a probability model in the form of a reputation system to summarize the observed information (FHI) and share the values of the parameters of the probability distributions as second-hand information(SHI). This shared information is soft data, requiring a proper way to incorporate it with the observed data into the trust model. The step of combining both sources of information is handled differently by different trust models. RFSN uses Dempster-Shafer belief theory model [48], solving it using the concept of belief discounting, and doing a reverse mapping from belief theory to continuous probability. In [49], a new Bayesian fusion algorithm to combine more than one trust component - data trust and communication trust to infer the overall trust between nodes is proposed. The trust value calculated between nodes based on their cooperation in routing messages to other nodes in the network is termed as Communication trust (CT). The trust value calculated based on the actual sensed data of the sensors in WSNs is known as Data trust (DT). As an extension to this work, authors proposed Recursive Bayesian Approach to Trust Management (RBTMWSN)[50] by introducing a new trust model and a Gaussian reputation system(GRSSN) for wireless sensor networks based on a sensed continuous data. In this work, Bayesian probabilistic approach based on the work done in modelling Expert Opinion[51] for mixing second-hand information from neighboring nodes with directly observed information is proposed. Opinions provided by knowledgeable sources are known as experts opinions. Such opinions are modulated by existing knowledge about the experts themselves, to provide a calibrated answer. It allows for the formal incorporation of informed knowledge into a statistical analysis. The probabilistic approach adopted is to consider the opinion given by the expert as soft data that is merged with the hard data according to the laws of probability[52]. In [53], authors proposed a Node Behavioral Strategies Banding Belief Theory of the Trust Evaluation (NBBTE) Algorithm. In this approach, at first, each node establishes the direct and indirect trust values of neighbor nodes by comprehensively considering various trust factors such as packet receive, send, strictness, delivery, consistency and availability, *etc*, and combining these factors together with network security grade, correlation of context time and rewards degree. Next, fuzzy set theory is used to decide the trustworthiness levels in accordance with the fuzzy subset grade of membership functions. Based on the levels of trustworthiness, the basic confidence function of D-S evidence theory[54] is accordingly formed. Finally, using the revised Dempster rules of combination, the integrated trust value of a node is obtained by integrating its trustworthiness of multiple neighbor nodes.

Current research challenge has been in designing an accurate and efficient trust and/or reputation model for distributed and heterogeneous environments[47]. When developing such models, different issues have to be taken into consideration. The problem to be solved here consists of deciding in a distributed environment which entity is the most reliable to interact with, in terms of trust and reputation. That is, having a system where different entities offer some services or goods and other ones are requesting those services, the former will always look for the best self profit, while the latter will demand the best services with respect to some quality characteristics, properties or attributes. Nevertheless, most of the times it is not feasible or realistic to assume the existence of service level agreements or the presence of a centralized entity or architecture supplying reliable information regarding the actual and current behavior of every service provider in the system. Hence, requesters have to determine on their own which service providers are the best ones according to certain criteria. Under these conditions, trust and/or reputation models are aimed to select the most trustworthy entity all over the system offering a certain service.

2.7. Trust aware routing

2.7.1. Definition

A trust aware routing protocol is a routing protocol in which a node incorporates in the routing decision its opinion about the behavior of a candidate router. This opinion is quantified and called the trust metric. Trust metric should reflect how much a router is expected to behave, for example, forward a packet when it receives it from a previous node.

Obtaining the trust metric is a problem by itself since it requires several operational tasks on observing nodes behavior, exchanging nodes' experience and opinions as well as modeling the acquired observations and exchanged knowledge to reflect nodes trust values. A system that provides these tasks to ultimately output a "rating" or a trust value on nodes is called a reputation system.

To appreciate the concept of trust behavior based routing, we provide in the next section some aspects that highlight the importance of trust aware routing.

2.7.2. Importance

Trust aware routing in WSN is important for both securing obtained information as well as protecting the network performance from degradation and network resources from unreasonable consumption.

Most WSN applications carry and deliver very critical and secret information like in military and health applications. A WSN network infected by misbehaving nodes can misroute packets to wrong destinations leading to misinformation or do not forward packets to their destination leading to loss of information. Such critical application can be very sensitive to these attacks. Having a trust aware routing protocol can protect data exchange, secure information delivery and maintain and protect the value of the communicated information.

Node misbehavior can cause performance degradation as well. For example, non forwarding attacks decrease the system throughput since packets will be retransmitted many times and they are not delivered. Denial of service attacks can increase the packet delay since some nodes acting as routers will be busy in responding to the attack and enforced to delay the processing of other packets. An infected WSN network can be partitioned into different parts that cannot communicate among each other due to non forwarding attacks. This leads to the demand of increasing the number of sensors or changing the node deployment to return network connectivity. This is very expensive, however, can be avoided if a good secure routing solution is adopted.

Network resources are also affected by misbehaving nodes. For example, Denial of service attacks affect resource availability, whether we consider an offended node as a resource for routing or we consider the availability of data itself. Also, this attack forces offended nodes to consume unnecessary energy on packet reception and processing.

As we can see, the information value and the network performance are directly affected by the security level provided by trust awareness of the routing operation in WSN.

2.8. Reputation systems

A reputation system is a type of cooperative filtering algorithm which attempts to determine ratings for a collection of entities that belong to the same community. Every entity rates other entities of interest based on a given collection of opinions that those entities hold about each other[5,60].

Reputation systems have recently received considerable attention in different fields such as distributed artificial intelligence, economics, evolutionary biology, etc. Most of the concepts in reputation systems depend on social networks analogy. As expected, reputation systems are complex in the sense that they do not have a single notion, but a single system will consist of multiple parts of notions. Thus, comparing reputation systems is, in fact, a very difficult problem. All known trials on such problem were based on qualitative approach. The work in [61] proposes an attempt on comparing reputation systems quantitatively based on game theory. The authors, thus, identify different notions of reputation systems like, contextualization, personalization, individual and group reputation, and direct and indirect reputation.

Reputation systems are often useful in large online communities in which users may frequently have the opportunity to interact with users with whom they have no prior experience. Such cases are clearly applicable to e-commerce applications and on line auctioning sites like eBay[62] and Epinions[63]. Another important field that derives the same concept of enforced interaction among entities that lack priori experience on each other is the field of ad hoc and wireless sensor network. This is because nodes in such networks need to route each others' packets. Thus, a trust relation should exist among themselves.

In the context of MANET and WSN [5, 11, 64], the reputation of a node is the amount of trust the other nodes grant to it regarding its cooperation and participation in forwarding packets. Hence, each node keeps track of each other's reputation according to the behavior it observes, and the reputation information may be exchanged between nodes to help each other to infer the accurate values. There is a trade-off between efficiency in using available information and robustness against misinformation. If ratings made by others are considered, the reputation system can be vulnerable to false accusations or false praise.

However, if only one's own experience is considered, the potential for learning from the experiences of others goes unused, which decreases efficiency.

Any reputation system in the context of MANET and WSN should, generally, exhibit three main functions [1, 65]:

- *Monitoring*: This function is responsible for observing the activities of the nodes of its interest set.
- *Rating:* A node will rate its interest set nodes based on the node's own observation, other nodes' observations that are exchanged among themselves, the history of the observed node and certain threshold values.
- *Response*: Once a node builds knowledge on others' reputations, it should be able to decide about different possible reactions it can take, like avoiding bad nodes or even punishing them.

2.9. Related work

2.9.1. SPINS - Security protocols for sensor networks

SPINS (Security Protocols for Sensor Networks) [24] is a set of security protocols that is optimized for WSN. It is mainly composed of two building blocks: (i) *SNEP (Secure Network Encryption Protocol)*: This protocol provides data confidentiality, two-party authentication and data freshness (ii) *μTESLA (micro version of Timed, Efficient, Streaming, Loss-tolerant Authentication protocol)*: This protocol provides authenticated streaming broadcast.

SNEP provides its features by semantic encryption; however, we can notice that these security services do not have a provision for secure routing. In other words, SNEP is an end to end security protocol and cannot prevent routing misbehavior.

On the other hand, μTESLA provides a secure broadcast communication, which is a common and important communication pattern in almost all WSN applications. μTESLA is developed to meet the special condition of WSN. For example, μTESLA authenticates initial packets using only symmetric keys instead of digital signature. μTESLA obtains routing security by authenticated routing that is achieved by deriving the operation on routing update packets and checking the correctness of the claiming parents by key disclosure.

2.9.2. INSENSE - Intrusion-tolerant routing in wireless sensor networks

INSENS (Intrusion-tolerant Routing in Wireless Sensor Networks)[66] constructs tree-structured routing for wireless sensor networks (WSNs). It aims to tolerate damage caused by an intruder who has compromised deployed sensor nodes and is intent on injecting, modifying, or blocking packets. INSENS incorporates distributed lightweight security mechanisms, including one-way hash chains and nested keyed message authentication codes to defend against routing attacks such as wormhole attack. Adapting to WSN characteristics, the design of INSENS also pushes complexity away from resource-poor sensor nodes towards resource-rich base stations.

2.9.3. SeFER - Secure, flexible, and efficient routing protocol

SeFER (Secure, flexible, and efficient routing protocol for sensor networks)[67] is based on random key pre-distribution mechanism. This mechanism aims to provide an easy way for managing the keys in WSN without using public key cryptography. The protocol assumes non symmetric communication architecture in which a tree of sensor nodes delivers information to a controller according to an inquiry sent into the network. Two nodes may communicate indirectly, but securely over a multiple hop path where each pair of nodes on this path shares a common key. The protocol provides the methods for nodes to securely share their keys and communicate directly so that the efficiency of communication is increased.

In fact, all previously mentioned protocols are crypto based solutions. They can successfully fight against attacks in which an intruder falsifies his identity to be a relay for the source such as sybil attack. However, other attacks like selective forwarding, blackhole and HELLO flooding are still possible especially when the attack is performed by an insider node or a node compromised by an intruder. Moreover, any misbehavior due to selfishness or faulty operational nodes cannot be prevented or even detected.

2.9.4. Watchdog and pathrater

Two extensions to the Dynamic Source Routing (DSR) protocol to mitigate the effects of routing misbehavior in ad-hoc networks were proposed in [6,7], namely the Watchdog and the Pathrater. The watchdog is the monitoring part that is designed to be responsible for detecting only non forwarding misbehavior. This is accomplished by overhearing the transmission of the next node. The node thus is assumed to be in a continuous promiscuous mode. When the attack is detected, the observing node informs the source of the concerned path. In this approach, each node maintains a buffer of recently sent packets; in case the packet is not forwarded on within a certain timeout or the overheard packet is different than the one stored in the buffer, the watchdog increments a failure counter for the node responsible for forwarding the packet. If the counter exceeds a certain threshold, the node is considered as misbehaving and the source is notified.

The pathrater is the component used for reputation. Ratings are kept about every node in the network based on its routing activity and they are updated periodically. Nodes select routes with the highest average node rating. Thus, nodes can avoid misbehaving nodes in their routes as a response. The pathrater combines knowledge of misbehaving nodes with link reliability data to select the route most likely to be reliable. Specifically, each node maintains a rating for every other node it knows about in the network and calculates a path metric by averaging the node ratings in the path, enabling thus the selection of the shortest path in case reliability information is unavailable. Negative path values indicate the existence of one or more misbehaving nodes in the path. If a node is marked as misbehaving due to temporary malfunction or incorrect accusation, a second-chance mechanism is considered, by slowly increasing the ratings of nodes that have negative values or by setting them to a non-negative value after a long-timeout. However, misbehaving nodes can still transmit their packets as there is no punishment mechanism adopted here. Moreover, no second hand information propagation view is considered which limits the cooperativeness among nodes.

2.9.5. CONFIDANT - Cooperation of nodes – Fairness in dynamic ad-hoc networks

In [3], the authors proposed CONFIDANT, a routing protocol for MANET with predetermined trust, and later improved it with an adaptive bayesian reputation and trust system and an enhanced passive acknowledge mechanism (PACK) in [68] and [69] respectively. It is a reputation based secure routing framework in which nodes monitor their neighborhood and detect different kinds of misbehavior by means of an enhanced PACK mechanism. The nodes use the second-hand information from others as a resource of rating, as well. The protocol is based on Bayesian estimation that aims to classify other nodes as misbehaving or normal. The observing node excludes misbehaving nodes from the network as a response, by both avoiding them for routing and denying them cooperation.

In this approach, Upon detection of the node's malice, its packets are not forwarded by normally behaving nodes, while it is avoided in case of a routing decision and deleted from a path cache. CONFIDANT architecture comprises 4 components residing on each node: the Monitor, the Reputation System, the Path Manager and the Trust Manager components. The Monitor component enables nodes to detect deviations of the next node on the source route by either listening to the transmission of the next node ("passive acknowledgement") or by observing route protocol behaviour. In order to convey warning information in case of identification of a bad behaviour, an ALARM message is sent to the Trust Manager component, where the source of the message is evaluated. The rating is updated only if there is sufficient evidence of malicious behaviour that is significant for a node and that has occurred a number of times, exceeding a threshold to rule out coincidences (e.g., collisions). Evidence could come either from a node's own experiences through the Monitor system or from the Trust Manager in the form of Alarm messages. Second-hand information is attributed with low significance (by means of a constant weighting factor w) with respect to the first-hand information, irrespective of its source node. Local rating lists and/or black lists are maintained at each node and potentially exchanged with friends. Black lists may be used in a route request, so as to avoid bad nodes along the way to the destination or to not handle a request originating from a malicious node and in forward packet requests, so as to avoid forwarding packets for nodes that have bad rating.

The protocol assumes a Dynamic Source Routing (DSR) operational routing protocol and lacks a provision on WSN constraints and conditions as it is designed for general ad hoc networks.

2.9.6. CORE - Collaborative reputation mechanism to enforce node cooperation in mobile ad hoc networks

Another famous reputation mechanism in literature is CORE protocol (Collaborative Reputation Mechanism to Enforce Node Cooperation in Mobile Ad Hoc Networks) [5]. It is a complete reputation mechanism that defines three different types of reputation: (i) Subjective Reputation - reputation observed locally by a node with regards to other nodes (direct observations), (ii) Indirect Reputation - reputation provided by nodes to other nodes which includes only the positive reports by others and (iii) Functional Reputation - also referred as task-specific behavior, which are weighted according to a combined reputation value that is used to make decisions about cooperation or gradual isolation of a node. That is, Subjective Reputation and Indirect Reputation are merged by means of a weighted combining formula in order to compute a final value of reputation concerning a specific evaluation criterion (e.g. packet forwarding) forming Functional Reputation, the last type of reputation considered. By combining different functional reputation values concerning different evaluation criteria, a global reputation value may be estimated. The subjective reputation is computed by giving more relevance to past observations than to recent ones. Subjective Reputation values are updated on the basis of a Watchdog mechanism, if misbehaviour is identified. Indirect Reputation values are updated by means of a reply message that contains a list of all entries that correctly behaved in the context of each function.

In this work, distribution of positive ratings is allowed so as to avoid potential denial of service attacks. In case reputation of an entity is negative, the execution of any requested operation will be denied by all other entities in the system. The system assumes a DSR routing in which nodes can be requesters or providers. The rating is done by comparing the expected result with the actually obtained result of a request. Here, nodes exchange only positive reputation information. The authors argue that this prevents a false-negative (badmouthing) attack, but do not address the issue of collusion to create false praise. In CORE, members have to contribute on a continuing basis (thereby enforcing node cooperation) to remain trusted or they will find their reputation deteriorating until they are excluded. CORE does not provide for a second-chance mechanism.

2.9.7. SORI - Secure and objective reputation-based incentive scheme for ad hoc networks

SORI Scheme for Ad Hoc Networks) [4] targets only the non forwarding attack. SORI monitors the number of forwarded packets from neighborhood and the number of forwarded packets to neighborhood. Reputation rating is then acquired by computing the ratio between the two numbers with a consideration for the confidence in the rating proportional to the number of packets that are initially requested for forwarding. Second hand information is delivered only to the immediate neighbors. This rating source; however, is weighted by what is called credibility, which is derived from the rating ratio. The delivery of the second hand information is achieved by hash-chain based authentication. SORI consists of three components, namely, neighbour monitoring (used to collect information about packet forwarding behaviour of neighbours), reputation propagation (employed so as to share information of other nodes with neighbours) and punishment (involved in the decision process of dropping packet action, taking into account the overall evaluation record of a node and a threshold so as to consider collision events). Reputation rating formation considers first-hand information weighted by a confidence value used to describe how confident a node is for its judgement on the reputation of another node and second-hand information weighted by the credibility of nodes which contribute to the calculation of reputation. Credibility of a node is defined on the basis of a node's behaviour as forwarder and not as a witness. Reputation rating itself is based on packet forwarding ratio of a node. SORI does not discriminate between selfish and misbehaving node. SORI does not comprise a second-chance / redemption mechanism. Finally, SORI, in order to tackle with impersonation threats, constructs an authentication mechanism based on a one-way-hash chain.

2.9.8. SAR - Security-aware routing

SAR [70] (Security-Aware Routing) is a protocol derived from AODV and based on authentication and a metric called the hierarchical trust value metric. The hierarchal trust values metric governs routing protocol behavior. This metric is embedded into control packets to reflect the minimum trust value required by the sender. Thus, a node that receives any packet can neither process it nor forward it unless it provides the required trust level presented in the packet. Moreover, this metric is also used as a criterion to select routes when many routes satisfying the required trust value are available.

2.9.9. TRANS - Trust routing for location aware sensor networks

TRANS (trust routing for location aware sensor networks) [72] is a geographic routing protocol (GPSR-based [8]) that provides security services using trust metric. It can be considered as a tight trust-based routing due to its specific targets and assumptions. It basically targets a misbehavior model in which an attacker selectively participates in routing signaling and control packets but drops consistently queries and data packets. The protocol also assumes static sensor networks in which a tight mapping can be done between the nodes' identities and their locations. TRANS assumes a location-centric architecture that helps it in isolating misbehavior and establishing trust routing in sensor networks. As a result of that, the protocol assumes a certain communication model in which a single or multiple sinks initiate communication requests with various locations. During that phase, insecure locations are identified and blacklisted. The trust metric used to judge on location security is calculated based on nodes' experience among each other regarding their identities, link availability and packet forwarding.

2.9.10. RGR - Resilient geographic routing

Resilient Geographic Routing (RGR) protocol [73] is also a trust-based routing protocol that relies on a modified routing operation in GPSR. The basic idea in RGR is to assign an initial trust value for each node. Then, this value is incremented or decremented depending on the forwarding activity of the monitored node using a step function. The source node selects probabilistically a subset among its neighbors to forward its packet. This subset is selected from the node's forwarding set that exhibits trust values greater than a threshold.

2.9.11. Robust reputation system for P2P and mobile ad-hoc networks

The main contribution in this work [68] is its proposal for a distributed reputation system that can handle false disseminated information. Every node maintains a reputation rating and a trust rating about every node that is of interest. The authors use a modified Bayesian approach so that they will accept only a second hand information set that is compatible with the current reputation rating. Also, Trust ratings are updated based on the compatibility of second-hand reputation information with prior reputation ratings. The work avoids exploitation of good behavior that can be incorrectly built over time by introducing a concept of re-evaluation and reputation fading.

2.9.12. RFSN - Reputation based framework for high integrity sensor networks

This work[2] proposes a reputation-based framework for sensor networks (RFSN) where nodes maintain reputation for other nodes and use it to evaluate their trustworthiness. The authors tried to focus on an abstract view that provides a scalable, diverse and a generalized approach hoping to tackle all types of misbehaviors resulting from malicious and faulty nodes. They also designed a system within this framework and employed a Bayesian formulation, using a beta distribution model for reputation representation. RFSN integrates tools from statistics and decision theory into a distributed and scalable framework. Bayesian formulation, specifically a beta reputation system is employed for the algorithm steps of reputation representation, updates, integration and trust evolution. This output metric of trust can be used by a node in several ways. For example, a data reading reported by a node can be weighted by the trust of the node when aggregating data from several nodes, thus reducing the impact of the faulty readings. Additionally, the evolution of trust over time provides an on-line tool to the end-user to detect compromised or faulty nodes. This can help the end-user to take appropriate countermeasures such as replacing the corrupted node or sensor.

The system starts the operation by monitoring. Monitoring mechanism follows the classic watchdog methodology in which a node is assumed to be in a promiscuous mode to overhear neighbors' packets. Monitoring behavioral events can result in either cooperative event, α, in which a node is behaving well or non cooperative behavior, β, in which a node misbehaves. The count of each type is injected into the beta distribution formula as the distribution parameters to calculate the node reputation R. This formula calculates node's reputation based on first hand information. The reputation is updated based on the new monitoring events, second hand information received and according to the age of the current reputation value. Any response action is based on selecting the most trusted node. The trust value of a node that is used for decision making is calculated as the statistical expectation of the reputation value.

2.9.13. DRBTS - Distributed reputation-based beacon trust system

In [74] authors propose a reputation based scheme called Distributed Reputation-based Beacon Trust System (DRBTS) for excluding malicious Beacon Nodes(BNs) that provide false location information. It is a distributed security protocol aimed at providing a method by which BNs can monitor each other and provide information so that the Sensor Nodes(SNs) can choose who to trust, based on a quorum voting approach. In order to trust a BN's information, a sensor must get votes for its trustworthiness from at least half of their common neighbor(s). In this approach, every BN monitors its 1-hop neighborhood for misbehaving BNs and accordingly updates the reputation of the corresponding BN in the Neighbor-Reputation-Table (NRT). The BNs then publish their NRT in their 1-hop neighborhood. BNs use this second-hand information published in NRT for updating the reputation of their neighbors after it qualifies a deviation test. On the other hand, the SNs use the NRT information to determine whether or not to use a given beacon's location information, based on a simple majority voting scheme.

Each BN is responsible for monitoring its neighborhood. When a sensor within its range asks for location information, it responds with its location, as do all other beacon nodes within the range of the requesting node. Due to the promiscuity of broadcast transmissions, a BN can overhear the responses of other BNs in its area. It can then determine its location using this claimed location of each BN and comparing them against its true location. If the difference is within a certain margin of error, then the corresponding BN is considered benign, and its reputation increases. If the difference is greater than the margin of error, then that BN is considered malicious and its reputation is decreased. This distributed model not only alleviates the burden on the base station to a great extent, but also minimizes the damage caused by the malicious nodes by enabling sensor nodes to make a decision on which beacon neighbors to trust, on the fly, when computing their location.

2.9.14. OCEAN - Observation based cooperation enforcement in ad hoc networks

OCEAN[75] approach to selfishness in ad-hoc networks is to disallow any second-hand information exchanges. Instead, a node makes routing decisions based solely on direct observations of its neighbouring nodes' interactions with it. OCEAN is designed on top of DSR protocol, may reside on each node in the network and hosts five components: Neighbour Watch (in order to observe the behaviour of the neighbours of a node), Route Ranker (estimating and maintaining ratings for each of the neighbouring nodes), Rank-based Routing (so as to avoid routes containing nodes in the faulty list), Malicious Traffic Rejection (rejecting all traffic from nodes it considers misleading so that a node is not able to relay its own traffic under the guise of forwarding it on somebody else's behalf) and Second Chance Mechanism (using a time-out based approach for removing a node from a faulty list after a fixed period of observed inactivity and assigning to it a neutral value). Once the rating of a node falls below a certain threshold, the node is added to the faulty list comprising all misbehaving nodes. In order to tackle selfish behaviour, the authors introduce a simple packet forwarding economy scheme, relying again only on direct observations of interactions with neighbours.

Due to the usage of only first-hand information, OCEAN is more resilient to rumour spreading. Finally, the authors rely on recent work on proof-of-effort mechanisms and mandate that a new identity will be accepted only if the owner shows reasonable effort in generating that identity.

2.9.15. TIBFIT - Trust index based fault tolerance for arbitrary data faults in sensor networks

In [76], authors propose a protocol called TIBFIT to diagnose and mask arbitrary node failures in an event-driven wireless sensor network. An event driven model of behaviour for sensing finds many applications in civilian, military as well as industrial scenarios. The goal of the proposed TIBFIT protocol involves event detection and location determination in the presence of faulty sensor nodes, coupled with diagnosis and isolation of faulty or malicious nodes. In this system model, sensor nodes are organized into clusters with rotating cluster heads. The nodes, including the cluster head, can fail in an arbitrary manner generating missed event reports, false reports, or wrong location reports. Correct nodes are also allowed to make occasional natural errors. The accuracy of the system is defined in terms of fraction of instances when an event occurrence is correctly detected, and its location determined within the given error bound. The approach followed by the protocol is to maintain state of the sensing nodes in terms of the fidelity of their previous sensing actions, and use this information in making decisions involving those sensing nodes. Sensor nodes report the occurrence and location of events to a data sink (cluster head), and remain silent otherwise. The data sink then decides on whether the event occurred and were based on the aggregated data. To determine the location of the event, the data sink must aggregate all reports from nodes within the detection radius. In this approach, a new parameter called *trust index* for this aggregation is introduced. Each node is assigned a trust index to indicate its track record in reporting past events correctly. The cluster head analyzes the event reports using the trust index and makes event decisions. The *Trust Index(TI)* of a node is a quantitative measure of the fidelity of previous event reports of that node as seen by the data sink. In a system comprised of sensing nodes, the data sink assigns and maintains a TI for each node in its domain, and does voting in a state-full manner. As the system runs over a longer time, more state is built up concerning the performance of the associated sensing nodes, and hence tolerance for faults also goes up accordingly. Authors claim that TIBFIT can tolerate faults in a network with more than 50% of its nodes compromised *after* it has built up adequate state of the nodes.

The main contributions of this paper are the following:

i. TIBFIT tolerates nodes that fail both naturally and maliciously, and makes decisions on event occurrence as well as location. Under several scenarios, accurate event determination and localization can be done even with more than 50% of the network compromised.

ii. No nodes are considered immune to failure, whether they are sensing nodes or the data sink.

iii. An adversary model is proposed with increasing levels of sophistication and demonstrated the effectiveness of the protocol in each case.

iv. The protocol is generic and can be applied to any data sensing and aggregation application in sensor networks.

2.9.16. PLUS - Parameterized and localized trust management scheme protocol

In [77] authors have proposed Parameterized and Localized trUst management Scheme (PLUS) for WSNs. The authors adopt a localized distributed approach and trust is calculated based on either direct observations or indirect observations. Whenever a node needs recommendation about another node, it will broadcast a request packet to its neighbors. This packet contains the identity of the evaluating node. In response all the nodes (except the node whose is going to be evaluated) send back a response packet to the requester. Once all the response packets are received, the requester will calculate the final trust value. If the node finds any misbehavior about the evaluated node, then the node will broadcast a exchange information packet to its neighbors. This packet contains information about identity of the node and error code. Based on the trust policy, the neighboring nodes send out its opinion: exchange Acknowledgement packet in case if they agree with the sender, otherwise neighbors will reply with exchange Argue packet indicating disagreement.

2.9.17. LARS - Locally aware reputation system

In [78], the authors propose LARS to mitigate misbehavior and enforce cooperation. Each node only keeps the reputation values of all its one-hop neighbours. The reputation values are updated on the basis of direct observations of the node's neighbours. If the reputation value of a node drops below an untrustworthy threshold, then it is considered misbehaving by the specific evaluator node. In such a case, the evaluator node will notify its neighbours about misbehaviour, by initiating a WARNING message. An uncooperative node is identified in the neighbourhood region, in case a WARNING message issued by a node is co-signed by m different one hop-neighbours, where $m-1$ is an upper bound on the number of nodes considered in the one-hop neighbourhood, in order to prevent false accusations and problems caused with inconsistent reputation values. Additionally, a fade factor has been introduced to give less weight to evidence received in the past. The misbehaving node is not excluded from the network for ever. After a time-out period, it is accepted, but with the reputation value unchanged so it would have to built its reputation by good cooperation.

2.9.18. TARF - A trust-aware routing framework for wireless sensor networks

In [79] authors propose a trust aware routing framework for WSNs called TARF to secure multi-hop routing in WSNs against intruders exploiting the replay of routing information. This approach identifies malicious nodes that misuse "stolen" identities to misdirect packets by their low trustworthiness, thus helping nodes circumvent those attackers in their routing paths. It incorporates the trustworthiness of nodes into routing decisions and allows a node to circumvent an adversary misdirecting considerable traffic with a forged identity attained through replaying. It significantly reduces negative impacts from these attackers. TARF is also energy efficient, highly scalable, and well adaptable.

In this approach, to route a data packet to the base station, a node only needs to decide to which neighbouring node it should forward the data packet considering both the trustworthiness and the energy efficiency. It maintains a neighbourhood table with trust level values and energy cost values for certain known neighbours. Two types of routing information that need to be exchanged in addition to data packet transmission are – (i) Broadcast messages from the base station about data delivery and (ii) Energy cost report messages from each node. Neither message needs acknowledgement. A broadcast message from the base station is flooded to the whole network. The other type of exchanged routing information is the energy cost report message from each node, which is broadcast to only its neighbours once. Any node receiving such an energy cost report message will not forward it. Each node has two modules – *Energy Watcher* and *Trust Manager* running on it in order to maintain a neighbourhood table with trust level values and energy cost values for certain known neighbours. *Energy Watcher* is responsible for recording the energy cost for each known neighbour, based on nodes observation of one-hop transmission to reach its neighbours and the energy cost report from those neighbours. A compromised node may falsely report an extremely low energy cost to lure its neighbours into selecting this compromised node as their next-hop node; however, these TARF-enabled neighbours eventually abandon that compromised next hop node based on its low trustworthiness as tracked by *Trust Manager*. *Trust Manager* is responsible for tracking trust level values of neighbours based on network loop discovery and broadcast messages from the base station about data delivery. At the beginning, each neighbour is given a neutral trust level. After any of those events occurs, the relevant neighbours' trust levels are updated. Occurrence of a loop degrades that node's next-hop node's trust level thereby gradually taking the trust level to a low value leading to the breaking of the loop by changing its next-hop selection. On the other hand, to detect the traffic misdirection by nodes exploiting the replay of routing information, *Trust Manager* computes the ratio of the number of successfully delivered packets which are forwarded by this node to the number of those forwarded data packets, denoted as Delivery Ratio. Once a node is able to decide its next hop neighbour according to its neighbourhood table, it sends out its energy report message - it broadcasts to all its neighbours its energy cost to deliver a packet from the node to the base station.

2.9.19. SensorTrust - A resilient trust model for wireless sensing systems

In[80], authors propose a resilient trust model, SensorTrust with a focus on data integrity for hierarchical WSNs. In this model, the aggregator maintains trust estimations for children nodes by integrating their long-term reputation and short-term risk and taking into consideration both communication robustness and data integrity. Long-term reputation, also called conventional reputation, refers to its average performance level in its whole past history, and short-term risk identifies to which degree its future behaviour is associated with its recent performance. Neither long-term reputation nor short-term risk alone could fully reflect current trustworthiness. On the one hand, a single fault could occasionally happen to even a trustworthy sensor node, but that doesn't necessarily mean the node is unreliable. That suggests the one-sidedness of short-term risk. On the other hand, long-term reputation treats the node's behaviour in each transaction equally. But in the real world, a node with good average performance level might begin to behave negatively during recent transactions. That could suggest that the sensor starts to malfunction. Since a node can behave maliciously regarding either wireless communication or data management, trustworthiness is evaluated from two aspects: communication robustness and data integrity. This model employs the Gaussian model to rate data integrity in a fine-grained style, and a flexible update protocol to adapt to different applications. In this model, to accurately identify the current trust level, past history and recent risk are synthesized in a real-time way. This model uses a SensorTrust value, which is a decimal number in [0,1], to represent trustworthiness level. The higher some node's SensorTrust value is, the more trustworthy that node is. Specifically, the SensorTrust value in terms of communication robustness is the estimated probability of a positive communication transaction; the SensorTrust value in terms of data integrity is the estimated probability of integrity of data. At the beginning, the aggregator assigns a SensorTrust value of 0 to its children nodes, since no evidence of trustworthiness is available. Each time a sensor node interacts with its associated aggregator, the aggregator evaluates the node's behavior by giving a rating number in [0,1] for this transaction in terms of communication robustness and data integrity respectively. This rating number reflects the aggregator's opinion of the current transaction: the higher the rating numbering is, the more positive the aggregator views the sensor node to be. The rating number together with its latest SensorTrust value will be used by the aggregator to update the node's SensorTrust value. With acceptable overhead, SensorTrust proves resilient against varied faults and attacks.

Considering the related work reported in the literature, it can be stated that there is a lack of standardization orientations when designing a trust and/or reputation model for distributed systems[46,47,55]. It has been found that approaches/schemes proposed in related research literature are based on quite different assumptions, while the trust/reputation framework considered varies significantly in many aspects. Some of the aspects in which these reported approaches differ can be listed as - Computation of trust/reputation considering only first hand information or both first-hand and second-hand information, Propagation of second-hand information considering only positive, negative or both types of recommendation, Degree of propagation, Adopted model for reputation value computation, Dishonest second-hand information provisioning, Identification of misbehaving nodes, Actions taken, Node re-integration in the system, etc. The proposed reputation systems use several debatable heuristics for the key steps of reputation updates and integration. Some systems maintain a statistical representation of the reputation by borrowing tools from the realms of game theory. These systems try to counter selfish routing misbehaviour of nodes by enforcing nodes to cooperate with each other. More recent reputation systems proposed in the domain of ad-hoc and sensor networks, formulate the problem in the realm of Bayesian analytics rather than game theory. Furthermore, most of the trust research focuses on communication behaviors without clearly indicating data integrity importance. Some reported recent approaches employ communication trust and data trust separately in their suggested trust models considering the fact that one of the main tasks of WSNs is data collection and moreover, different applications have their own specific requirements regarding communication trustworthiness and data trustworthiness.

3. Reputation system overview

In this section, an overview of the proposed reputation system will be presented. The section will start by describing the general framework. This is followed by a brief description of our customized reputation system that fits into the framework guidelines.

3.1. Reputation system framework

The conceptual operation of the reputation system is based on building a trust relation between different members of the community as they learn about each other. Thus, irrespective of why a node needs to build such relations, any reputation system must have two basic components, i.e. monitoring component to allow nodes to learn about each other and rating component to build the trust relations among nodes. However, the purpose of these trust relations will determine the specifications of each component and may imply a new component responsible for further actions based on the trust relations.

Our reputation system is fully distributed in the sense that each node implements all modules with the full functionality. Moreover, at the initial deployment stage, all nodes start with default and equal reputation values. This implies that all nodes have the same trust relation among each other. However, these initial reputation values are not the ones that imply a full trust. This is because our system assumes an always-suspicious environment in which all nodes are always '*suspects*'. A node can increase its reputation by good behavior or, otherwise, it decreases.

Since the purpose of the reputation system in this work is to provide trust aware routing in WSN, there are three basic components in our system. They are as follows.

3.1.1. Monitoring component

The reputation system operation starts by the execution of monitoring component. Monitoring component is responsible for collecting behavior information by direct observation of neighbor's activities. In this work, we are concerned only in routing activities and, more specifically, in packet forwarding, i.e. monitoring whether a router is forwarding a packet or not. After a monitoring node detects some misbehavior, it reports its observation as a quantity to the rating component.

3.1.2. Rating component

This component is responsible for evaluating the reputation of an observed node. Assume that node A wants to evaluate a reputation value for a node B that may or may not be directly monitored by A. Then, the reputation value of B evaluated by A is a number that reflects how good or bad node B behaves from the perspective of node A considering:

- Monitoring results of all types of routing activities.
- Monitoring results obtained by direct observations from A as first hand information, if any.

- Monitoring results gathered from other nodes observing B and shared with A as second hand information, if any.

Once a reputation value for B is formed by A, A will decide about a certain level of trust relationship with B. Notice that, according to these specifications of the rating module, it is not necessary that A and B are neighbors to each other. Thus, A can have a trust relation with any node in the network. This, in fact, helps in generalizing the framework to allow the use of various routing protocols that differ in the obtainable level of hop information. For example, with DSR, in which multi-hop information can be collected, this module provides the ability to build trust relations among all nodes from source to destination. Also, on the other extreme, the module can work with geographic routing protocols like GPSR and GEAR with one hop information.

3.1.3. Response component

Once node A gets reputation knowledge about node B and decides a trust relation with it, A may or may not respond to B's behavior. Since our system treats the secure routing purpose, A should respond in a proper manner. Among different possible reactions provided in many reputation systems [3, 5, 72], our system framework assumes three main response approaches with regard to node A.

- *Defensive approach*: Here, node A just avoids using node B as a router. This avoidance can be gradual as the reputation value of node B decreases. However, B can still use A or any node to forward its packets.
- *Offensive approach*: In this approach, node A avoids B as in the previous approach. In addition to that, A takes further actions by punishing node B. However, node B still has the right to defend itself and is treated normally if it can prove a good behavior.
- *Dismissal approach*: in this approach, node A totally ignores node B as if it is not in the network. So, A does not receive any packet coming through B and does not forward to it. Moreover, B will never rejoin the network as seen by node A.

The previous approaches show possible single responses that can be taken by a single node. However, by the assumption of nodes cooperation, these approaches can extend to more than one node or possibly to the whole network by the propagation of second hand information or some sort of alarms.

With these three components of our framework, the following block diagram in figure 1 illustrates our reputation system operation and inter-components relationships.

3.2. Customized reputation system – SNARE overview

This section describes our new customized reputation system that fits into the general framework described earlier. We called our system: *Sensor Node Attached Reputation Evaluator* (SNARE) system[82][83].

SNARE is a collection of protocols and algorithms that interacts directly with the network layer. The system consists of three main components; i.e. monitoring component, rating component and response component.

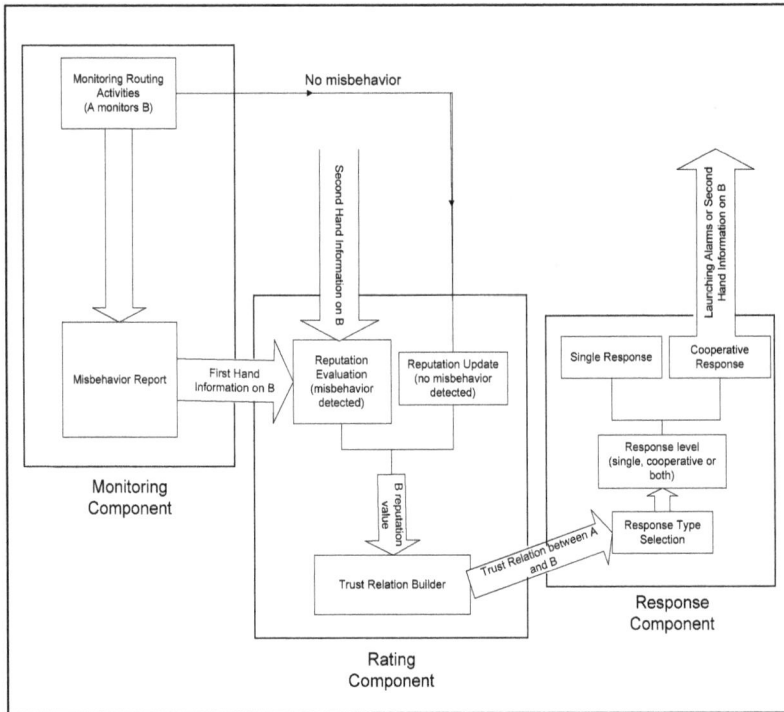

Figure 1. SNARE Reputation system framework

The monitoring component, EMPIRE(Efficient Monitoring Procedure In REputation system)[84], observes packet forwarding events. A monitoring node will not be in a continuous monitoring mode of operation, rather, it will monitor the neighborhood periodically and probabilistically to save resources. When a misbehaving event is detected, it is counted and stored until an update time, T_{update} or T_{ON} is due, then a report is sent to the rating component.

The rating component, CRATER(Cautious Rating for Trust Enabled Routing)[85], evaluates the amount of risk an observed node would provide for routing operation. The risk value is a quantity that represents the previous misbehaving activities that a malicious node (a node that drops packet) obtained. This value is used as an expectation for how much risk would be suffered by selecting that malicious node as a router. It is calculated based on the first hand information and the second hand information. The first hand information is achieved by the direct observation done by the node of concern. Risk values are updated based on the first hand information every time a new misbehavior report is received from the monitoring component. Moreover, if an observed node shows an idle behavior during a certain period, its risk value is reduced. A monitor also updates the risk values of its neighbors by second hand information received periodically from some announcers.

In this work, our system adopts the defensive response approach of the proposed framework. Thus, depending on the trust relations, a node will try to avoid malicious nodes based on the routing decision made by the developed routing protocol - Geographic, Energy, Trust Aware Routing protocol (GETAR). GETAR incorporates the trust information along with distance and energy information to choose the best next hop for the routing operation. The detailed description of this enhanced protocol GETAR is presented in section 4.

3.3. System assumptions

In order to understand how our system works and how simulations have been carried out, it is essential to formally identify the general assumptions on system requirements and boundaries. We will look at system assumptions from different perspectives.

3.3.1. WSN Perspective

In this work, we consider a WSN with a total number of nodes deployed in a random topology or in a grid topology inside a square area. It is assumed that the nodes communicate via bidirectional links so that the nodes can monitor packet forwarding. Moreover, all nodes have equivalent power transmission capabilities, i.e. all have equivalent transmission range. It is also assumed that the consumed power during the simulation time does not impact the transmission range of nodes. This assumption is made to keep the focus of our work on security issues and not on power control. The transmission and reception power are set to 1 Watt whereas the processing power is considered to be 1 milli-Watt per transmission, reception or monitoring operation. Finally, in this work, we assume a static WSN. Mobile WSN can be an interesting subject of a future research work.

3.3.2. Communication model perspective

The system adopts a general communication model in which each node in the system can initiate a routing operation. Thus, any node can be a source. Moreover, any node can be a destination for that node. The selection of the source-destination pair is done randomly. The reason of adopting this model is to study a very general case and not limiting our scope to particular scenarios. Other more realistic scenarios are important to consider. However, such scenarios should also account for the application specifications.

3.3.3. Security perspective

The existence of the reputation system does not imply a complete solution for all security problems. Our proposed solution tries to solve a particular security problem that is related to nodal behavior in the routing operation, as has been discussed earlier. Thus, some reasonable assumptions are made to make the work more focused on our problem:

- The system assumes always-suspicious nodes. This means that a node can not be fully trusted. Every node is assumed to have a minimum risk that can be encountered if that node is used as a router.
- The system assumes a crypto system for any setup requirements. This system is dependent on the routing protocol and, hence, it would imply different implementations that are left to the desire of the operator.
- The system assumes collusion-free attacks. The design of the system, however, can be easily modified to handle collusion based attacks since we adopt modular design. Changes need to be done in the rating component.
- The system treats only one type of behavior related attacks, i.e. non forwarding attack. Although the reputation system can be applied to any other attack, we concentrate here on non-forwarding attack. This is because we are not interested in Intrusion detection Systems (IDS), and we want to maintain the focus of the work on reputation system evaluation.
- The system assumes honesty in treating information exchange about nodes energy levels or risk values. Honesty can be accounted for in the rating component.

3.4. Monitoring component: Efficient monitoring procedure in reputation systems (EMPIRE)

In the context of reputation systems, monitoring is the function that is responsible for observing the activities of the nodes of its interest set, for example, the set of its neighbors.

Monitoring operation can be considered as the most expensive part in terms of resource usage for WSN. That is because it requires a node to track the events occurring around it by overhearing packet transmissions, which consumes lots of energy. Moreover, the computations and allocations of such events may consume a considerable amount of processing power and memory space, which are also important to conserve in WSN. As a result, a node has to monitor the behavior of its neighbors in an efficient manner that can provide a better possible way of resource conservation, while being able to reach to a good conclusion about the neighbors' behaviors so that it will take a proper action based on what it has observed.   Thus, an efficient monitoring mechanism should guarantee a satisfactory level of capturing neighborhood activities, while trying to minimize power consumption, memory usage, processing activities, communication overhead, etc.

A new monitoring strategy that is called *Efficient Monitoring Procedure In REputation system* (EMPIRE) to solve the problem of efficient monitoring in WSN is proposed in [84]. Monitoring efficiency is realized here by the association between the nodal monitoring activity (NMA) and various performance measures. NMA is determined by the frequency of monitoring actions that a node takes to collect direct observation information. Reducing the frequency of monitoring, i.e. reducing NMA, will affect the quantity and/or the quality of the obtained information. Thus, the performance measures will be affected. However, on the other hand, this reduction implies a saving in node's resources such as power, processing and memory. EMPIRE provides a probabilistic approach to reduce nodal monitoring activities (NMA), while keeping the performance of the system, from the behavior and trust awareness perspective, at a desirable level.

In this procedure as depicted in figure 2, every sensor node is alternating between two nodal monitoring activity states, i.e. ON state and OFF state. A node that is in ON state is a node that performs monitoring activities such as overhearing packets, checking the headers for validation, storing packets to validate events, etc. On the other hand, an OFF node is a node that does not do any monitoring activity. Notice that ON and OFF states are associated with the nodal monitoring activity. Thus, an OFF node may still receive, send and process data not related to monitoring issues. As explained earlier, the objectives of this procedure are realized through the frequency of nodal monitoring activity, NMA. Since nodes alternate between ON and OFF states, reducing NMA is determined by how much a node will stay in each of these states. Thus, when a node stays longer in ON state, its NMA will increase and when it stays longer in OFF state, NMA will decrease. The basic phenomenon of EMPIRE is to allow each node to enter a certain state probabilistically, stay there for a deterministic duration and then, at the end of that duration, it probabilistically leaves its state to the other one or stay for another epoch.

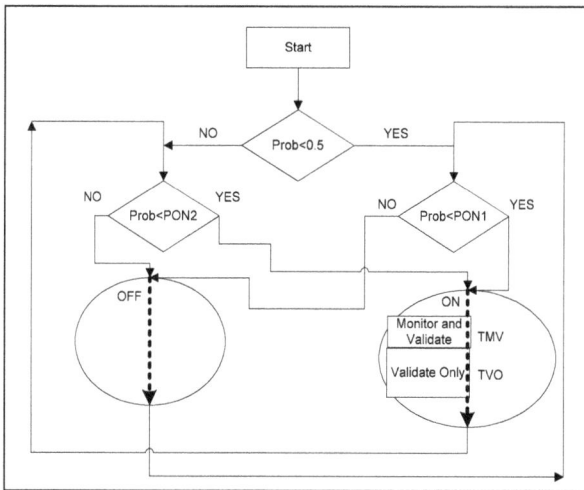

Figure 2. EMPIRE algorithm block diagram

In a cooperative monitoring environment, a node does not need to have a high NMA by continuously monitoring its neighbors' activities as long as there are a sufficient set of nodes that can monitor the same activities. So, if an activity can be monitored by two or more nodes who can share their knowledge among each other, then it is enough to have only one monitor active at a time. Then, upon some scheduling approach, the active node sleeps and another one gets awake. However, this scheduling problem is very complex and depends on different conditions like the network topology, network deployment, nodes mobility, etc.

Thus, in our EMPIRE solution, we are trying to induce a condition-independent and probabilistic "virtual scheduling" among nodes to overcome that problem. It is very important here to emphasize the node cooperation assumption. Node cooperation implies that a node will be willing to inform other nodes about its findings from its NMA. This is known as indirect reputation knowledge sharing or second hand information propagation. With this assumption, nodes will still be able to capture the events it loses during its OFF state. EMPIRE is based on a distributed and probabilistic monitoring approach. The main goal of EMPIRE is to provide good monitoring operation that satisfies the security requirements, while using the least possible nodal monitoring activity. This way, a node will also be able to conserve its resources. Our simulation results show that EMPIRE can satisfy various levels of monitoring requirements with different possible choices of nodal monitoring activity levels. Moreover, EMPIRE is safe in the sense that it can differentiate between malicious and non malicious nodes regardless of the choice of the nodal monitoring activity.

A detailed discussion and analysis of the EMPIRE procedure, simulation setup, performance measures, and simulation results can be found in [84].

3.5. Rating component: Cautious rating for trust enabled routing (CRATER)

In this section, a new rating approach for reputation systems in WSN called CRATER[85] is presented. CRATER evaluates nodes reputation by a risk representation. This risk value is computed based on FHI, SHI and idle behavior (NBP). The mathematical modeling of CRATER assumes a set of conditions that we define as cautious assumptions in which a node is very cautious in dealing with other's information.

In reputation systems, after a node gathers some information regarding the behavior of other nodes of interest, it needs to evaluate or rate these nodes. This is done by the rating function or the rating component of the system. Rating function is based on the node's own observation, other nodes' observations that are exchanged among themselves and the history of the observed node.

The rating component of a reputation system is a very critical part since it is responsible for providing the reputation of nodes. Thus, it can be considered as the heart of any reputation system.

To illustrate the rating operation, assume that node A wants to evaluate a reputation value for a node B that may or may not be directly monitored by A. Then, the reputation value of B evaluated by A is a number that reflects how good or bad node B behaves from the perspective of node A, considering:

- Monitoring results of all types of routing activities.
- Monitoring results obtained by direct observations from A as first hand information, FHI.
- Monitoring results gathered from other nodes observing B and shared with A as second hand information, SHI.

Once a reputation value for B is formed by A, A will decide about a certain level of trust relationship with B

An important issue in rating is the reputation update. Since rating is related to node behavior, the reputation of a node should be a dynamic metric that changes with time. This change would be due to new FHI observations, new SHI, or other defined aspects like, for example, to "forgive" some idle malicious nodes.

A new rating technique called *Cautious RAting for Trust Enabled Routing* (CRATER) is presented in [85]. Basically, this technique identifies three rating factors: FHI, SHI and Neutral Behavior period (NBP) during which a node is not doing any activity. The new contribution in CRATER is its mathematical approach that is used to rate nodes based on what we call cautious assumptions, which are very true in many applications in WSN.

3.5.1. Cautious assumptions

The rating methodology proposed in CRATER assumes what we call "the cautious assumptions". These assumptions are:

- Pessimistic start: The default status of a node joining the WSN network is to be untrustworthy. However, its reputation, or what we will call later the risk value, will not be at the extreme level.
- Unreliable SHI: A node tries to be as much independent from SHI as possible to avoid dishonesty issues.
- Rejecting good news: Announcing "good news" about other nodes in SHI can be a trial from the announcer to relieve itself from routing duties and put the burden on the others or it can be thought as collusion between the announcer and an attacker. Thus, nodes are not interested in hearing good news. On the other hand, "bad news" is very much welcomed. The differentiations between these good or bad announcements are realized by a threshold.
- Local interest: This means that a node is only interested in rating its immediate neighbors.

3.5.2. Rating factors in CRATER

In CREATER, each node rates its neighbor by assigning a risk value to the corresponding monitored node. The risk value of node j assigned by node i, $r_{i,j}$ is defined as a quantity that represents how much risk the node i will encounter when it uses node j as a next hop to route its packets. This value ranges from 0 to 1 where 0 represents the minimum risk and 1 represents the maximum risk. The reputation of node j as per node i is then computed as:

$$rep_{i,j}=1-r_{i,j} \tag{1}$$

The CRATER operation is based on rating the nodes on the risk notion. Each node evaluates the risk values of its neighbors and takes the proper action based on the values it obtains. The risk values are affected by three factors:

- FHI: The direct observation of the neighbor's behavior, this will be referred to as first hand information, FHI.
- SHI: The opinion of other nodes regarding the neighbor of concern. This will be called second hand information, SHI.
- Neutral Behavior Periods (NBP): these are the periods during which a neighbor is observed doing nothing. That is, a neighbor does not receive anything to be tested for forwarding.

Each node in the system continuously and periodically updates the risk values of its neighbors based on the information collected during these update periods .

The general algorithm that a node i follows to rate its neighbor j is:

- Node i monitors node j for the duration of the update period, T_{update}.
- At the end of each update period, do the following:
    - Calculate $r_{i,j,FHI}$ using the new FHI.
    - Update the old risk value, $r_{i,j,old}$ using the new calculated $r_{i,j,FHI}$ to get $r_{i,j}$.
    - Calculate the $r_{i,j,SHI}$ using the SHI.
    - Update $r_{i,j}$ using the $r_{i,j,SHI}$
    - Update $r_{i,j}$ if neutral behavior periods are realized.

When node j is observed by i for n consecutive update periods to be idle in its behavior, node i will give node j a chance to be more trusted by reducing its current risk value. A node is considered to be in idle behavior if it does not perform any routing operation. The reduction procedure follows exactly the same methodology explained in rating based on FHI when $r_{i,j,FHI}=0$. The only difference here is that in the case of neutral behavior the update is done after we observe such behavior during n consecutive update periods whereas it is done immediately after an update period in the case of $r_{i,j,FHI}=0$. The choice of n is a design parameter that depends on how much a network is tolerable against attacks. High values of n mean that we are not willing to forgive malicious nodes quickly.

A detailed discussion and analysis of the CRATER approach, simulation setup, performance measures, and simulation results, can be found in [85].

3.6. Reputation systems-independent scale for trust on routing (RESISTOR)

Reputation systems are very complicated systems to evaluate or compare. This is because each system has its own components' implementation methods, like monitoring strategy, rating approach and response mechanism. All these components affect the efficiency of the reputation system individually as well as a complete system. Therefore, it is important to come up with a simple mechanism that can evaluate and analyze a reputation system. Such a mechanism must be:

- Independent of the reputation system: This means that the inputs of the formulae or equations used in this mechanism should not use the specific parameters that determine how the individual component of the reputation system is working.

- Representative for the effect of each individual component: This means that the mechanism should provide parameters that reflect the role of individual components in the reputation system.

In this work, we propose a simple but strong, independent and representative scale to evaluate reputation systems called *REputaion Systems-Independent Scale for Trust On Routing* (RESISTOR)[85].

3.6.1. The resistance concept

RESISTOR is an evaluation procedure that is used to evaluate the performance of reputation systems that are designed to provide trust aware routing. The basic idea behind RESISTOR is to utilize some of the objectives of a reputation system in an analytical way to evaluate the performance of the system.

Any reputation system that is concerned with trustworthy routing has two main objectives:

- Recognizing the malicious nodes by ultimately reaching to their theoretical reputation values or risk values as in the context of CRATER.
- Reducing the flow of packets into the malicious nodes so that they will not have a chance to drop packets, or do any other type of attacks.

Having these two objectives, we introduce the resistance metric. We define, generally, the resistance between node i and a malicious node j in the direction from i to j; $RES_{i,j}$, as a ratio of the risk value $r_{i,j}$ to the number of packets that flow from node i to j, i.e. $P_{i,j}$.

Please notice here that the concept of resistance is only associated with malicious nodes. Thus, if $r_{i,j}$ is high, the resistance value will be high, reflecting that the reputation system is performing well since we are "resisting" a malicious node. Similarly, if $P_{i,j}$ is small, the resistance value gets high, inferring that the reputation system is performing well, too. This is because we expect to pass few packets to a malicious node, ideally zero packets.

The resistance concept is analogous to the resistance phenomenon in electric circuits. We can think of the risk value of a malicious node j as seen by i as the voltage difference between j and i and the packet flow from i to j as the current flow. The resistance, then, increases as the voltage, $r_{i,j}$ increases and the current $P_{i,j}$ decreases similar to Ohm's law; R=V/I. Following this analogy, we have:

$$RES_{i,j} = \frac{r_{i,j}}{P_{i,j}} \qquad (2)$$

Thus, a good reputation system must provide high resistance. A perfect reputation system should provide an infinite resistance since $P_{i,j}=0$. A detailed discussion and evaluation of CRATER using RESISTOR approach, simulation setup, and simulation results, can be found in [85].

4. Response component : Geographic, energy and trust aware routing (GETAR) protocol

In this section, an enhanced routing protocol that aims to provide a secure packet delivery service guarantee by incorporating the trust awareness concept into the routing decision is presented. Our proposed protocol is called Geographic, Energy and Trust Aware Routing (GETAR) which is an enhanced version of the Geographic and Energy Aware Routing (GEAR) protocol [9]. GEAR is basically a geographic routing protocol in which the next hop is selected based on two metrics: the distance between the next hop and the destination and the remaining energy level the next hop owns. The new contribution in GETAR is to add a third metric in the next-hop selection process, i.e. the risk value of a node that is computed by the rating component, CRATER in our case.

After a node monitors its neighborhood using EMPIRE and rate them based on CRATER, the node should make the proper response that leads to a proper routing decision. Assume that node A computed a risk value for a malicious neighboring node, B. Then, node A may or may not respond to B's behavior. Since our system treats the secure routing purpose, A should respond in a proper manner. Among different possible reactions provided in many reputation systems [3, 5, 72], we can identify three main response approaches:

- *Defensive approach*: Here, node A just avoids using node B as a router. This avoidance can be gradual as the risk value of node B increases. However, B can still use A or any node to forward its packets.
- *Offensive approach*: In this approach, node A avoids B as in the previous approach. In addition to that, A takes further actions by punishing node B. However, node B still has the right to defend itself and to be treated normally if it can prove a good behavior.
- *Dismissal approach*: In this approach, node A totally ignores node B as if it is not in the network. So, A does not receive any packet coming through B and does not forward to it. Moreover, B will never rejoin the network as seen by node A.

In this work, the defensive approach where malicious nodes are simply avoided without any further actions against them is adopted.

4.1. The original protocol: GEAR

4.1.1. GEAR description

Geographic and Energy Aware Routing [9] (GEAR) is a geographic routing protocol in which the routing decision accounts for the geographic location of a selected node with respect to the destination. It is also considered as a location based routing protocol because nodes are assumed to be interested in communicating with other nodes that reside in certain geographic locations regardless of their identities. The protocol implements greedy forwarding approach based on distance to destination and energy consumption considerations. In fact, the protocol tries to fairly consider energy balancing among the neighbors of a packet forwarder node.

In GEAR, the routing mechanism involves two phases:

- Forwarding the packet to a target region R with a greedy algorithm that tries to balance energy.
- Disseminating the packet within the target region by recursive forwarding.

Forwarding: Forwarding operation in GEAR can be summarized by the following steps:

- Each node N maintains a state value h(N,R) which is called the *learned cost* to region R. A node infrequently updates its h(N,R) value to its neighbors. Thus, every node N has state value knowledge for each neighbor N_i.
- A Source N picks a neighbor N_{min} with the minimum learned value to the region R.
- If N does not have the learned cost of a neighbor N_i, N estimates the learned cost by using the *estimated cost* function $c(N_i,R)$. The function combines the distance d from N_i to R and the consumed energy value e at N_i, as follows:

$$h(N_i,R) \approx c(N_i, R) = \alpha\, d\, (N_i, R) + (1-\alpha)\, e\, (N_i) \qquad (3)$$

where $d(N_i, R)$ is the distance from Ni to the center of R normalized (divided) by the largest such distances among all other candidates. $e(N_i)$ is the so far consumed energy at node N_i normalized by the largest consumed energy among all candidates. α is a tunable weight parameter that varies from 0 to 1 and indicates the routing decision preference. So, if α is close to one, the decision will be biased by the distance. If α is close to zero, the decision will be biased by the consumed energy levels.

- After selecting the N_{min} for routing, N updates its learned cost value to the destination region R as follows:

$$h(N, R) = h\, (N_{min}, R) + c(N, N_{min}) \qquad (4)$$

where the latter term is the cost of transmitting a packet from N to N_{min} considering the same approach in equation (3).

As we can see, from equation (3), when all nodes are equal in energy, the routing decision will be simply the greedy approach as in GPSR [8]. In case all nodes are equidistance from the destination, the selected node will be the one that consumed the least energy among others. This guarantees a fair selection of the node in terms of energy balancing.

Dissemination: Once a packet reaches the center node C_i of the destination region R, the protocol switches to the dessimination phase as follows:

- C_i splits the region R to sub regions R_i, for example four sub regions.
- C_i, then, sends four copies of the packet to the centroids of each sub region R_i.
- Each center node in different sub regions repeats the operation of splitting and forwarding until the center node finds that it is the only node in its sub region.

In our proposed protocol, this phase is avoided and we restrict the operation to forwarding with the cost functions since there is actually no routing decision to be made in the dissemination phase as suggested by GEAR.

Void Regions Problem : If a node wants to forward a packet and it finds out that the learned costs of all its neighbors are greater than its own learned cost, the node should select itself. However, the node's transmission range does not cover the destination. In this case the node is said to be in a void region. GEAR escapes this void region as follows:

- Assume that a source node S wants to transmit a packet to a destination T.
- S selects a next hop, C, that is in a void region, i.e. $h(C,T) < h(N_i,T)$ where N_i is a neighbor to C.
- C forwards the packet to a node, call it B, based on a predefined ordering, e.g. node ID. Then it updates its cost function $h(C,T)$ to be $h(C,T)=h(B,T)+c(C,B)$.
- Now, $h(C,R) > h(B,R)$
- Later, when node S wants to transmit a new packet to T, it will forward it to B instead of C (see the figure 3).

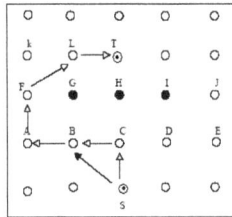

Figure 3. Escaping void regions in GEAR

4.2. The enhanced protocol: GETAR

4.2.1. Basic idea

GETAR is a geographic and energy aware routing protocol that has the additional feature of trust awareness. The trust awareness is achieved by the rating functionality of a running reputation system that will feed the routing protocol with the trust metric that will be the risk values, $r_{i,j}$. The risk value $r_{i,j}$, as discussed earlier, is a quantity that reflects, to some extent, the expectation that a node j will not forward the packet received from node i, assuming non forwarding attack.

The risk value metric, along with distance and energy metrics, are used to compute the learned cost function for each neighbor. The concerned node, then, makes the routing decision by selecting the neighbor of lower cost as in normal GEAR.

As we can see, GETAR is a modification extension to the GEAR protocol to account for some security issues. In GEAR, the choice of a next hop router to the desired destination is made locally by each node based on the learned cost function obtained using equations (3) and (4).

It should be clear that the main idea behind this cost function in GEAR is to provide a tunable preference to the distance or energy consumption as routing metrics based on the value of α. It is important to notice that the two metrics are considered to be routing resources for the node as well as the network. The new contribution in GETAR is to add in the cost function the risk value as the trust metric to account for trust awareness and which is also considered to be a routing resource.

To illustrate the idea, let's make an analogy between energy and trust. From the energy perspective, a node will prefer to select the next hop that has the least consumed energy level according to GEAR. This local decision and selection is the best effort that the node can do to cooperate in the routing operation and simultaneously conserve the total network energy. Similarly, from the security perspective, a node will prefer to select the next hop that is least risky among others in neighborhood. Such a selection will guarantee the safest decision that the node can do to cooperate in packet delivery. However, the node here tries to maintain trust as a resource.

4.2.2. Forwarding in GETAR

GETAR forwards the packets and makes routing decision following the same procedure in GEAR. However, the major difference is in calculating the estimated cost function that is used to learn the cost to different destinations. In GETAR, the estimated cost function that a node i evaluates for every neighbor j is given by:

$$t(j,R) = \beta r(j) + (1 - \beta)[c(j,R)] \tag{5}$$

where t(j,R) is the *trust-aware* cost of using the node j by node i as a router to the center of R. r(j) is the *risk function* that evaluates the risk value of using j as a router. β is a tunable parameter to prefer trust as opposed to other resources.

Using equation (3), we can rewrite equation (5) as:

$$t(j,R) = \beta r(j) + (1 - \beta)[\alpha d(j,R) + (1 - \alpha)e(j)] \tag{6}$$

If we are concerned about trust more than other resources, β should be close to 1. When β equals 1, the trust-aware cost will consider only the trust part of equation (6) and the next hop will be the most trusted one. Setting β to zero, however, turns the protocol to pure GEAR without any security considerations from the routing protocol perspective.

4.2.3. The risk function r(.)

There can be several ways to represent the risk function evaluated by a node i for using node j as a router. In this work, however, the risk function r(.) is nothing but the risk value $r_{i,j}$. Thus equation (6) is rewritten as:

$$t(j,R) = \beta(r_{i,j}) + (1 - \beta)[\alpha d(j,R) + (1 - \alpha)e(j)] \tag{7}$$

4.2.4. Dissemination and voids in GETAR

In GETAR, the routing operation involves only packet forwarding phase and does not implement dissemination. This is because in the dissemination phase in GEAR, the packets are intended to be forwarded to all nodes in the target region. However, when we consider trust awareness, a misbehaving node should not be given a chance to have the packet since it will not forward the packet. Thus, GETAR continues to forward packets based on the routing decisions made by the learned cost function.

Regarding the problem of void regions, there is no change in the escaping operation proposed by GEAR. The only difference in GETAR is that the reason of being in a void region can be also related to the existence of misbehaving nodes in the proximity of the node of interest.

4.3. Simulation objectives and setup

4.3.1. Objectives

In this work, we are studying the effect of incorporating trust aware metric in routing decision in GETAR. The simulation work aims to analyze the following issues:

- The efficiency of GETAR in terms of packet delivery. Therefore, we are analyzing how our proposed protocol will improve the packet delivery, decrease the impact of attacks on dropping packets and decrease the number of packet retransmission due to malicious dropping.
- The efficiency of GETAR in terms of energy conserving. This issue is related to the hypothesis that GETAR will reduce the retransmission due to malicious behavior. Thus, we expect that the power that could have been used for retransmission will be saved with GETAR.
- Studying the impact of malicious nodes population on GETAR performance.
- The trade-off between trust awareness and energy balancing.

4.3.2. Assumptions

- As mentioned earlier in this section, the risk value of a node is assumed to be abstractly calculated by the monitoring and rating components of a reputation system. This risk value is assumed to be constant during the simulation duration. This assumption is valid if we consider that the update period of the risk values is greater than the simulation time, or the updated values during the simulation time are not very far from the starting values. This is valid as long as we assume that the rating and monitoring component have a moderate or slow pace. Moreover, our focus in this work is to study the impact of injecting trust into the routing decision during a period that holds this trust metric unchanged.
- We assume that all nodes are able to locate themselves in the (x,y) coordinates and that sender nodes are able to locate their destinations.
- We assume that nodes will announce their energy and location information honestly. Handling false updates is beyond the focus of this work.
- Attackers are assumed to follow GETAR protocol. They are also allowed to initiate packet transmission sessions. This is because this work does not consider an offensive-

response to malicious behavior. A future work would include that issue, i.e. how to punish malicious nodes from the routing perspective.

4.3.3. Simulation setup

In this simulation work, we used the parameters in table 1. In our simulation, we tested one type of attacks; i.e. non forwarding attack. Moreover, a malicious node in this attack will drop all packets that it receives with probability=1. For this type of attacks, we experiment four different percentages of attackers of the total number of nodes; i.e. 10%, 30%, 50% and 70%.

All experiments are performed by varying the value of the trust awareness parameter β in GETAR cost function. Then, the outputs are used to compare the behavior of the performance metric versus the change in β values.

Parameter	Value	Parameter	Value
Number of nodes	100 nodes	Queuing model	M/M/1
Network dimensions	square 90 units * 90 units	Simulation platform	Event driven simulation using Java programming language
Transmission range	15 units	Simulation duration	100 seconds
Network Deployment	Random topology	Retransmission Timeout	Explicit retransmission request
Power consumption	1 unit per reception and 1 unit per sending operation	Retransmission trials	Unlimited
Mean arrival rate	1 pps	Update Strategy	Periodic, every 5 seconds.
Mean service rate	500 pps	α	0.5 (GEAR parameter)
Outsider attackers deployment	Random	Communication discipline	random source to random destination.
Escaping void	using GEAR part and then distance	Void failure: max number of hops	100

Table 1. Simulation parameters fo GETAR experiments

4.3.4. Performance Measures

- Delivery ratio: This is defined as the ratio between the number of packets delivered successfully to their destinations to the total number of generated packets; i.e:

$$\text{delivery ratio} = \frac{\text{number of successful packets}}{\text{total number of packets}} \qquad (8)$$

The objective of this metric is to show the effect of injecting the trust knowledge into the routing decision on improving the success of the routing operation. The metric is studied under the effect of increasing the trust awareness feature by increasing the β parameter of GETAR.

- Outsider attacks' drop ratio: This is defined as the ratio between the number of packets dropped due to outsider malicious nodes to the total number of generated packets; i.e:

$$\text{Outsider attacks drop ratio} = \frac{\text{number of dropped packets by malicious nodes}}{\text{total number of packets}} \qquad (9)$$

- Retransmission ratio: This is defined as the ratio between the number of retransmitted packets to the total number of generated packets; i.e

$$\text{retransmit ratio} = \frac{\text{number of retransmissions}}{\text{total packets}} \qquad (10)$$

Retransmitted packets include all possible causes, i.e. outsider drops or congestion drops due to voids or exceeding time out. However, if a decrease in this ratio shows up with an increase in β, this proves that most of these retransmissions are due to attacks. Moreover, this ratio indicates the ratio of power spent for packet retransmission to the total network consumed power. Thus, a decrease in this ratio will indicate a saving in power consumption.

- Coefficient of variation of node consumed power (COV): This metric is obtained by dividing the standard deviation of the consumed power per node by the average consumed power per node. A large value of this metric indicates that there is large variation around the mean value. This can be then viewed as a non balancing effect of energy consumption. Small values of this metric indicate that almost all nodes are consuming an amount of power that is around the mean value. This means that there is a better energy balancing among nodes. The metric is computed mathematically as:

$$\text{COV of consumed power} = \frac{\sigma(\text{consumed power})}{\mu(\text{consumed power})} \qquad (11)$$

where σ is the standard deviation and μ is the mean.

4.3.5. Simulation results and analysis

4.3.5.1. Delivery ratio

Figure 4 shows the delivery ratio versus β assuming a non forwarding attack. We simulate different scenarios of percentages of attackers from the total population of nodes. The maximum percentage of attackers is set to 70% as a very pessimistic case to see how GETAR would work with such extreme unacceptable scenarios. However, the practical cases of less percentages are also presented. For each scenario, we can notice that the delivery ratio increases as β increases until a knee point at which the delivery ratio remains almost unchanged. This agrees with the expectation that higher values of β will make GETAR more trust aware and, hence, the developed routes will include fewer attackers. At around β=0.4, all curves saturate at their corresponding maximum possible delivery ratio. This is an interesting result as it indicates that the effect of β is fully utilized for the trust awareness issues at 0.4. This means that increasing β beyond that value is not efficient in terms of trust-awareness. Moreover, as β increases, it will mask the GEAR part of the cost function. Thus, the minimum β that guarantees the maximum achievable delivery ratio is the best choice from the perspective of trust awareness.

Another point to be noticed in this figure is that when β is equal to zero, the delivery ratio is very low (e.g. 0.34 with 10% attackers), while we should expect values around 0.9 since the attackers should drop 10% of the traffic. The reason of this low delivery ratio can be related to GETAR cost function propagation. When a node selects a malicious node as a router, it may get stuck with this router for several transactions before it switches to another router based on energy and distance information. As a result, such low delivery ratio is expected.

The figure also shows the effect of the percentage of the malicious nodes (attackers) in the network on the delivery ratio. As expected, the more the attacker percentage, the less the delivery ratio is. Moreover, the improvement of the delivery ratio by increasing the value of β becomes more significant as attacker percentage increases. For example, with 10% attackers, the ratio increases from 0.34 at β=0 to 0.85 at β=0.4, whereas it improves from 0.1 at β =0 to 0.3 at β=0.4 with 70% attackers. Thus, with 70% attackers, one may decide to keep β<0.4 to give a preference for normal GEAR operation since the delivery ratio is not improving significantly.

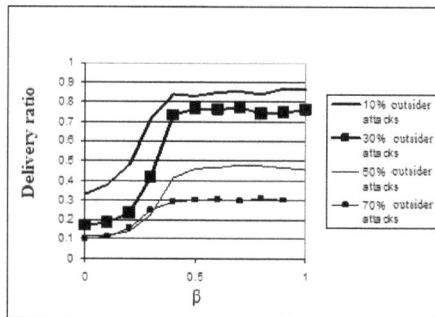

Figure 4. Comparison of the delivery ratio for different attackers' percentage

4.3.5.2. Outsider attacks' drop ratio

Figure 5 provides the relationship between the drop ratio and β parameter. For each scenario of attack percentage, the drop ratio decreases as β increases. The same analysis provided for figure 4 is also valid here.

If we compare this figure with figure 4, we can notice that they almost complement each other. This would be very true if we consider the total drops in the drop ratio to include, in addition to the attack related drops, other drops due to network congestion. However, in our simulation, we are interested only in the attack-related drops. Since this figure is almost complementing figure 4, it is very evident that most of the drops are due to attacks.

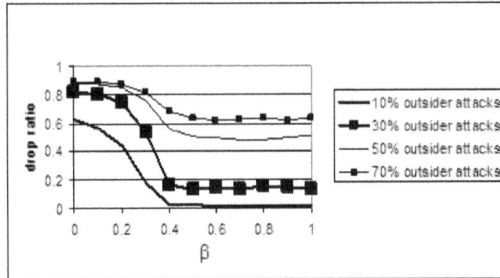

Figure 5. Drop ratio for different attackers percentages

4.3.5.3. Retransmission ratio

The retransmission ratio accounts for two types of retransmitted packets, i.e. packets dropped due to attacks and packets that are not delivered due to path congestion. In figure 6, we can notice two different behaviors of the curves in two regions separated by certain values of β <0.5 for different scenarios. In the first regions for β <0.5, we notice that as β increases, the retransmission ratio increases. This is because when β gets higher values, more packets will suffer longer delays to avoid malicious nodes. Thus, retransmission due to congestion will increase. Also, as we are still below β=0.4, the drops due to attacks are still significant according to figure 5. As a result, an increase in β will cause more retransmissions.

Once we exceed a certain value of β, like 0.4 in case of 30% attackers, most of the packets will have the same routes with the same delays and, as a result, the retransmissions due to congestion will remain almost constant. However, since the drop ratio is decreased dramatically as has been discussed in figure 5, the retransmission ratio will now be affected only by the drop ratio. Thus, the retransmission ratio decreases, also dramatically.

An increase in retransmission ratio gives an indication of the wasted power. That is, the more the retransmission ratio is, the more power is wasted. Thus, an important objective here is to reduce the retransmission ratio as much as possible. However, this fact is very much affected by the percentage of attackers and routing metric preference. For example, assume we have a 10% attackers scenario. It is very obvious that the best choice of β is 0.4 where we have 0 retransmissions or, equivalently, 0 wasted power. However, with 70% attackers, the minimum "wasted power" can be achieved with 0.123 retransmission ratio in two different regions at β <0.3 and β>0.4. In such a situation, if we are more concerned about the energy as a routing metric, it is better to choose β= 0.2 or 0.1. However, if the preference is given for trust awareness, β should be 0.5.

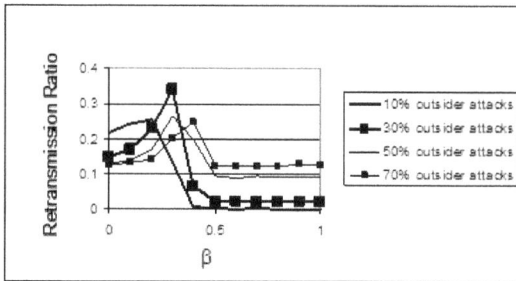

Figure 6. Comparison of the retransmission ratio for different percentages of attackers

4.3.5.4. Coefficient of variation of node consumed power

The importance of the consumed power coefficient of variation metric in figure 7 is to show the impact of trust aware routing decision on energy balancing proposed by normal GEAR. We can see that as β gets higher values, the consumed power coefficient of variation increases until a knee point like $\beta=0.5$ in the case of 10% attackers. After that, this metric remains almost unchanged.

Before the knee point, an increase in β will cause the routing decision to select a trusted node with less consideration for energy. This is because high value of β will mask the GEAR part of the cost function. As a result, trusted nodes that are in the proximity of attackers will suffer heavy routing duties whereas other nodes will be balanced with their neighbors. As a result, we will have a larger variation of power consumption as β increases. However, after the knee point, the increase of β will have the same masking effect on the GEAR part of the cost function. Thus, the routing decisions will not change as well.

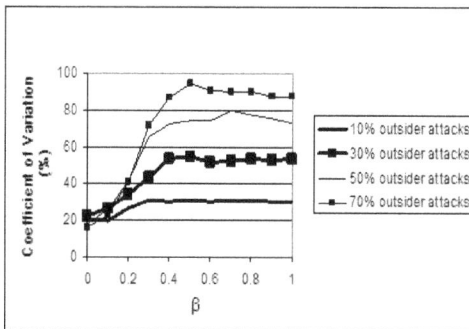

Figure 7. Comparison of the coefficient of variation for different percentages of attackers

This section proposed an enhanced trust aware routing protocol, GETAR, for WSN. The suggested protocol promises to provide trust awareness as well as energy efficiency as it is based on an enhancement of GEAR protocol. This way, GETAR abides by the constrained energy usage in WSN while providing its security service.

5. Comparison with previous work and main contributions

5.1. Routing approach

We provided a simulation based performance analysis for the efficiency of our proposed GETAR routing protocol. Simulation results proved the following points : GETAR improves the delivery ratio, decreases the drop and retransmission ratio and saves the retransmission power when compared with the previous work. The improvement in a performance metric can be achieved at different values of β parameter starting at a minimum value of β at a knee point in the curve. This value can be an optimum choice that guarantees best delivery ratio and better energy balancing. Energy balancing is negatively impacted by an increase in trust awareness. Thus, trade off considerations should be taken carefully in order to design an appropriate value of β. This will be subjected to the application preference between security and energy.

In SAR[70], The routing operation needs to encounter a trusted route setup phase, which contributes some initial delay, especially with the crypto-based authentication required at the route setup. The trust metric used in SAR does not reflect nodes' behavior; rather, they represent a "rank" that a node exhibits based on its identity and various security service provision. Thus, a trusted node in SAR is a node that has the appropriate rank that meets the routing requirements. To rank a node is a problem by itself and requires crypto mechanisms. Our protocol, GETAR, is much simpler in that it assigns trust values to nodes based on nodes behavior. The routing decision rules in SAR are governed by the source, which makes the protocol less flexible. The routing decision is not to select the next hop but to decide to participate in the trusted route. As a result, selfish behavior is not addressed well in SAR. WSN constraints of power consumption are not treated. In fact, SAR targets ad hoc networks with an assumption of more relaxed conditions as compared to WSN.

In TRANS[72], the trust, in fact, is associated with locations rather than nodes. The problem is that a location can be infected by a single node. The detour, then, will be around a larger area rather than a single node. "Innocent" nodes located in proximity of an infected location might be also isolated. If not, they are also exposed to heavy routing duties that may induce selfish behavior. TRANS is limited by single or multiple sink communication models. This assumption is necessary for the efficient operation of the protocol. Our proposed protocol, however, is more generic and can be applied to TRANS model or even for peer-to-peer model. TRANS discusses approaches to decrease energy consumption due to the security provision overhead. However, the protocol does not provide energy efficient techniques in the routing operation itself since it relies on GPSR.

The RGR[73] protocol has no provision for energy efficiency as it relies on GPSR. The protocol totally relies on trust-based forwarding. If a node is completely surrounded by misbehaving nodes, there is no other mechanism proposed to select a next hop since all nodes will be eliminated from the node's forwarding list. RGR is a multi-path trust-based routing. Although multi-path is important for reliable services, we believe that it can be energy consuming which we try to conserve in our work using GETAR.

In RFSN[2], the monitoring mechanism uses a normal watchdog mechanism that assumes a promiscuous mode operation for every node. This is not suitable for the WSN conditions in terms of energy scarcity as discussed earlier. The system does not show a practical solution implementation of monitoring and rating phases. From an implementation point of view, the study should provide an example of how monitoring and rating will be done under some application assumptions. The work does not propose a response methodology, for example, a routing algorithm. Instead, it leaves it as an open issue. Therefore, the work lacks performance figures that can show the efficiency and security gain and benefits in routing operation that can be obtained in adopting this solution.

Main Contributions of our work are the following :

- *Energy awareness*: Our protocol relies on an enhanced operation of GEAR which has energy awareness, whereas RGR, TRANS and SAR do not.
- *Identity-independent trust*: As opposed to SAR and TRANS, our trust metric is behavior-dependent and not identity-dependent. Thus, to obtain trust metrics, we do not require a crypto system to validate nodes identities.
- *Source-sink-Independent routing decision*: In our protocol, routing decision is performed completely based on the individual node vision of the vicinity conditions, whereas in TRANS and SAR, the routing decision is governed by the sink or source requirements.
- *Applicability to different communication models*: Our proposed protocol can be applied to any communication model and architecture

5.2. Rating approach

The rating component of a reputation system deals with combining the first-hand and second-hand information meaningfully into a representative value. Moreover, it is responsible for updating such values as the behavior of nodes are evolving.

In literature, some rating approaches use a single value, called reputation, like CORE [5] and DRBTS [74]. This is similar to our approach in CRATER where we use a single value called the risk value, $r_{i,j}$. Other rating systems like RFSN[2] and CONFIDANT[3] use two separate values, to represent the node reputation.

Some rating approaches updates the node reputation using both first-hand and secondhand information. In CRATER, we use this approach and we also introduce the neutral behavior period as another rating factor. Some other approaches like OCEAN (Observation-based Cooperation Enforcement in Ad Hoc Networks) [75] use just FHI.

In CRATER, SHI is accepted based on the cautious assumptions and the collected SHI by a node i about a node j is averaged to calculate a single $r_{i,j,SHI}$. No validation check or honesty consideration is performed. However, some rating methods use validity and credibility tests for the gathered SHI. One method is to use a deviation test proposed in [64, 74].

Rating functions and mathematical modeling vary depending on the target applications. However, Beta distribution has been the most popular among researchers in reputation and trust-based systems. It was first introduced in the field by Josang and Ismail [65]. Since then, many researchers have used the beta distribution including Ganeriwal and Srivastava [2] and Buchegger and Boudec [64]. In CRATER, however, we are using a simpler approach similar to the exponential average weighting. This is similar to the approach proposed in DRBTS [74].

When the weighing approach is used, an important issue in maintaining and updating reputation is how past and current information is weighted. For example, CORE tends to give more weight to the past observations assuming that a current observation should have a lower impact on a "greatly built history". On the other hand, RFSN tends to give more weight to recent observations based on the issue of aging. Aging means that we give higher weights to recent observations such that if you behave well you will survive more. As a result, malicious node will be enforced to reduce their attack to survive. In CRATER, we adopt the aging approach with some detailed modifications.

Up to our knowledge, there is no simple and global technique that can independently and efficiently evaluate reputation systems or rating components in the context of WSN and ad hoc networks as compared to our proposed technique, RESISTOR. However, the work in [61] proposes an attempt on comparing reputation systems quantitatively based on game theory. The authors, thus, identify different notions of reputation systems like, contextualization, personalization, individual and group reputation, and, direct and indirect reputation. But, it is more complicated than RESISTOR. Moreover, RESISTOR can be used as an indicator to understand the flaws or plus points in the rating system.

6. Conclusion

The problem of secure routing in WSN is an important area of research that has various aspects of considerations. One important direction under this area is to provide security mechanisms against behavioral related attacks. In this chapter, a comprehensive treatment of the Reputation system based Trust-Enabled Routing framework for wireless sensor networks is provided. We have highlighted the importance of Trust-Aware Routing considering the different network aspects and special conditions of WSN. We have provided a comprehensive review and an in-depth discussion of different Reputation system based Trust-Aware routing approaches highlighting their pros and cons. In our proposed work, we investigated reputation based systems as a promising solution for behavioral related routing security problems. The work developed a new reputation system called SNARE (Sensor Node Attached Reputation Evaluator)[82] that is designed to meet WSN conditions and constraints. This system is divided into three components; i.e. monitoring, rating and response components. Each component is designed with the features that make it possible to apply and then optimize for WSN applications and conditions. In the response component, an enhanced trust aware routing protocol was proposed, called GETAR(Geographic, Energy & Trust Aware Routing). Simulation results showed that this enhanced protocol performs well in terms of increasing packet delivery ratio with tradeoffs in terms of energy balancing. Energy balancing raises an issue of optimization, as well.

As future work, some of the interesting issues to be analyzed to build a robust reputation system are - accurate and efficient trust/reputation modeling and management specific to Wireless Sensor Networks environments, performance of the examined cooperation enforcement with respect to network throughput realized, communication overhead introduced, time required for obtaining accurate reputation ratings/detecting misbehaving nodes, robustness against spurious ratings under a common reference scenario.

Author details

A. R. Naseer
Principal and Professor of Computer Science & Engineering, Jyothishmathi Institute of Technology & Science (JITS), Affiliated to Jawarharlal Nehru Technological University(JNTU) Hyderabad, India

7. References

[1] A. R. Naseer, I.K. Maarouf , and M. Ashraf, "Routing Security in Wireless Sensor Networks", Book Chapter published in Handbook of Research on Wireless Security, Publisher: Idea Group Reference, USA, 2008.

[2] S. Ganeriwal and M. Srivastava, "Reputation-based framework for high integrity sensor networks", Proceedings of the 2nd ACM workshop on Security of ad hoc and sensor networks, Washington DC, USA, pp. 66-77, October 2004.

[3] S. Buchegger and J.-Y. Le Boudec, "Performance Analysis of the CONFIDANT Protocol: Cooperation of Nodes — Fairness In Dynamic Ad-hoc Networks", Proc. IEEE/ACM Symp. Mobile Ad Hoc Net. and Comp., Lausanne, Switzerland, June 2002.

[4] Q. He, D. Wu, and P. Khosla, "SORI: A Secure and Objective Reputation-Based Incentive Scheme for Ad Hoc Networks", WCNC 2004, Atlanta, GA, Mar. 2004.

[5] P. Michiardi and R. Molva, "Core: A Collaborative Reputation Mechanism to Enforce Node Cooperation in Mobile Ad Hoc Networks", Commun. and Multimedia Security 2002 Conf. Portoroz, Slovenia, Sept. 26–27 2002.

[6] S. Marti et al., "Mitigating Routing Misbehavior in Mobile Ad Hoc Networks", Proc. MOBICOM 2000, 2000, pp. 255–65.

[7] Buchegger, S.; Le Boudee, J.-Y, "Self-policing mobile ad hoc networks by reputation systems", Communications Magazine, IEEE Volume 43, Issue 7, July 2005 Page(s):101 – 107.

[8] B. Karp and H. T. Kung. "GPSR: Greedy Perimeter Stateless Routing for Wireless Networks", Mobicom 2000.

[9] Y. Yu, R. Govindan, and D. Estrin, "Geographical and energy aware routing: A recursive data dissemination protocol for wireless sensor networks", Technical Report UCLA/CSD-TR-01-0023, May 2001.

[10] Yang Xio, et al. "Security and Routing in wireless Networks", Chapter 4. Nova Science Publishers, Inc. 2005.

[11] Djenouri, D.; Khelladi, L.; Badache, A.N. "A survey of security issues in mobile ad hoc and sensor networks", Communications Surveys & Tutorials, IEEE.Volume 7, Issue 4, Fourth Quarter 2005 Page(s):2 – 28.

[12] I.F. Akyildiz, W. Su, Y. Sankarasubramaniam, E. Cayirci,"Wireless sensor networks: a survey", Computer Networks 38 (2002) 393–422

[13] A. Cerpa, J. Elson, M. Hamilton, J. Zhao, "Habitat monitoring: application driver for wireless communications technology", ACM SIGCOMM'2000, Costa Rica, April 2001.

[14] C. Jaikaeo, C. Srisathapornphat, C. Shen, "Diagnosis of sensor networks", IEEE International Conference on Communications ICC'01, Helsinki, Finland, June 2001.

[15] B. Warneke, B. Liebowitz, K.S.J.Pister, "Smart dust: communicating with a cubic millimeter computer",IEEE Computer (January2001) 2–9.

[16] http://www.fao.org/sd/EIdirect/EIre0074.htm

[17] J.M. Kahn, R.H. Katz, K.S.J. Pister, "Next century challenges: mobile networking for smart dust", Proceedings of the ACM MobiCom'99, Washington, USA, 1999, pp. 271–278.

[18] N. Noury, T. Herve, V. Rialle, G. Virone, E. Mercier, G. Morey, A. Moro, T. Porcheron, "Monitoring behavior in home using a smart fall sensor", IEEE-EMBS Special Topic Conference on Microtechnologies in Medicine and Biology, October 2000, pp. 607–610.

[19] D. Estrin, R. Govindan, J. Heidemann, "Embedding the Internet", Communication ACM 43 (2000) 38–41.

[20] N. Priyantha, A. Chakraborty, H. Balakrishnan, "The cricket location-support system", Proceedings of ACM MobiCom'00, August 2000, pp. 32–43.

[21] Karl, Holger, Willig, Andreas, "Protocols and architectures for wireless sensor networks", Wiley, c2006.

[22] Slijepcevic, S.; Potkonjak, M.; Tsiatsis, V.; Zimbeck, S.; Srivastava, M.B."On communication security in wireless ad-hoc sensor networks. Enabling Technologies: Infrastructure for Collaborative Enterprises", WET ICE 2002, Proceedings. Eleventh IEEE International Workshops on 10-12 June 2002 Page(s):139 – 144.

[23] Krishnamachari, Bhaskar, "networking wireless sensors", Cambridge University Press, 2005.

[24] A. Perrig, R. Szewczyk, V. Wen, D. Culler, and D. Tygar, " SPINS: Security Protocols for Sensor Networks", Proceedings of Seventh Annual International Conference on Mobile Computing and Networks MOBICOM 2001.

[25] G.J. Pottie, W.J. Kaiser, "Wireless integrated network sensors", Communications of the ACM 43 (5) (2000) 551–558.

[26] J.M. Rabaey, M.J. Ammer, J.L. da Silva Jr., D. Patel, S. Roundy, "PicoRadio supports ad hoc ultra-low power wireless networking", IEEE Computer Magazine (2000) 42–48.

[27] Kemal Akkaya, Mohamed Younis, "A Survey on Routing Protocols for Wireless Sensor Networks", Department of Computer Science and Electrical Engineering. University of Maryland.

[28] W. Heinzelman, J. Kulik, and H. Balakrishnan, "Adaptive protocols for information dissemination in wireless sensor networks", Proceedings of the 5th Annual ACM/IEEE

International Conference on Mobile Computing and Networking (MobiCom'99), Seattle, WA, August 1999.

[29] C. Intanagonwiwat, R. Govindan and D. Estrin, "Directed diffusion: A scalable and robust communication paradigm for sensor networks", Proceedings of the 6th Annual ACM/IEEE International Conference on Mobile Computing and Networking (MobiCom'00), Boston, MA, August 2000.

[30] D. Braginsky and D. Estrin, "Rumor Routing Algorithm for Sensor Networks", Proceedings of the First Workshop on Sensor Networks and Applications (WSNA), Atlanta, GA, October 2002.

[31] W. Heinzelman, A. Chandrakasan, and H. Balakrishnan, "Energy-efficient communication protocol for wireless sensor networks", in the Proceeding of the Hawaii International Conference System Sciences, Hawaii, January 2000.

[32] A. Manjeshwar and D. P. Agrawal, "TEEN : A Protocol for Enhanced Efficiency in Wireless Sensor Networks", in the Proceedings of the 1st International Workshop on Parallel and Distributed Computing Issues in Wireless Networks and Mobile Computing, San Francisco, CA, April 2001.

[33] T. He et al., "SPEED: A stateless protocol for real-time communication in sensor networks", in the Proceedings of International Conference on Distributed Computing Systems, Providence, RI, May 2003.

[34] C. Perkins, "Ad Hoc Networks", Addison-Wesley, Reading, MA, 2000.

[35] Chris Karlof and David Wagner, "Secure routing in wireless sensor networks: Attacks and countermeasures", Elsevier's Ad Hoc Network Journal, Special Issue on Sensor Network Applications and Protocols, September 2003.

[36] T. Grandison and M. Sloman, "A survey of trust in internet applications," *IEEE Comm. Surveys & Tutorials*, vol. 3, no. 4, 2000.

[37] S. Marsh, "Formalising Trust as a Computational Concept," in *Departmet of Computer Science and Mathematics*, vol. PhD: University of Stirling, 1994, pp. 184.

[38] P. Resnick, R. Zeckhauser, E. Friedman, and K. Kuwabara, "Reputation systems: Facilitating trust in internet interactions," *Comm. of the ACM*, vol. 43, no. 12, pp. 45–48, 2000.

[39] Li H, Singhal M. A Secure routing protocol for wireless ad hoc networks. *39th Hawaii International Conference on system Sciences*, Kauai, 2006.

[40] Rezgui A, Eltoweissy M, TARP: a trust-aware routing protocol for sensor-actuator networks. *IEEE International Conference on Mobile Ad Hoc and Sensor Systems*, Pisa, Italy, 2007.

[41] Hur J, Lee Y, Yoon H, Choi D, Jin S. Trust evaluation model for wireless sensor networks. *Advanced Communication Technology Conference*, Phoenix Park, Korea, 2005; 491–496.

[42] Crosby GV, Pissinou N. Cluster-based reputation and trust for wireless sensor networks. *Consumer Communications and Networking Conference*, Las Vegas, NV, USA, 2007.

[43] Lewis N, Foukia N., Using trust for key distribution and route selection in wireless sensor networks. *IEEE Globecom*, Washington DC, USA, 2007.

[44] Jing, Q.; Tang, L.Y.; Chen, Z. Trust Management in Wireless Sensor Networks. *J. Softw.* **2008**, *19*, 1716-1730.

[45] Hur J, Lee Y, Yoon H, Choi D, Jin S. Trust evaluation model for wireless sensor networks. *Advanced Communication Technology Conference*, Phoenix Park, Korea, 2005; 491–496.

[46] G. Theodorakopoulos and J. S. Baras, "On trust models and trust evaluation metrics for ad hoc networks," *IEEE J. on Selected Areas in Comm.*, vol. 24, no. 2, pp. 318–328, Feb. 2006.

[47] Marmol, F.G.; Perez, G.M. Towards Pre-standardization of Trust and Reputation Models for Distributed and Heterogeneous Systems. *Comput. Stand. Interfaces* **2010**, *32*, 185-196.

[48] G. Shafer, "A mathematical theory of evidence,", Princeton University, 1976.

[49] M. Momani, S. Challa, and K. Aboura, "Modelling Trust in Wireless Sensor Networks from the Sensor Reliability Prospective," presented at International Joint Conferences on Computer, Information, and Systems Sciences, and Engineering (CIS2E 06) University of Bridgeport, USA, 2006.

[50] M. Momani, K. Aboura and S. Challa, "RBATMWSN: Recursive Bayesian Approach to Trust Management in Wireless Sensor Networks", in *The Third International Conference on Intelligent Sensors, Sensor Networks and Information*, Melbourne, Australia, 2007.

[51] P. A. Morris, "Bayesian expert resolution," in *Department of Engineering-Economic Systems*, vol.Ph.D: Stanford University, 1971.

[52] D. V. Lindley and N. D. Singpurwalla, "Reliability (and fault tree) analysis using expert opinions. ," *Journal of the American Statistical Association*, vol. 81, pp. 87-90, 1986.

[53] Feng, Renjian; Xu, Xiaofeng; Zhou, Xiang; Wan, Jiangwen. 2011. "A Trust Evaluation Algorithm for Wireless Sensor Networks Based on Node Behaviors and D-S Evidence Theory." *Sensors* 11, no. 2: 1345-1360, 2011

[54] Dempster, A. Upper and Lower Probabilities Induced by Multivalued Mapping. *Ann. Math. Stat.* **1967**, *38*, 325-339.

[55] Lopez, J.; Roman, R.; Agudo, I.; Fernandez, C.G. Trust Management Systems for Wireless Sensor Networks: Best Practices. *Comput. Commun.* **2010**, *33*, 1086-1093.

[56] Li, J.L.; Gu, L.Z.; Yang, Y.X. A New Trust Management Model for P2P Networks with Time Self-Decay and Subjective Expect. *J. Electron. Inf. Technol.* **2009**, *31*, 2786-2790.

[57] Li, L.; Fan, L.; Hui, H. Behavior-Driven Role-Based Trust Management. *J. Softw.* **2009**, *20*, 2298-2306

[58] Lewis N, Foukia N., Using trust for key distribution and route selection in wireless sensor networks. *IEEE Globecom*, Washington DC, USA, 2007.

[59] Atakli IM, Hu H, Chen Y, Ku WS, Su Z. Malicious node detection in wireless sensor networks using weighted trust evaluation. *Spring Simulation Multiconference*, Ottawa, Canada, 2008.

[60] http://en.wikipedia.org/wiki/Reputation_systems

[61] L. Mui, A. Halberstadt, and M. Mohtashemi, "Notions of Reputation in Multi-Agents Systems: A Review", *Proc. First Int'l Joint Conf. Autonomous Agents and Multi-Agent Systems*, pp. 280-287, July 2002.

[62] http://www.ebay.com

[63] http://www.epinions.com

[64] S. Buchegger and J.-Y. L. Boudec, "A Robust Reputation System for Mobile Ad-hoc Networks", EPFL IC, Tech. Rep. IC/2003/50, July 2003.

[65] Audun Josang, Roslan Ismail, "The Beta Reputation System", 15th Bled Electronic Commerce Conference, e-Reality: Constructing the e-Economy. Bled, Slovenia, June 2002.

[66] Deng, R. Han, and S. Mishra, "Insens: Intrusion-tolerant Routing in Wireless Sensor Networks", 23rd IEEE Int'l. Conf. Distributed Comp. Sys. (ICDCS 2003), May 2003.

[67] Oniz, C.C. Tasci, S.E. Savas, E. Ercetin, O. Levi, A, "SeFER: secure, flexible and efficient routing protocol for distributed sensor networks", Proceedings of the Second European, Workshop on Wireless Sensor Networks, 2005. Publication Date: 31 Jan.-2 Feb. 2005

[68] S. Buchegger and J.-Y. Le Boudec. A Robust Reputation System for Peer-to-Peer and Mobile Ad-hoc Networks. *Proceedings of P2PEcon 2004*, Harvard University, Cambridge MA, U.S.A., June 2004.

[69] S. Buchegger, C. Tissieres and J.-Y. Le Boudec. A Test-Bed for Misbehavior Detection in Mobile Ad-hoc Networks - How Much Can Watchdogs Really Do?. *Proceedings of IEEE WMCSA 2004*, English Lake District, UK, December 2004.

[70] S. Yi, R Naldurg, and R. Kravets, "Security-aware Ad-hoc Routing for Wireless Networks", ACM Wksp. Mobile Ad Hoc Networks, Mobihoc, 2001.

[71] S. Buchegger and J.-Y. Le Boudec, "A Robust Reputation System for Peer-to-Peer and Mobile Ad Hoc Networks", P2PEcon, Harvard Univ., Cambridge, MA, June 2004.

[72] S. Tanachaiwiwat, P. Dave, R. Bhindwale, and A. Helmy, "Location-centric isolation of misbehavior and trust routing in energy-constrained sensor networks", April 2004.

[73] Nael AbuGhazaleh, Kyoung Don Kang and Ke Liu. "Towards Resilient Geographic Routing in WSNs", MSWiM'05, October 10–13, 2005.

[74] A. Srinivasan, J. Teitelbaum and J. Wu, "DRBTS: Distributed Reputation based Beacon Trust System", 2nd IEEE International Symposium on Dependable, Autonomic and Secure Computing (DASC'06), Indianapolis, USA, pp. 277–283, 2006.

[75] S. Bansal and M. Baker, "Observation-based Cooperation Enforcement in Ad Hoc Networks", http://arxiv.org/pdf/cs.NI/0307012, July 2003.

[76] M. Krasniewski, P. Varadharajan, B. Rabeler, S. Bagchi, and Y. Hu, "TIBFIT: Trust Index Based Fault Tolerance for Arbitrary Data Faults in Sensor Networks", Proceedings of the International Conference on Dependable Systems and Networks (DSN'05), (Yokohama, Japan), June 2005.

[77] Z. Yao, D. Kim, and Y. Doh. PLUS: Parameterized and localized trust management scheme for sensor networks security. In *Proc. of the 3rd IEEE Int. Conf. on Mobile Ad-hoc and Sensor Systems*, pages 437–446, Vancouver, Canada, Oct. 2006.

[78] Hu, J., Burmester, M., 2006. "LARS: a locally aware reputation system for mobile ad-hoc networks", in *44th annual ACM Southeast Regional Conference, 2006*.

[79] G. Zhan, W. Shi, and J. Deng, "Tarf: A trust-aware routing framework for wireless sensor networks," in *Proceeding of the 7th European Conference on Wireless Sensor Networks (EWSN'10)*, 2010.

[80] Zhan, G., Shi, W., Deng, J., "Sensortrust - a Resilient trust model for WSNs", SenSys 2009, Proceedings of the 7th International Conference on Embedded Networked Sensor Systems (2009)

[81] M. Momani and S. Challa, "GTRSSN: Gaussian Trust and Reputation System for Sensor Networks", in *International Joint Conferences on Computer, Information, and Systems Sciences, and Engineering (CISSE '07)*, University of Bridgeport 2007.

[82] I. K. Maarouf and A. R. Naseer, "SNARE : Sensor Node Attached Reputation Evaluator", in Proceedings of IEEE/ACM 2nd International CONEXT conference, Dec. 4-7, 2006, Lisboa, Portugal.

[83] I. K. Maarouf and A. R. Naseer, "WSNodeRater: An optimized Reputation System Framework for Security Aware Energy Efficient Geographic Routing in WSNs", in Proceedings of ACS/IEEE International Conference on Computer Systems and Applications, AICCSA '2007, May 13-16, 2007 Amman, Jordan.

[84] A. R. Naseer, I.K. Maarouf, U. Baroudi, , "Efficient Monitoring Approach for Reputation System based Trust-aware Routing in Wireless Sensor Networks", International Journal of IET Communications – Wireless Adhoc Networks, May 2009, Volume 3, Issue 5, pp. 846-858, ISSN 1751- 8628

[85] I.K. Maarouf, U. Baroudi, A. R. Naseer, "Cautious Rating for Trust-enabled Routing in Wireless Sensor Networks", EURASIP International Journal on Wireless Communications and Networking, 2010, Volume 2, Article ID 718318, 16 pages, ISSN: 1687-1472.

Chapter 11

Asymmetric Encryption in Wireless Sensor Networks

Gustavo S. Quirino, Admilson R. L. Ribeiro and Edward David Moreno

Additional information is available at the end of the chapter

1. Introduction

A Wireless Sensor Network (WSN) is composed of autonomous devices called sensor nodes that generally have low computational power, limited data transmission and power constraints. A WSN consists of sensor nodes that capturing information from an environment, processing data and transmitting them via radio signals. WSNs are increasingly present in our days and can be found in environmental area (climatic measurements, presence of smoke), in health area (measurement of vital signs, temperature), home automation (motion sensor and image sensor) and other areas. Generally, WSNs have no fixed structure, and in many cases there is no monitoring station of sensor nodes during the operational life of the network, so a WSN must have mechanisms for self-configuration and adaptation in case of failure, inclusion or exclusion of a sensor node.

Security requirements of WSNs are similar to conventional computer networks, therefore parameters such as confidentiality, integrity, availability and authenticity must be taken into account in creation of a network environment. Due to limitations of WSNs, not all security solutions designed for conventional computer networks can be implemented directly in WSN. For a long time, it was believed that the public key cryptography was not suitable for WSNs because it was required high processing power, but through studies of encryption algorithms based on curves was verified the feasibility of that technique in WSN.

The cryptographic algorithm RSA is currently the most used among the asymmetric algorithms, working from the difficulty of factoring large prime numbers. Standardized by NIST[1], this algorithm is widely used in transactions on the Internet. The algorithms Elliptic / Hyperelliptic Curve Cryptography (ECC / HECC) were created in 80s, and are based on the difficulty of solving the discrete logarithm problem on elliptic curves and hyperelliptic respectively. Despite its complexity the algorithm based on elliptic and hyperelliptic curves have been extensively studied in academia. Recently, the public key algorithm called

---

[1] U.S. Agency for technology that has a partnership with industry to develop and apply technology, measurements and standards. Further information: www.nist.gov

Multivariate Quadratic Almost Group (MQQ) was proposed in academia. Experiments performed in the FPGA and PC platforms showed that MQQ is faster than algorithms such as RSA and ECC [1, 2]. Algorithms involved in this study are asymmetric, but each one works with a specific encryption mode.

Many studies have evaluated performance of cryptographic algorithms in WSNs, but there is no standardization in the performance analysis. As stated by Margi [3] studies on performance evaluation of cryptographic algorithms for WSNs are often quite different in terms of methodology, platform, metrics and focus of analysis, what difficult a direct comparison among the obtained results. Thus, this chapter describes a theoretical study of cryptographic such as RSA, ECC, HECC and MQQ as well as the performance analysis of these algorithms in WSN.

2. Wireless sensor network

Sensor nodes are electronic devices that have as main components units of storage, processing, sensing and transmission. Usually, these devices have low computational power, nevertheless play an important role in ubiquitous computing, because they have function of collecting data in a given environment, passing them through a wireless network. According to [4] WSNs can be seen as a special type of MANET (Mobile Ad hoc NETwork) that tend to run a collaborative basis where the elements (sensor nodes) provide data that are processed (or consumed) by special nodes called sink nodes.

The operation area of a WSN is very large and can be used in environmental monitoring, control temperature and humidity, vehicle traffic control, monitoring of human body organs, among others. Figure 1 illustrates a scenario of WSNs in the medical area where patients that are being monitored can be in a hospital, at home, or anywhere else performing an activity routine. Sensing data are sent to health professionals through the Internet.

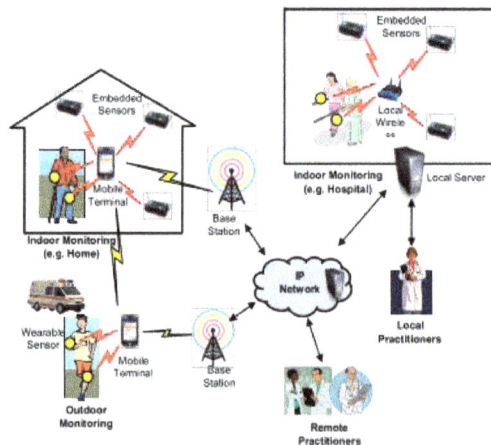

Figure 1. Scenario of Wireless Sensor Network [5]

Some application areas of WSNs require security in the information transport, such as the scenario illustrated by figure 1, where sensor nodes implanted in the human body reporting to a hospital. According to [5] in the health field, authentication and access control are the main challenges of a mobile dynamic network topology with limited resources. Besides medical area, several other areas also need security in their transmissions as: industry, asset security and military applications. A U.S. security agency called DARPA[2] has been developed numerous studies involving security in WSNs for military purposes.

2.1. Sensor devices

Sensor devices are basically formed by a computational part responsible for storing and transmitting data, and a sensing portion which can be formed by one or more sensors, such as acoustic, seismic, infrared video camera, heat temperature and pressure [4]. In general two modulation formats are available: Frequency-Shift-Keyed (FSK) operating at 433 and 868-915 MHz and direct sequence spread spectrum (DSSS) operating at 2.4 GHz band that transmit 802.15.4 and ZigBee standards. The reach of the radios varies from 10 to 100 meters. The antenna configuration can cause transmission rates from 19.2kbps to 240kbps [6]. Currently, the sensor nodes can vary between mode activity, inactivity (idle) and low consumption (sleep) in order to save energy. The energy issue is important because most of sensor nodes are powered by batteries. Nowadays, the main sensor nodes available are LOTUS, IRIS, MICAz, Mica2, and TELOSB CRICKET.

Figure 2 illustrates the actual format of sensor nodes such as IRIS TelosB and MicaZ. During the development of this work were not found Brazilian companies that commercialize sensor nodes. A budget held in the Chinese company Mensic[3] in jan/2012 showed that a Micaz cost U$ 114.00 and a TelosB U$ 160.00, excluding import duties also should be minimum purchases of U$ 1000.00.

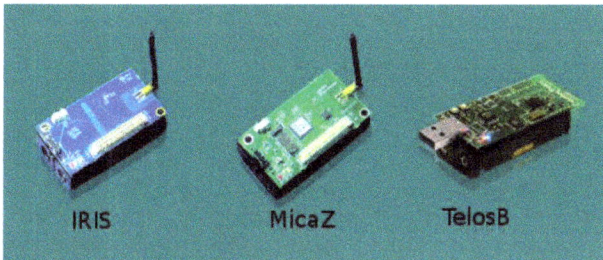

Figure 2. Sensors [6]

2.2. Application environments

WSNs can be applied in various areas. According to Loureiro [4], WSNs can be used in following situations:

- **Environment -** Monitoring of environmental variables such as buildings, residences and external locations such as oceans, volcanoes, deserts, etc..
- **Traffic -** Monitoring of vehicle traffic on highways, railroads, rivers, oceans, etc..

---

[2] Advanced Research Projects Agency. More information at: http://www.darpa.mil
[3] Chinese company specializing in electronic devices. Further information: http://www.memsic.com/

- **Security** - To provide security in homes, shopping centers, farms, among others.
- **Military** - To detect the presence of enemies, explosions, presence of hazardous materials as poison gas and radiation.

2.3. Security vulnerabilities

In most of applications, sensor devices are spread over large areas, what difficult a individual control of network components. Moreover, wireless communication allows an attacker can trigger attacks without having physical access to the device, so according to Shi and Perrig [7] attacks on WSNs can be divided into three main types: (1) Attack of authentication and confidentiality: Consists of attacks change, repetition or modification packages. (2) Availability network Attack: Generally known as DoS attacks or negation of service, this attack involves the application of techniques that make the network unavailable. (3)Attack on integrity: this type of attack the attacker's goal is to inject false data on the network, keeping the network available, but traveling fictitious data. Table 1 described by Wang [8] illustrates the most common types of attacks in WSN considering the network layer in which they operate.

Layer	Types of Attacks
Physical Layer	*jamming* ou ataque de interferï£¡ncia
Link Layer	*collision, exhaustion, unfairness*
Network Layer	*spoofed routing information and selective forwarding, sinkhole, sybil, wormhole, Hello flood, Ack Flooding,*
Transport Layer	*Flooding De-synchronization*
Application Layer	*Malicious Node*

Table 1. Possible attacks on a Wireless Sensor Network [8]

At the physical layer can occur the following attacks: jamming and tampering. The attack jamming consists in the interference of radio frequency signal that sensor nodes use to communicate. The tampering attack occurs due to physical vulnerability of sensor nodes spread over large areas, therefore susceptible to capture, breaking the circuit, setting modification or even replacement of a network node by a malicious sensor node [9]. At link layer attacks can be of the collision, when two sensor nodes attempt to transmit while at the same frequency, in this case the packet is discarded and must be retransmitted [10]. The attacker may cause intentional collisions by a malicious sensor node. Repeated collisions can lead to exhaustion of resources, making it unavailable sensor nodes. Also in the link layer unfairness attack is a type of DoS when the adversary causes degradation of real-time applications run on other sensor nodes by intermittent interruption of the transmission of their frames.

Denial of Service (DoS) attacks consist of flooding the receiver with no other requests for communication can be performed during the attack, leaving the involved nodes unavailable for new connections.

In the network layer attacks can occur of type Spoofed Routing Information, where the attacker modifies routing table information. The routes make false packets do not reach the correct destination, or even make the referral to consume more resources than normal [11]. The Selective Forwarding attack is the involvement of a sensor node by an attacker who causes some messages to be routed and other discarded [11]. In the Sinkhole attack the attacker causes a compromised sensor node is seen as most efficient route to the sink of the network, thus the neighboring nodes will always use the attacker to send their data [12][11][10].

The Sybil attack happens when a malicious node takes over a network identity. According to Douceur [13] this attack was originally intended for distributed systems of redundant data storage, but it is also effective against routing algorithms, data aggregation, and resource allocation, among others. The Wormhole attack consists in a low latency link between two sensor nodes of a network through which an attacker generates messages with court order to exhaust the resources of the devices [11]. In the Hello Flood attack the attacker can use a high power transmitter to fool a large number of sensor nodes, making them believe they are close [11].

Subsequently the attacker sends a fake shortest path to base station, and all nodes receiving Hello packets, try to convey through the attacking node. However, these nodes are out of radio range of the malicious node. Some routing information algorithms use state of sensor nodes. The Acknowledgment Spoofing attack consists in spreading false information about the states of neighboring sensor nodes performed by a malicious sensor node in order to prevent packets from reaching their destinations [11].

In the transport layer, Flooding attack consists in the flood of requests to new connections in order to exhaust the resources of memory and prevent the closure of legitimate requirements of provisions. The De-synchronization attack refers to the interruption of an existing connection [10]. In this attack the attacker captures messages forcing the sender to resend them expending energy unnecessarily.

There are also attacks that exploit vulnerabilities in authentication and data confidentiality. The attack consists of setting replication of a malicious node assumes the identity of a network node. This false node can forward packets in corrupt or false routes. If the attacker has physical access to network, it can copy cryptographic keys and use them in false messages. Also the attacker can deploy the malicious node at strategic locations in order to divide the WSN.

Preserving privacy in data transmission in WSN is challenging, since this type of network allows remote access. Moreover, a single adversary can monitor multiple networks simultaneously [14]. Eavesdropping and passive monitoring are the most common and easiest attack to data privacy. In this type of attack the spy monitors the data transfer and can access its contents if no encryption mechanism implemented in the network being monitored. The traffic analysis is usually applied in conjunction with the attack of listening and passive monitoring. It consists of the preliminary analysis of network traffic to identify nodes that are generating data exchange that interest to the attacker. Finally, the camouflage attack, wherein the malicious attacker deploys a node in the network forwards packets to sensor nodes being monitored.

Through this analysis one can see that there is a range of attacks for WSNs in all layers of the TCP / IP protocol stack. Furthermore, it is apparent that a common point in most attacks is the exploitation of low computing capacity of sensor nodes, as are injected false data and routes are always altered in order to occupy the lower transmission capacity of the sensor nodes, or eliminate its reserve energy. Others attacks yet unidentified may occur in WSN, and protect the network from these threats can be a difficult task.

2.4. Defense mechanisms

Different types of WSN applications require different security requirements. In an environment of temperature monitoring, where researchers collect data for research, it may be that safety requirements are not very important, but the monitoring of radiation near a nuclear power plant requires authenticity assurance, confidentiality, availability and integrity. Various architectures have been developed to provide security in WSNs, among them are: SPINS, TinySec MiniSec and besides these the IEEE 802.15.4 include a security framework to meet the services of data integrity, confidentiality and authenticity [3].

SPINS (Security Protocols for Sensor Networks) developed by Perrig [15] consists of a set of security protocols that acts through encryption and message authentication codes. The TinySec was designed and implemented in the TinyOS operating system to be a mechanism for providing confidentiality, integrity and authenticity of the data link layer. It uses the CBC mode of operation that may be combined with various block ciphers as RC5 and skipjack [16]. The MiniSec is a protocol layer of security to WSN using OCB (Offset Codebook) mode for operating the block cipher, which eliminates the need of adding filler to the clear text blocks [17]. The standard IEEE 802.15.4 provides integrity, access control, confidentiality and replay protection in the link layer. The cryptographic algorithm used in this standard is AES [18].

According to Loureiro [4], a WSN tends to be autonomous and requires a high degree of cooperation to perform the tasks defined for the network. This means that traditional distributing algorithms, such as communication protocols and election of leader, should be reviewed for this type of environment before being used directly. Taking account also the limited computational power and especially of limited energy of devices is possible to deduce that not everything that works efficiently in traditional computer networks can be used in WSNs The computational limitations of a device restricting the choice of cryptographic algorithms and protocols safety. Furthermore, the lifetime of the batteries using techniques preclude the complex of security because it drastically decreases the life span of the network. [18].

Encryption is the security solution most applicable in computing. In recent years asymmetric algorithms have been extensively studied in embedded systems with low computational power. The next section discusses concepts of cryptography, and the description of the algorithms RSA, ECC, HECC and MQQ.

3. Concepts of cryptography

Data encryption emerged before the invention of computer. Diplomats, enthusiasts and mainly militaries contributed to the evolution of this art that consists in distort the information that is being transported, so that only the authorized recipient can decipher it. In this regard, a cryptographic algorithm can be set as a function that converts encrypted message in clear messages and vice versa, making use of a cryptographic key.

Most cryptographic algorithms are public, according to Tanembaum [19] keeping the algorithm public gets rid of the creator from eager cryptologist to decode the system in order to publish articles, and that after five years of their exposure and no decoding was successful, the algorithm is assumed to be solid. Secrecy is the key that has the function to parameterize the cryptographic function, ie only with the key can encrypt or decrypt a message. Another important factor is that the key have the ability to change the output of the algorithm, so every change of key cryptographic algorithm generates a new encrypted message. The key size is critical in a project, because the longer the key, more work will be crypto analyst to try to decipher the message. In general, keys have sizes of 64, 128 or 256 bits and may be higher or lower, according to security needs.

Currently, in addition to confidentiality, encryption also operates in the fields of integrity of authentication and is described below:

- **Confidentiality:** ensuring that only the sender and receiver have the ability to understand the message being exchanged.
- **Integrity:** Ability to check if a message was altered during transmission.
- **Authentication:** Medium to prove the identity of an individual communication.

According to Boyle and Newe [20] encryption is the standard method for defending a WSN of most possible attacks, and the various levels of encryption implicate variations in overhead in the form of growth in the size of the package data, code size, processor usage, memory, etc.. The choice of a cryptographic algorithm to get efficient for a WSN is a large debate among researchers. According to Chen [21] the cryptographic methods used in WSN should meet the constraints of computational devices, and go through evaluation before being implanted.

3.1. Classes of cryptographic algorithms

Traditionally users of encryption algorithms used simple, but currently the goal is to make the algorithm so complex that without the key is practically impossible to extract some information through a cryptanalysis. The classes of cryptographic algorithms say about it as an encryption key is changed and also the quantity of keys involved in the application of the method. Most existing cryptographic algorithms can be classified as symmetric or asymmetric.

3.1.1. Symmetric encryption

Symmetric encryption or secret key cryptography is the use of only a key, both in the encryption and decryption of data. By the year 1976 this was the only known method for the use of encryption, but to be effective you need a secure channel for communication in which a cryptographic key can be changed.

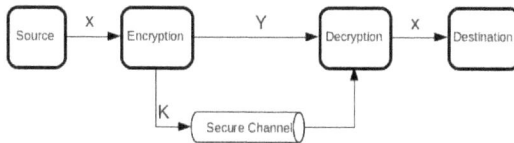

Figure 3. Symmetric Cryptography [22]

Figure 3 illustrates a communication through symmetric encryption. The text is encrypted X and Y become the message through the encryption algorithm and key k. The message Y is sent to the receiver, which uses the key k to decrypt it, turning it on again in the text X. Also according to figure 3 you can see that the key k is transported by a secure channel, for the possession of it, a potential attacker could easily make the reading the original text. AES and DES are two examples of algorithms that are part of the class symmetrical.

3.1.2. Asymmetric encryption

The public key cryptography or asymmetric cryptography came up with a radical change of paradigms. According to Stallings [22] public key algorithms are based on mathematical functions, instead of permutation and substitution. Besides the single most important thing is that the public key cryptography is asymmetric, involving the use of two different keys, in contrast to the conventional symmetric encryption, which uses only one key. The use of two keys has profound consequences in the areas of confidentiality, key distribution and authentication. The main distinguishing feature of asymmetric encryption is that it allows the establishment of a secure communication between individuals, without the requirement of the previous share a single cryptographic key.

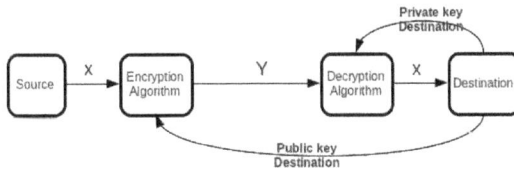

Figure 4. Asymmetric Cryptography [22]

In this class of cryptographic algorithms are used two different keys for encryption and decryption: a public key and its corresponding private key. In this model, in accordance with figure 4, the receiver releases its public key to the sender can encrypt the message, but only the private key of the receiver, which is kept secret is able to decrypt it.

3.1.3. Symmetric x asymmetric cryptography

The IEEE 802.15.4 standard of 2011 defines parameters for low-range personal area networks (LR-WPANs). The first version of this standard was launched in 2003, and the second one [20] was appointed to be the standard communication protocol for WSNs. The encryption mechanism specified in IEEE 802.15.4 standard is based on encryption symmetric key. But according to Sen [23] recent studies have shown that it is possible to implement public key encryption using the right selection of algorithms and associated parameters, and optimization techniques for low power. In some cases the public-key cryptography efficiently obtained similar or even greater than symmetric key encryption using keys smaller. According to Struik [24] is already proven that public-key algorithms developed are suitable for hardware in WSNs.

3.2. RSA algorithm

In the introductory paper about RSA, the authors [25] proposed a method to implement a public key cryptosystem whose security is based on the difficulty to be factoring large prime numbers. Through this technique it is possible to encrypt data and to create digital signatures. It was so successful that today is the RSA public key algorithm used most in the world. The encryption scheme uses RSA and signature of the fact that:

$$m^{ed} \equiv m(mod\,n) \tag{1}$$

for m integer. The encryption and decryption schemes are presented in Algorithms 1 and 2. The decryption works because $c^d \equiv (m^e)^d \equiv m(mod n)$. The safety lies in the difficulty of computing a clear text m from a ciphertext $cm^e mod n$ and the public parameters n (e).

Algorithm 1: RSA Encryption

Input: RSA public key (n,e), Plain text m $\in$ [0, n-1]
Output: Cipher text c
begin
  | 1. Compute c = m^e mod n **2.** Return c.
end

Algorithm 2: Decryption RSA

Input: Public key (n,e),Private key d, Cipher text c
Output: Plain text m
begin
  | 1. Compute m = c^d mod n
  | 2. Return m.
end

3.3. Algorithms based on curves

The main idea of the algorithms based on curves is to build a set of points of an elliptic curve for which the discrete logarithm problem is intractable. According to Blake [26] cryptosystems based on elliptic curves is an interesting technology because they reach the same level of security systems such as RSA, using minor keys, and thus consuming less memory and processor resources. This characteristic makes them ideal for use in smart cards and other environments where features such as storage, time and energy are limited.

The scenario of using public key cryptographic algorithms are changing, because according to Koc [27] in terms of public key encryption algorithm RSA continues to lead the number of implementations, but the number of applications that are using algorithms elliptic curves is increasing considerably thanks to the standardization performed by NIST. The algorithms based on curves are standardized according to the ANSI X9.62, FIPS 186-2, IEEE 1363-2000 and ISO / IEC 15946-2. According to Amin [28] public key encryption includes algorithms for key agreement, encryption and digital signatures. Among the algorithms that operate in key agreement, it can mention the Elliptic Curve Diffie-Hellman (ECDH), data encryption on the Elliptic Curve Integrated Encryption Standard (ECIES) and generating the digital signature Elliptic Curve Digital Signature Algorithm (ECDSA).

3.3.1. ECC algorithm

In the mid-80 [29] and [30] proposed a method of cryptography based on elliptic curves ECC . According to creators of the ECC [4], an elliptic curve is a plane curve defined by the following equation:

$$y^2 = x^3 + ax + b \tag{2}$$

---

[4] Elliptic curve cryptography. More information on the site of the workshop on Elliptic Curve Cryptography which is in issue. Site: http://ecc2011.loria.fr/index.html

The efficiency of this algorithm is based on finding a discrete logarithm of a random element that is part of an elliptic curve. To get an idea of the applicability of the algorithms based on elliptic curves on devices with computational constraints [31] argue that the efficiency of ECC cryptographic algorithm with key sizes of approximately 160 bits is the same obtained using the RSA algorithm with 1024 bit key. Algorithms several features are based on elliptic curves, including key management, encryption and digital signature. Key management algorithms are used to share secret keys, encryption algorithms enable a confidential communication and digital signature algorithms authenticate a participant communication as well as validate the integrity of the message.

The procedures of decryption and encryption through elliptic curve analogous to ElGamal encryption scheme are described in the algorithms 3 and 4. The pure text m is first represented as a point M, and then encrypted by the addition to kQ, where k is an integer chosen randomly, and Q is the public key.

Algorithm 3: *ElGamal elliptic curve encryption*

Input: Parameters field of elliptic curve (p, E, P, n), Public key Q, Plain text m
Output: Cipher text (C_1, C_2)
begin
> **1.** Represent the message m as a point M in E (F_p)
> **2.** Select $k \in R^{[1,n-1]}$.
> **3.** Compute $C_1 = kP$
> **4.** Compute $C_2 = M + kQ$.
> **5.** Return (C_1, C_2)
end

Algorithm 4: *ElGamal elliptic curve decryption*

Entrada: Parameters field of elliptic curve (p, E, P, n), Private key d, Cipher text (C_1, C_2)
Saída: Plain text m
início
> **1.** Compute $M = C_2 - dC_1$, and m from M.
> **2.** Return (m).
fim

The transmitter transmits the points $C_1 = kPeC_2 = M + kQ$ to receiver who uses his private key d to compute:

$$dC_1 = d(kP) = k(dP) = kQ, \tag{3}$$

and then calculating $M = C_2 - kQ$. An attacker who wants to read of M need to calculate kQ. This model algorithm have been extensively studied since according to Amin [28] in recent years the ECC has attracted attention as a security solution for wireless networks, because the use of small keys and low computational overhead.

3.3.2. HECC algorithm

The HECC was created in 1988 by Koblitz [32] as a generalization of elliptic curves. According to Batina [33] the unique difference between ECC and HECC is at average level that in this case consists of different sequences of operations. The HECC uses more complex operations, but works with smaller operands. According to Chatterjee [31] the hierarchy of operations in the HECC and ECC algorithms can be divided into three levels. The first level is the scalar multiplication on the second level are point operations group / splitter and the third level, finite field operations. The authors further inform that the main difference between the ECC and HECC is in the operations group, as different from the ECC, the points on the curve hiperelliptic not form a group. HECC is more complex than the ECC, but uses small numbers.

According to [27] a hyperelliptic curve is a special type of non-singular, projective curve. For our purposes, a hyperelliptic curve, of genus $g \geqslant 1$ over k is the set of points $(X, Y) \in k^2$ that satisfy

$$y^2 + h(X)Y = f(X) \tag{4}$$

where h and f are polynomials in k[X] with $deg(f) = 2g + 1, deg(h) \leq g$, together with a point "at infinity", P_∞. An elliptic curve is just a hyperelliptic curve of genus 1.

3.4. Multivariate Quadratic Quasigroup (MQQ)

The cryptographic algorithms presented above have their security based on computationally intractable mathematical problems: computational efficiency of calculating the discrete logarithm and integer factorization [1]. In 2008, it was proposed a new scheme called multivariate quadratic public key near group (MQQ) [34]. This algorithm is based on multivariate polynomial transformations of nearly quadratic and groups having the following properties [1, 34].

- Highly parallelizable unlike other algorithms that are essentially sequential.
- The encryption speed is comparable to other cryptosystems public key based on multivariate quadratic.
- The decryption speed is typical of a symmetric block cipher.
- Post-Quantum Algorithm

According to Ahlawat [34, 35] MQQ gives a new direction for the cryptography field and can be used to develop new cryptosystems the public key as well as improve existing cryptographic schemes. Furthermore according to El-Hadely and Maia [2, 34] experiments showed that the hardware MQQ can be as fast as a typical symmetric block cipher, being several orders of magnitude faster than algorithms such as RSA, DH and ECC.

A generic description for the scheme is a typical system MQQ multivariate quadratic $T \circ P' \circ S : \{0,1\}^n \to \{0,1\}^n$ where T and S are two nonsingular linear transformations and P'is a multivariate mapping bijetivo quadratic over $\{0,1\}^n$. The mapping $P' : \{0,1\}^n \to \{0,1\}^n$ is defined in the algorithm 5.

The algorithm for encryption with the public key is the direct application of the set of n multivariate polynomials $P = \{P_i(x_1, ..., x_n) | i = 1, ..., n\}$ on the vector $x = (x_1, ..., x_n)$, or is $y = P(x)$. Can be represented as $y = P(x) \equiv y \equiv A.X$. The algorithm 6 is described a decryption using the private key $(T, S, *_1, ..., *_8)$.

Algorithm 5: Non-linear mapping P′

Input: A vector $x = (f_1, ..., f_n)$ of n linear Boolean functions of n variable. We implicitly suppose that a multivariate quadratic quasigroup * is previously defined, and that n = 32k, k = 5,6,7,8 is also previously determined.

Output: : 8 linear expressions $P'_i(x_1, ..., x_n)$, i = 1,...,8 and n - 8 multivariate quadratic polynomials $P'_i(x_1, ..., x_n)$, i = 9, ..., n.

begin
> 1. Represent a vector $x = (f_1, ..., f_n)$ of n linear Boolean functions of n variables $x_1, ..., x_n$ as a string $x = (X_1, ..., X_{n/s})$ where X_i are vector of dimension 8;
> 2. Compute $Y = Y_1, ..., Y_{n/8}$, where $Y_1 = X_1, Y_{j+1} = X_j * X_{j+1}$,for even j=2,4,...and $Y_{j+1} = X_{j+1} * X_j$, for odd j=3,5,...
> 3. Output (y)

end

Algorithm 6: Decryption Algorithm MQQ and sign with private key T, S, *1,...,*8

Input: Vector $y = y_1, ..., y_n$
Output: Vector $x = x_1, ..., x_n$ such that P(x) = y
begin
> 1. $y' = T_{-1}(y)$.
> 2. $W = y'_1, y'_2, y'_3, y'_4, y'_5, y'_6, y'_{11}, y'_{16}, y'_{21}, y'_{26}, y'_{31}, y'_{36}, y'_{41}$.
> 3. $Z = Z_1, Z_2, Z_3, Z_4, Z_5, Z_6, Z_7, Z_8, Z_9, Z_{10}, Z_{11}, Z_{12}, Z_{13} = Dob^{-1}(W)$.
> 4. $y'_1 \longleftarrow Z_1, y'_2 \longleftarrow Z_2, y'_3 \longleftarrow Z_3, y'_4 \longleftarrow Z_4, y'_5 \longleftarrow Z_5, y'_6 \longleftarrow Z_6, y'_{11} \longleftarrow Z_7, y'_{16} \longleftarrow Z_8, y'_{21} \longleftarrow Z_9, y'_{26} \longleftarrow Z_{10}, y'_{31} \longleftarrow Z_{12}, y'_{41} \longleftarrow Z_{13}$.
> 5. $y' = Y_1, ..., Y_k$ onde Y_i are vectors of dimension 5.
> 6. Considering $*_i$, i = 1,...,8, obter $x' = X_1, ..., X_k$, such that, $X_1 = Y_1, X_2 = X_{1\setminus 1}Y_2, X_3 = X_{2\setminus 2}Y_3, X_1 = X_{i-1\setminus 3+((i+2)mod6)}Y_i$
> 7. $x = S^{-1}(x')$

end

---

4. Performance evaluation

Some authors [36] believed that HECC would be less efficient than ECC due to complex structure of the group's operations, but it was reported that there was a detailed analysis of the efficiency of these cryptosystems in embedded systems. The work done by [37] and [38] confirmed the superiority in efficiency of ECC compared to RSA.

[37] showed that the ECC 160 bits is two times better than RSA 1024 bits considering code size and power consumption. [37]performed the tests in 8051 and AVR platforms. [38] pointed out that ECC 160 bits uses four times less energy than RSA 1024 bits in Mica2dot platform.

Only [36] and [31] presented a general analysis, comparison of ECC and HECC, which showed a trend of superiority of HECC on embedded systems. [31] showed that in the encryption, the HECC reaches the same time of ECC using smaller keys. Regarding the time decrypting, the HECC always performed better. The scalar multiplication of HECC is two times faster than ECC. [31] used the platform jdk1.6 in his assessment.

[39] conducted a comparison of ECC and HECC over the computational time of point multiplication on a platform with limited resources. The results showed that the ECC consumed 210ms and the HECC consumed 546ms.

[40] have implemented ECC and HECC on diferent embedded platforms with high practical relevance, namely ARM, ColdFire, and PowerPC. Table 2 show that for the boards at hand they could achieve the best timings for the HECC implementation on the PowerPC. One scalar multiplication for HECC took 117 ms and 84.9 ms for genus-2 and genus-3 curves, respectively. The scalar multiplication for ECC can be performed fastest on the PowerPC at 50MHz resulting in 106.3 ms.

Group Order		ECC	HECC		
			$g = 2$	$g = 3$	$g = 4$
$\approx 2^{160}$	ARM @ 50Mhz	469.96	446.46	515.46	316.6
	ColdFire @ 90Mhz	152.1	155.6	219.4	123.6
	PowerPC @ 50Mhz	106.3	117	141.4	84.9

Table 2. Timings of the scalar multiplication of ECC and HECC on diferent embedded platforms (in ms). [40]

[41] conducted tests with authentication protocols based on RSA and HECC algorithms, comparing the computational time in the Palm III and Wireless Tolkit platforms. The results showed that the protocol based on HECC is 1.37 times faster in key generation and 1.38 times faster with respect to signal generation.

According to Gligoroski [42] in software, digital signature performed by MQQ is 300 to 7000 times faster than the signature of RSA and ECC algorithms. Already in hardware, the superiority of MQQ can reach 10,000 times. The speed of 59 bytes of authentication is compared by the authors [42] and the results are shown in Table 3. The results of the performance evaluation showed that MQQSIGN is at least 325 times faster than RSA and ECC.

[34] evaluated the time of encryption and decryption of algorithms RSA and MQQ in the MicaZ and TelosB platforms. The MQQ showed the time 825.1 ms to encrypt and 116.6 ms to decrypt in TelosB and 445ms to decrypt in MicaZ . Still according to [34] the MQQ 160bits is 909 times faster in the encryption and 5470 times faster in the decryption when compared to the RSA 8bits.

Algorithm	Signing of 59 bytes
RSA 1024	2,230,848
ECC 160	1,284,800
MQQSIGN 160	3,440
-	-
RSA 2048	14,815,324
ECC 224	2,108,556
MQQSIGN 224	4,160

Table 3. Performance comparison of RSA, ECC and MQQ in CPU cycles. [42]

According to [2] that implemented in FPGA a 160bit instance of the newly published public key scheme MQQ, the results of their implementation and the Table 4 show that in hardware, MQQ public key algorithm in encryption and decryption (that means also in verication and signing) can be as fast as a typical block cipher and is several orders of magnitude faster than most popular public key algorithms like RSA, DH or ECC.

Algorithm	1024-bit RSA	160-bit MQQ	128-bit AES
Throughput	40Kbps	44.27Gbps / 399.04Mbps	7.78Gbps

Table 4. Synthesis Results for 160bit MQQ realized in Virtex-5 chip xc5vfx70t-2-δ1136 [2]

5. Conclusion and future work

It is natural that the spread of ubiquitous computing to increase the number of devices with low computing power scattered all over the planet. The security of data transmissions from these devices should be improved in a preventative manner to avoid possible attacks. Regarding WSNs, RSA public key algorithm is the most commonly used is standardized, and achieves efficiency relatively good. The algorithm based on elliptic curves have been extensively studied in academia as an alternative to RSA, and the results show that it is possible to achieve good results with smaller keys. The algorithm MQQ was discovered recently and showed significant results when compared to RSA and ECC, taking as parameters authenticity and digital signature. This algorithm is post-quantum, and may even be a good solution when the quantum computation is standardized. Despite the satisfactory results of MQQ front of RSA and ECC algorithms, there is not a work about performance evaluation specific to encryption and decryption of data.

Author details

Gustavo S. Quirino, Admilson R. L. Ribeiro and Edward David Moreno
Universidade Federal de Sergipe, Brasil

6. References

[1] D. Gligoroski, S. Markovski, and S.J. Knapskog. A public key block cipher based on multivariate quadratic quasigroups. *Arxiv preprint arXiv:0808.0247*, 2008.

[2] M. El-Hadedy, D. Gligoroski, and S.J. Knapskog. High performance implementation of a public key block cipher-mqq, for fpga platforms. In *Reconfigurable Computing and FPGAs, 2008. ReConFig'08. International Conference on*, pages 427–432. IEEE, 2008.

[3] Marcos Simplicio Margi, MS Jr, and Tereza C. M. B. Carvalho Barreto. Segurança em Redes de Sensores Sem Fio. In *Simpósio Brasileiro em Segurança da Informação*, pages 149–194, 2009.

[4] A.A.F. Loureiro, J.M.S. Nogueira, L.B. Ruiz, R.A. de Freitas Mini, E.F. Nakamura, and C.M.S. Figueiredo. Redes de sensores sem fio. In *Simpósio Brasileiro de Redes de Computadores (SBRC)*, pages 179–226, 2003.

[5] Xuan Hung Le, Ravi Sankar, Murad Khalid, and Sungyoung Lee. Public Key Cryptography - based Security Scheme for Wireless Sensor Networks in Healthcare. In *4th International Conference on Ubiquitous Information Management and Communication*, pages 1–7, 2010.

[6] MEMSIC. Mote Processor Radio Mote Interface Boards User Manual - Document Part Number: 7430-0021-09 Rev A, 2012

[7] E. Shi and A. Perrig. Designing secure sensor networks. *Wireless Communication Magazine*, 11(6):38–43, 2004.

[8] Y. Wang, G. Attebury, and B. Ramamurthy. A survey of security issues in wireless sensor networks. *IEEE Communications Surveys and Tutorials*, 8(2):2–23, 2006.

[9] X. Wang, W. Gu, S. Chellappan, Dong Xuan, and Ten H. Laii. Search-based physical attacks in sensor networks: Modeling and defense. Technical report, Department of Computer Science and Engineering, Ohio State University, 2005.

[10] A.D. Wood and J.A. Stankovic. Denial of service in sensor networks. *IEEE Computer*, 35(10):54–62, 2002.

[11] C. Karlof and D. Wagner. Secure routing in wireless sensor networks: Attacks and countermeasures. In *1st IEEE International Workshop on Sensor Network Protocols and Applications*, pages 113–127. IEEE, 2003.

[12] J. Newsome, E. Shi, D. Song, and A. Perrig. The sybil attack in sensor networks: Analysis and defenses. In *3rd International Symposium on Information Processing in Sensor Networks*, pages 259–268. ACM Press, 2004.

[13] J. Douceur. The sybil attack. In *1st International Workshop on Peer-to-Peer Systems - IPTPS*, 2002.

[14] H. Chan and A. Perrig. Security and privacy in sensor networks. *IEEE Computer Magazine*, pages 103–6105, 2003.

[15] A. Perrig, R. Szewczyk, JD Tygar, V. Wen, and D.E. Culler. Spins: Security protocols for sensor networks. *Wireless networks*, 8(5):521–534, 2002.

[16] C. Karlof, N. Sastry, and D. Wagner. Tinysec: a link layer security architecture for wireless sensor networks. In *Proceedings of the 2nd international conference on Embedded networked sensor systems*, pages 162–175. ACM, 2004.

[17] M. Luk, G. Mezzour, A. Perrig, and V. Gligor. Minisec: a secure sensor network communication architecture. In *Information Processing in Sensor Networks, 2007. IPSN 2007. 6th International Symposium on*, pages 479–488. IEEE, 2007.

[18] IEEE. Ieee 802.15.4 - wireless medium access control (mac)and physical layer (phy) specifications for low-rate wireless personal area networks (lr-wpans). Technical report, Park Avenue, New York, USA: IEEE, 2011.

[19] Andrew S. Tanembaum. *Redes de computadores*. Elsevier, Rio de Janeiro, 4ž edition edition, 2003.

[20] David Boyle and Thomas Newe. Securing Wireless Sensor Networks: Security Architectures. *Journal of Networks*, 3(1):65–77, January 2008.

[21] Xiangqian Chen, Kia Makki, Kang Yen, and Niki Pissinou. Sensor network security: a survey. *IEEE Communications Surveys & Tutorials*, 11(2):52–73, 2009.

[22] W. Stallings. *Network and internetwork security: principles and practice*. Prentice-Hall, Inc., 1995.

[23] Jaydip Sen. A Survey on Wireless Sensor Network Security. *International Journal of Communication Networks and Information Security (IJCNIS)*, 1(2):55–78, 2009.

[24] RenÃľ Struik. Cryptography for highly constrained networks. In *NIST - CETA Workshop 2011*, 2011.

[25] R.L. Rivest, A. Shamir, and L. Adleman. A method for obtaining digital signatures and public-key cryptosystems. *Communications of the ACM*, 21(2):120–126, 1978.

[26] I. Blake, G. Seroussi, and N. smart. *Elliptic Curves in Cryptography*. Cambridge, 1999.

[27] C.K. Koc, Nigel Boston, and Matthew Darnall. *About Cryptographic Engineering - Elliptic and Hyperelliptic Curve Cryptography*. Springer, 2009.

[28] F Amin, A H Jahangir, and H Rasifard. Analysis of Public-Key Cryptography for Wireless Sensor Networks Security. *World Academy of Science, Engineering and Technology*, 31(July):530–535, 2008.

[29] N. Koblitz. Elliptic curve cryptosystems. *Mathematics of computation*, 48(177):203–209, 1987.

[30] V. Miller. Use of elliptic curves in cryptography. In *Advances in Cryptology - CRYPTO Proceedings*, pages 417–426. Springer, 1986.

[31] Kakali Chatterjee, Asok De, and Daya Gupta. Software Implementation of Curve based Cryptography for Constrained Devices. *International Journal of Computer Applications*, 24(5):18–23, 2011.

[32] N. Koblitz. A family of jacobians suitable for discrete log cryptosystems. In *Advances in Cryptology - Crypto88*, pages 94–99. Springer, 1990.

[33] L. Batina, N. Mentens, B. Preneel, and I. Verbauwhede. Flexible hardware architectures for curve-based cryptography. In *Circuits and Systems, 2006. ISCAS 2006. Proceedings. 2006 IEEE International Symposium on*, pages 4–pp. IEEE, 2008.

[34] R.J.M. Maia. Análise da viabilidade da implementação de algoritmos pós-quânticos baseados em quase-grupos multivariados quadráticos em plataformas de processamento limitadas. Master's thesis, USP, 2010.

[35] R. Ahlawat, K. Gupta, and S.K. Pal. From mq to mqq cryptography: Weaknesses new solutions. In *Western European Workshop on Research in Cryptology*, 2009.

[36] Thomas Wollinger, Jan Pelzl, Volker Wittelsberger, Christof Paar, Gökay Saldamli, and Çetin K. Koç. Elliptic and hyperelliptic curves on embedded μP. *ACM Transactions on Embedded Computing Systems*, 3(3):509–533, August 2004.

[37] Nils Gura, Arun Patel, Arvinderpal Wander, Hans Eberle, and Sheueling Chang Shantz. Comparing Elliptic Curve Cryptography and RSA on 8-bit CPUs. In *In Proceedings of the 2004 Workshop on Cryptographic Hardware and Embedded Systems (CUES 2004), Boston Marriott Cambridge Cambridge (Boston)August*, 2004.

[38] Shish Ahmad, Mohd. Rizwan beg, and Qamar Abbas. Energy Saving Secure Framework for Sensor Network using Elliptic Curve Cryptography. *IJCA Special Issue on ?Mobile Ad-hoc Networks?*, pages 167–172, 2010.

[39] Lejla Batina, Nele Mentens, Kazuo Sakiyama, Bart Preneel, and Ingrid Verbauwhede. Public-Key Cryptography on the Top of a Needle. In *2007 IEEE International Symposium on Circuits and Systems*, pages 1831–1834. Ieee, May 2007.

[40] T. Wollinger. *Software and hardware implementation of hyperelliptic curve cryptosystems*. Europ. Univ.-Verl., 2004.

[41] S. Prasanna Ganesan. An Authentication Protocol For Mobile Devices Using Hyperelliptic Curve Cryptography. *International J. of Recent Trends in Engineering and Technology*, 3(2):2–4, 2010.

[42] D. Gligoroski, S.J. Knapskog, S. Markovski, R.S. Ødegård, R.E. Jensen, L. Perret, and J.C. Faugère. The digital signature scheme mqq-sig. *Arxiv preprint arXiv:1010.3163*, 2010.

Localization & Positioning

Distributed Range-Free Localization of Wireless Sensor Networks via Nonlinear Dynamics

Shuai Li and Yangming Li

Additional information is available at the end of the chapter

1. Introduction

In the past years, the development in micro electromechanical systems (MEMS), radio frequency (RF), integrated circuit (IC), etc., greatly enhanced the advancement of wireless sensor networks (WSNs). As an ubiquitous sensing technology, WSNs find more and more applications, such as structural monitoring [34], precision agriculture [3], gas-leak localization [14], volcano monitoring [33], robot navigation [4, 15], health monitoring [20], to name a few. For most existing applications of WSNs, the location information is crucial. For example, in the structural monitoring application, we can conclude that the structure is out of condition if fault is detected by one or more sensors in the network of sensors mounted everywhere on the structure. However, we are unable to accurately report the faulty position without localization capability of the WSN. In contrast to other type of networks, e.g., Internet, a prominent difference is that WSNs are location-based networks. Therefore, the design of localization hardware and localization algorithms is an important procedure in the development of a WSN system.

There are mainly two classes of localization approaches for WSNs: one is pre-localization and the other one is self-localization. The pre-localization method measures the position of sensors in the deployment stage. After the deployment and position measurement, the position is stored in the memory of the sensor. For this method, any movement of the sensors will result in errors in the location information. Differently, the self-localization method computes the locations of each sensor based on real-time measurements and therefore is robust to the variance of the environment. With GPS devices embedded, sensors are enabled with self-localization capability. However, the relatively high cost of GPS devices often makes it not practical to apply GPS to all sensors in a network. Instead, the strategy with a portion of sensors equipped with GPS as beacons and using triangulation or trilateration to iteratively determine the positions of blind sensors based on the distance or angle measurements between neighboring sensors provides a less expensive way for self-localization [13, 16, 26]. Although many GPS devices are saved, as a tradeoff, the sensors are required to have the ability to measure the distance or the relative angle to its neighbor, which may result in costs for extra hardware. Without introducing extra hardware, received signal strength (RSS) based distance measurement method [17, 30], relying on the estimated distance according to the signal strength received from the neighboring sensor, provides a promising direction for self-localization. Another promising self-localization method is range-free localization, which even does not require the information on the signal strength received from the neighbor but the connectivity information, i.e., a sensor only need to know who is its neighbor. This technology implies that localization can be a by-product of communication since connectivity information can be obtained in communication. For example, if Sensor A can communicate with Sensor B, then we conclude they are connected. Due to this promising property, range-free localization is becoming more and more popular in both practice and research. In this chapter, we investigate the range-free localization of WSNs.

Dynamic models gained great success in realtime signal processing [28], robotics [12, 22], online optimization [29], etc.. In this chapter, we overview our previous work on dynamic model based range-free localization [10, 11, 25]. Particularly, we will examine two dynamic models for the real time localization of WSNs. The models are described by nonlinear ordinary differential equations (ODEs). The state value of the ODEs converges to the expected position estimation of sensors. Both of the two models find feasible solutions to the formulated optimization problem. Particularly, the second model, by exploiting heuristic information, has a tendency to converge to better solutions in the sense of localization error. The real time processing ability of the models allows possible movement of the sensor nodes, which often happens in mobile sensor networks [23]. Besides the real time localization capability, another prominent feature of the proposed models is that both of them are completely distributed, i.e., each sensor in the network only need to exchange information with its neighbor and thus no message passing is needed in the network. This advantage makes the proposed algorithms scalable to large scale networks involving thousands of sensors or more.

The remainder of this paper is organized as follows. In Section 2, some preliminaries on range-free localization of WSNs are presented. In Section 3, we formulate the localization problem from an optimization perspective. Two dynamic models are presented in Section 4 to solve the formulated optimization problems. In Section 5, simulations are performed to demonstrate the effectiveness of our method. Section 6 concludes this paper.

2. Preliminaries

In this chapter, we assume that all sensors are equipped with communication modules and the locations of beacon sensors are known. Fig. 1 sketches the connectivity topology of a WSN consisting of beacon sensors and blind sensors. In the network, The beacon sensors are those with known locations. The locations can be obtained either by GPS or by pre-deployment. The blind sensors are those without pre-known positions. A sensor can communicate with other sensors within the signal coverage area. The communication links and sensors therefore form a network with sensors as nodes and communication links as edges.

The signal strength at a given distance from the emitter varies due to propagation conditions, material coverage, antenna configurations and battery conditions [31] and the calculated distance according to the received signal strength often has a large error [8, 18]. Nevertheless, the nominal maximum range, which is measured under ideal conditions in open environments without obstacles along the signal propagation route, without material coverage, with a proper configuration of the antenna and with a full power of the battery, etc.,

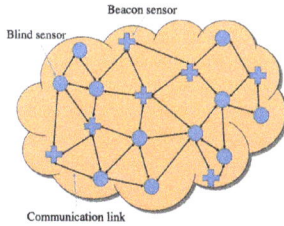

Figure 1. Schematic of a WSN topology in two dimensional space.

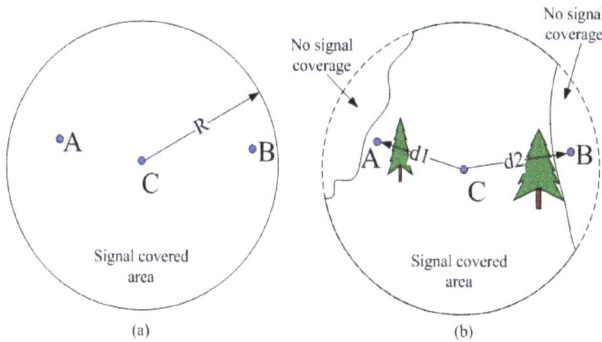

Figure 2. Range-free localization in environments with obstacles.

gives an upper bound on the distance between the emitter and receiver pair. As depicted in Fig. 2. Subfigure (a) depicts the ideal open environment, where the communication radius, denoted by R in the figure, equals the nominal maximum range. In this situation, both the point A and the point B are within the communication range of the sensor located at point C and therefore the distance from the sensor to A and that to B are both less than R. In the situation with the presence of obstacles (shown as trees in the subfigure (b)), the signal covered area shrinks and some positions, such as the point B in the subfigure (b), even with a distance less than R to the sensor, cannot be covered by the signal. Therefore, $d1$, which is the distance from the sensor to point A, is less than R if the sensor located at A can detect the signal.

3. Problem formulation

In this section, we present the mathematical formulation of the problem.

3.1. Nonlinear inequality problem formulation

As discussed in Section 2, the position of beacon sensors are known and the distance between two neighbor sensor (in the sense of communication) is less than R, which is the nominal maximum communication range. In equation, we have

$$(x_i - x_j)^T(x_i - x_j) \le R^2 \quad \text{for } i \in \mathbb{N}(j) \tag{1a}$$

$$x_k = \bar{x}_k \quad \text{for } k \in \mathbb{B} \tag{1b}$$

where $\mathbb{B}$ is the beacon sensor set, x_i, x_j represent the position of the ith sensor and the jth sensor, respectively, R is the maximum communication range of sensors, $\mathbb{N}(j)$ denotes the jth sensor's neighbor set, which includes all sensors connected to it via communication, $\mathbb{B}$ is the beacon sensor set, $\bar{x}_k$ is the true position of the kth beacon sensor.

Remark 1. *There is no explicit objective function but inequality and equality constraints in problem (1). The solution to this problem is generally not unique. We are more concerned with finding a feasible solution in real time instead of finding all the feasible solutions. Based on this consideration, we explore finding a feasible solution to problem (1) in real time via a dynamic model.*

3.2. Optimization problem formulation

To find a feasible solution of problem (1) numerically, we first transform the problem into an optimization problem and employ dynamic evolutions to solve it.

The solution of problem (1) is identical to the one of the following normal optimization with an explicit objective function,

$$\text{minimize} \quad \sum_{i=1}^{n} \sum_{j \in \mathbb{N}(i)} w_{ij} \max\{(x_i - x_j)^T(x_i - x_j) - R^2, 0\}$$

$$\text{subject to} \quad x_k = \bar{x}_k \quad \text{for } k \in \mathbb{B} \tag{2}$$

where n denotes the number of sensors, $w_{ij} > 0$ is the weight of the connection between the ith and the jth sensor. Note that the problem (2) is a non-smooth optimization problem due to the presence of the function $\max(\cdot)$.

4. Solving the problem via nonlinear dynamic evolution

In this section, we present two ODE models, both of which are able to solve the range-free localization problem (2). As the solution to the problem is generally not unique. Property employment of heuristic information may improve the solution performance. Based on the feasible solution obtained by the first ODE model, the second ODE model proposed in this chapter indeed realizes the improvement in performance.

4.1. Model I

The partial sub-gradient relative to x_i of the objective function switches between $4 \sum_{j \in \mathbb{N}(i)} (x_i - x_j)$ and 0 at the critical point $(x_i - x_j)^T(x_i - x_j) - R^2 = 0$. For smooth arbitration, we use the following dynamic evolution to find a feasible solution of the optimization problem (2),

$$\dot{x}_i = -\epsilon_1 \sum_{j \in \mathbb{N}(i)} w_{ij} I_{ij}(x_i - x_j) \tag{3}$$

where x_i is the position estimation of the blind sensor labeled i, which is initialized randomly, $\epsilon_1 > 0$ is a scaling factor, w_{ij} is a positive weight, I_{ij} is an indicator function defined as follows:

$$I_{ij} = \begin{cases} 1 & \text{if } (x_i - x_j)^T (x_i - x_j) - R^2 > 0 \\ 0 & \text{if } (x_i - x_j)^T (x_i - x_j) - R^2 \leq 0 \end{cases} \tag{4}$$

In the ODE model, each blind sensor is associated with a dynamic module. The modules interact with their neighbor modules and all the modules together perform the localization task and solve the problem (1). The dynamic evolution of x_i in the system (3) depends on its neighbor values x_j for $j \in \mathbb{N}(i)$. In detail, the neighbor x_j has an action $-\epsilon_1 I_{ij}(x_i - x_j)$ on x_i. This action term is analogous to a force pointing from x_i to x_j and pulling x_i to x_j with an amplitude ϵ_1 or 0 respectively when $\|x_i - x_j\| > R$ or $\|x_i - x_j\| \leq R$. This negative feedback mechanism guides position estimations of neighbor sensors to aggregate to within the maximum range R.

Notably, the ODE model (3) is a distributed one. Communication only happens between neighboring sensors. No routing or cross-hop communication is required for the implementation of the ODE model. The distributed nature of the model thoroughly reduces the communication burden and makes the method scalable to a network with a large number of sensors involved.

About the ODE model I (3), we have the following theorem,

Theorem 1 ([25]). *The ODE model I (3) with $\epsilon_1 > 0$, w_{ij} for all possible i and j, asymptotically converges to a feasible solution x_i^* (for all i in the blind sensor set) of problem (1).*

The proof of this theorem is based on Lyapunov stability theory. Interested readers are refereed to our previous work [10, 25] for a detailed proof. This theorem reveals that the ultimate output of the ODE model I is a feasible solution to problem (1).

4.2. Model II

Section 4.1 provides an ODE model to find a feasible solution of the problem. The model presented in this part is also a dynamic ODE model. Different from Model I, which is initialized randomly and does not use any heuristic information, Model II is initialized with the ultimate output of Model I with R replaced by $R - \delta$ in (4) with $\delta \ll R$ and takes advantages of heuristic information to result in sensor position estimations with inclination to uniformly distribution. We first define the following optimization problem to incorporate heuristic information,

$$\text{minimize} \quad \sum_{i=1}^{n} \sum_{j \in \mathbb{N}(i)} (x_i - x_j)^T (x_i - x_j)$$

$$- c_0 \sum_{i=1}^{n} \sum_{j \in \mathbb{N}(i)} \log \left(R^2 - (x_i - x_j)^T (x_i - x_j) \right) \tag{5a}$$

$$x_k = \bar{x}_k \quad \text{for } k \in \mathbb{B} \tag{5b}$$

where $\mathbb{B}$ is the beacon sensor set, x_i is initialized with the ultimate output of (3) with R replaced by $R - \delta$ in (4). $c_0 > 0$ is a coefficient. Note that the first term in (5a) contributes to the equal distribution in space. In (5a) the terms involving x_i write $2\sum_{j\in\mathbb{N}(i)}(x_i - x_j)^T(x_i - x_j)$. The minimization of $2\sum_{j\in\mathbb{N}(i)}(x_i - x_j)^T(x_i - x_j)$ in terms of x_i tends to adapt x_i to the center formed by all x_j for $j \in \mathbb{N}(i)$. The second term in (5a) is essentially a barrier term and approaches to infinitely large when the solution tends to violate the inequality constraints given in (1). This term works to restrict the solution in the feasible set.

We use the following gradient based dynamics to solve (5):

$$\dot{x}_i = -\epsilon_2 \sum_{j\in\mathbb{N}(i)} \left(1 + \frac{c_0}{R^2 - (x_i - x_j)^T(x_i - x_j)}\right)(x_i - x_j)$$

$$x_k = \bar{x}_k \quad \text{for } k \in \mathbb{B}$$

$$x_i(0) = x_i' \tag{6}$$

where x_i is the position estimation of the ith blind sensor, x_i' is the ultimate output of Model I (3) with R replaced by $R - \delta$, i.e., the solution of x_i obtained by solving (2) with R replaced by $R - \delta$ in (4). The expression $x_i(0) = x_i'$ means that x_i is initialized with x_i'. $\epsilon_2 > 0$ is a scaling factor and $c_0 > 0$ is a positive constant.

The ODE model (6) is a distributed one since the update of x_i in (6) only depends on x_j for $j \in \mathbb{N}(i)$, i.e., the position estimations of the neighbor sensors. Therefore, communication only happens between neighbor sensors.

About the initialization of the ODE model, we have the following remark,

Remark 2. *The ODE model (6) is initialized with the ultimate output of the ODE model I (3) with R replaced by $R - \delta$ in (4) with $\delta \ll R$. The goal is to ensure the ultimate output of Model I strictly locates inside the open set formed by (1), which is necessary for the barrier term in (5) to restrict the solution always stays inside the feasible region.*

According to Theorem 1, the ODE model I with R replaced by $R - \delta$ ultimately converges to a solution in the following set,

$$(x_i - x_j)^T(x_i - x_j) \leq (R - \delta)^2 \quad \text{for } i \in \mathbb{N}(j) \tag{7a}$$

$$x_k = \bar{x}_k \quad \text{for } k \in \mathbb{B} \tag{7b}$$

with which we conclude that $(x_i - x_j)^T(x_i - x_j) \leq (R - \delta)^2 < R^2$ for $i \in \mathbb{N}(j)$. With the effect of the barrier term $\frac{c_0}{R^2 - (x_i - x_j)^T(x_i - x_j)}(x_i - x_j)$ in the model II (6), the ultimate solution of (6) with an initialization inside the feasible set will still stay inside this set. We have the following theorem to state this point rigourously,

Theorem 2 ([11]). *The ODE model II (6) with $\epsilon_2 > 0$, $c_0 > 0$, initialized with x_i', which is the ultimate output of the ODE model I (3) with R replaced by $R - \delta$ in (4) with $\delta \ll R$, stays in the open set constructed by (1).*

5. Simulations

In this section, simulations are used to verify the two ODE models in both the one dimensional space and the two dimensional space.

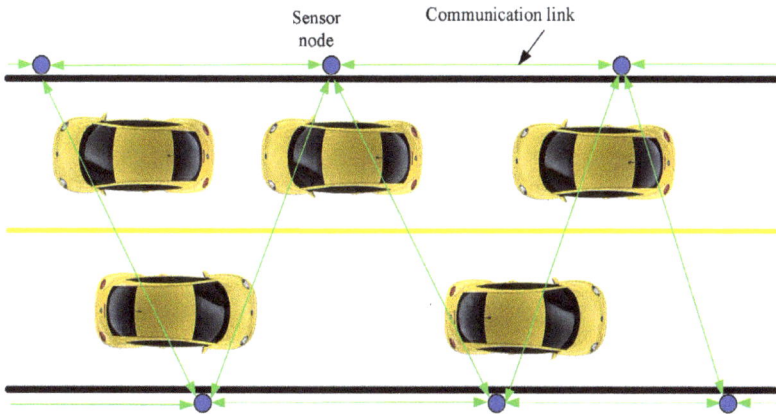

Figure 3. A schematic description of WSNs for highway monitoring.

5.1. Range-free localization in one dimensional spaces

In this part, we investigate the range-free localization of sensors in a network deployed in a one dimensional topology.

5.1.1. Background

There are a bunch of applications which deploys sensors along an one dimensional line. For example, WSNs for highway monitoring [1, 9, 21] are often deployed along the highway direction and thus form a one dimensional deployment topology, as sketched in Fig. 3. Other applications, such as WSNs for bridge health monitoring [32] and WSNs along a tunnel [5] for traffic safety, can also be put into the category of one dimensional localization problem.

5.1.2. Simulation setup and simulation results

We consider a wireless sensor network with one dimensional deployment. There are 4 beacon sensors deployed at 0m, 166.6667m, 333.3333m and 500.0000m, and 16 blind sensors deployed at 26.6011m, 56.1963m, 83.3216, 119.9182m, 147.6692m, 176.9903m, 208.3049m, 238.5405m, 263.6398m, 290.4771m, 320.4868m, 355.1442m, 384.0493m, 407.3632m, 440.0192m and 470.5006m, respectively. The communication range of sensors is 50m.

For the dynamic models, the state values of Model I are randomly initialized. We choose $\epsilon_1 = 10^5$, $\epsilon_2 = 20 \times 10^5$ as the scaling parameters, the coefficient $c_0 = 5$. The shrinking constant δ is chosen as 5. Fig. 4 plots the transient behavior of the position estimation by Model I. From this figure, we can clearly see that the estimation converges with time. For Model II, it is initialized with the output of Model I by replacing R with $R - \delta$. As $R \approx R - \delta$ in this simulation, Fig. 4 and Fig. 5, which shows the transient to obtain the initial position estimation for Model II, demonstrate similar behaviors. Fig. 6 shows the transient of the position estimation by Model II. By comparing the final values and the initial values in Fig. 6, it can be found that the values tends to equal distances between neighbors. The position estimation results are shown in Fig. 7. It can be observed that both models result in estimations meeting the nonlinear inequalities (neighbor sensors are within a distance of 50m). However, the result by Model I may break

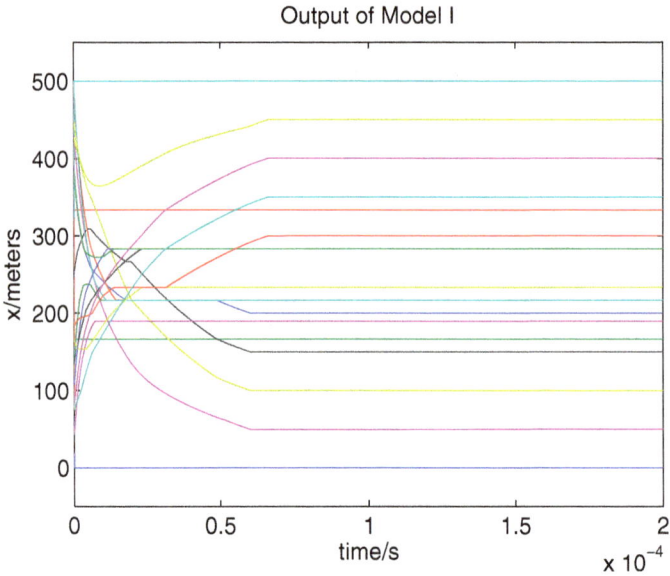

Figure 4. Transient of the position estimation by Model I in the one dimensional localization problem.

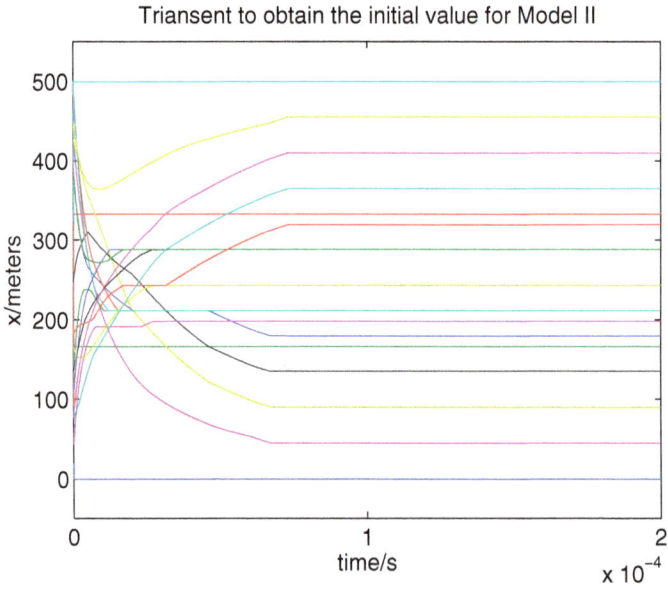

Figure 5. Transient to obtain the initial position estimation for Model II in the one dimensional localization problem.

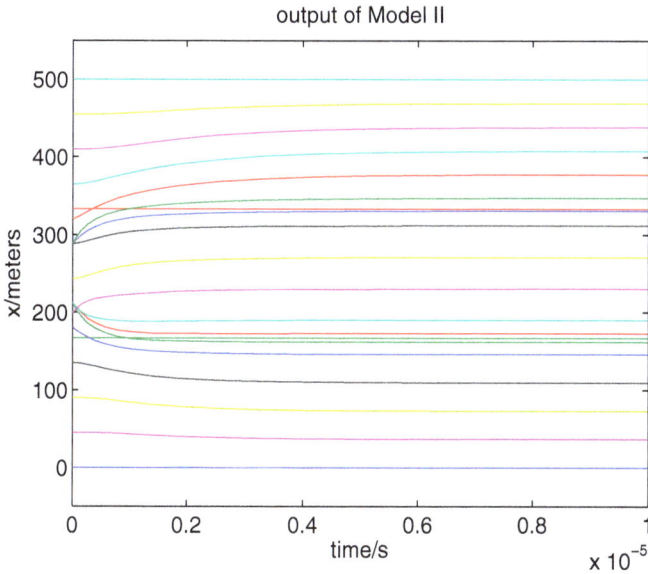

Figure 6. Transient of the position estimation by Model II in the one dimensional localization problem.

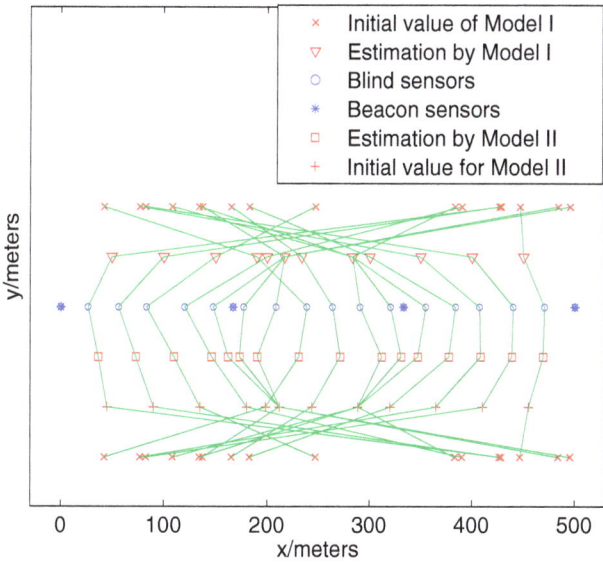

Figure 7. Position estimation results in the one dimensional localization problem.

the real order, i.e., as shown in Fig. 7 the sixth sensor from the left actually locates to the left of the eighth one while the estimated position of the sixth sensor by Model I is to the right of the eighth sensor. However, the performance is improved by using Model II and the estimation results follows the real order. For better comparisons of Model I and Model II in the sense of estimation error, we use the Root-mean-square error $E1$ defined as $E1 = \sqrt{\frac{\sum_{i=1}^{n}||\hat{x}_i - x_i||^2}{n}}$ (where $\hat{x}_i$ and x_i represent the estimated value and the true value of the ith sensor's position), and the maximum absolute error $E2$ defined as $E2 = \max_{i=1,2,...,n}\{||\hat{x}_i - x_i||\}$ to evaluate the estimation error. Ten independent simulations with random initializations are performed and the estimation errors are calculated for all runs. As shown in Table 1, the error $E1$ is around 40 and the error $E2$ is around 80 for Model I with the simulation setup. In contrast, the estimation errors for Model II are much lower, which are about 10 for $E1$ and 26 for $E2$ in the ten simulation runs. This result demonstrates the advantage of Model II over Model I for position estimation of sensors by introducing heuristic information. Also note that there are only 4 beacon sensors in contrast to 16 blind sensors, meaning that the ratio of beacon sensors to blind sensors is 25%. For such a low beacon vs. blind sensor ratio, the estimation errors $E1$ and $E2$ for both Model I and Model II, especially for Model II, as shown in Table 1, are acceptable for rough estimations of sensor positions in applications.

♯	Estimation Error of Model I		Estimation Error of Model II	
	E1	E2	E1	E2
1	41.2948	84.0870	9.9419	23.7890
2	44.5053	84.4876	10.2123	27.2300
3	43.7765	80.0258	13.2497	26.1105
4	44.3971	83.3457	11.1692	26.9130
5	40.9815	83.3549	12.4231	28.3951
6	40.7992	79.4815	10.8819	25.6199
7	39.5168	86.9493	11.4397	26.0199
8	37.4373	65.4550	10.4166	24.0404
9	42.9898	83.3865	11.4330	26.6570
10	44.0201	78.4496	13.2752	28.7782

Table 1. Estimation errors for Model I and Model II in different simulation runs of the one dimensional localization problem.

5.2. Range-free localization in two dimensional Spaces

In this part, we investigate the range-free localization of sensors in a network deployed in a two dimensional topology.

5.2.1. Background

Most existing literatures deal with the general localization problem in two dimensional space. The one dimensional sensor localization problem investigated in the last section falls into this category by fixing the value of sensor positions along one dimension. Localization in applications, such as wildlife monitoring [7], WSN aided robot navigation [6, 19] and animal tracking [27], can be abstracted as two dimensional localization problems.

5.2.2. Simulation setup and simulation results

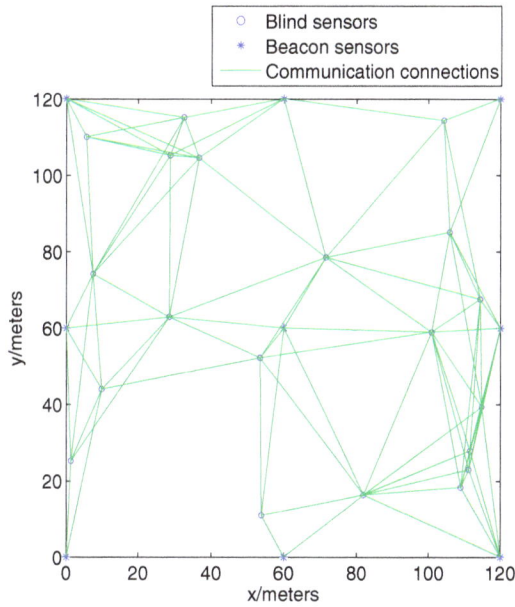

Figure 8. True positions of sensors in a typical simulation run of the two dimensional WSN localization problem.

In the simulation, we consider a $100 \times 100m^2$ square area with 9 beacon sensors uniformly deployed (the beacon sensors are deployed along the perimeter and at the center, with relative coordinates $[0,0]$m, $[60,0]$m, $[120,0]$m, $[0,60]$, $[60,60]$, $[120,60]$, $[0,120]$, $[60,120]$, $[120,120]$ respectively.) and 20 blind sensors randomly deployed (see Fig. 8 for the deployment of sensors in a typical simulation run). The maximum communication range of sensors are chosen as $R = 50$m.

As to the dynamic models, we choose the scaling factors $\epsilon_1 = \epsilon_2 = 10^5$, the connection weight w_{ij} equals 5 for connections with a beacon sensor and 1 otherwise for Model I, the relaxation parameter $delta = 0.5$ for Model II and the coefficient $c_0 = 1$ for Model II.

Fig. 9 and Fig. 10 show the time histories of the position estimations by Model I along x-direction and that along y-direction respectively. From the figures we can observe that after a short period of transient, the estimation results converge to constant values. The time histories of the position estimations along x-direction and y-direction estimations by Model II are plotted in Fig. 11 and Fig. 12, respectively. Compared to the transient of Model I, the change of state values in Fig. 11 and Fig. 12 are much milder. The adjustment of values refine the initial estimation with the tendency to even distributions under the communication connectivity constraints. With time elapses, the estimation results by Model II converge. It is worth noting that the ultimate values by Model II shown in Fig 11 and Fig. 12 are not exactly uniformly distributed. This is due to the compromise of the heuristic information driving to even

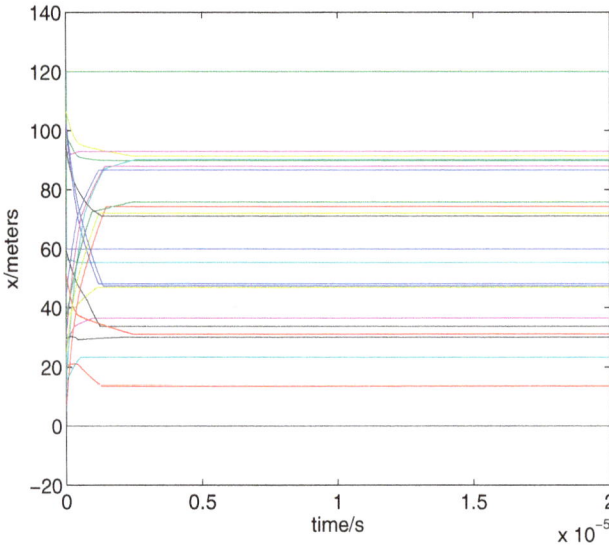

Figure 9. Time history of the position estimation in x-direction by Model I in the two dimensional WSN localization problem.

distribution in space and the inequality constraints imposed by communication connectivity. The final position estimations by Model I and Model II are shown in Fig. 13 and Fig. 14, respectively. It can be observed that the estimated results shown in both figures are within the area covered by the circle centered at the true position with a radius $R = 50$m, which verifies the effectiveness of Model I and Model II in modeling the communication connectivity constraint. On the other hand, it is clear that the estimation results shown in Fig. 14 outperforms the results shown in Fig. 13 thank to introducing heuristic information in Model II. For better comparisons of Model I and Model II for estimating sensor locations in two dimensional scenario, the Root-mean-square error $E1$ defined as $E1 = \sqrt{\frac{\sum_{i=1}^{n}||\hat{x}_i - x_i||^2}{n}}$ and the maximum absolute error $E2$ defined as $E2 = \max_{i=1,2,...,n}\{||\hat{x}_i - x_i||\}$, both of which are the same as the definition in the one dimensional case, are used to evaluate the estimation error. Ten independent simulations with random initializations are performed and the estimation errors are calculated for all runs. As shown in Table 2, the error $E1$ is around 20 and the error $E2$ is around $r0$ for Model I with the simulation setup. In contrast, the estimation errors for Model II are about 10 for $E1$ and 20 for $E2$ in the ten simulation runs, which are much lower than the results obtained by Model I and again verifies the advantage of Model II in position estimation.

5.3. Discussions

In the above two subsections, we considered the range-free localization problem in one dimensional case and two dimensional case respectively. In some applications of WSNs, higher dimensional cases [24] (see Fig. 15 for the sketch of a typical three dimensional one

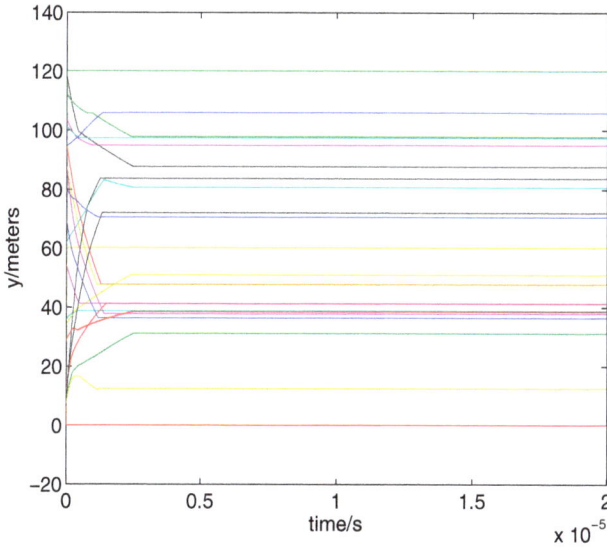

Figure 10. Time history of the position estimation in y-direction by Model I in the two dimensional WSN localization problem.

♯	Estimation Error of Model I		Estimation Error of Model II	
	E1	E2	E1	E2
1	19.2071	34.9071	10.9093	24.0850
2	20.6657	42.7582	10.2647	18.3666
3	21.2873	37.9441	10.2106	18.2110
4	20.0403	41.3617	9.9603	16.8761
5	20.6692	39.0069	9.1838	17.2634
6	18.1865	40.7181	9.5004	21.8774
7	21.0053	37.7784	10.2448	18.6334
8	23.1289	44.2787	12.3591	25.5854
9	23.5922	40.3665	10.6922	21.0049
10	19.5115	41.8718	11.6586	24.0909

Table 2. Estimation errors for Model I and Model II in different simulation runs of the one dimensional localization problem.

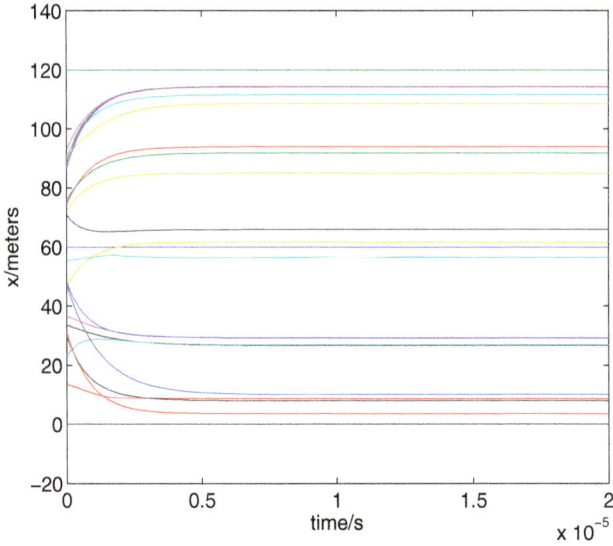

Figure 11. Time history of the position estimation in x-direction by Model II in the two dimensional WSN localization problem.

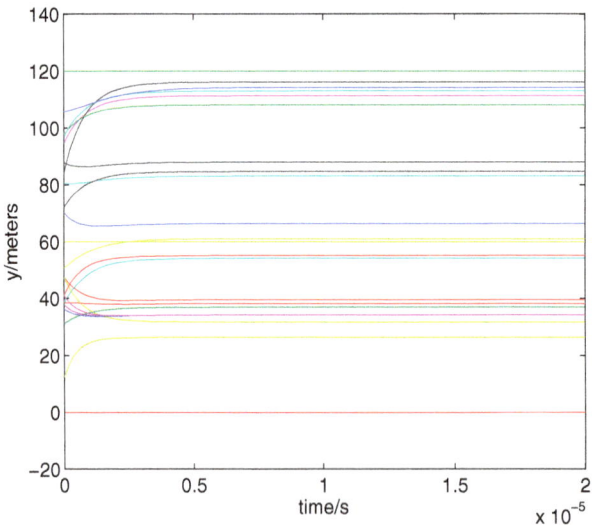

Figure 12. Time history of the position estimation in y-direction by Model II in the two dimensional WSN localization problem.

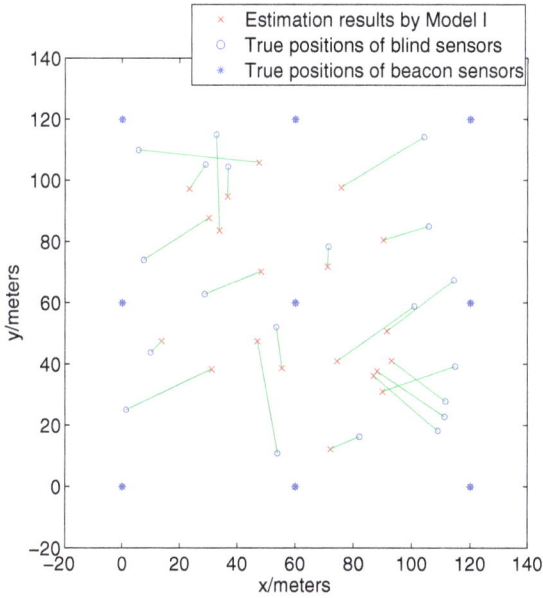

Figure 13. Position estimation results by Model I in the two dimensional WSN localization problem.

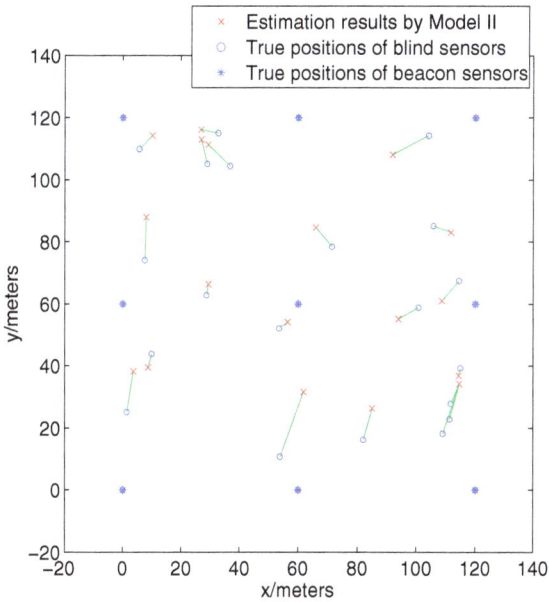

Figure 14. Position estimation results by Model II in the two dimensional WSN localization problem.

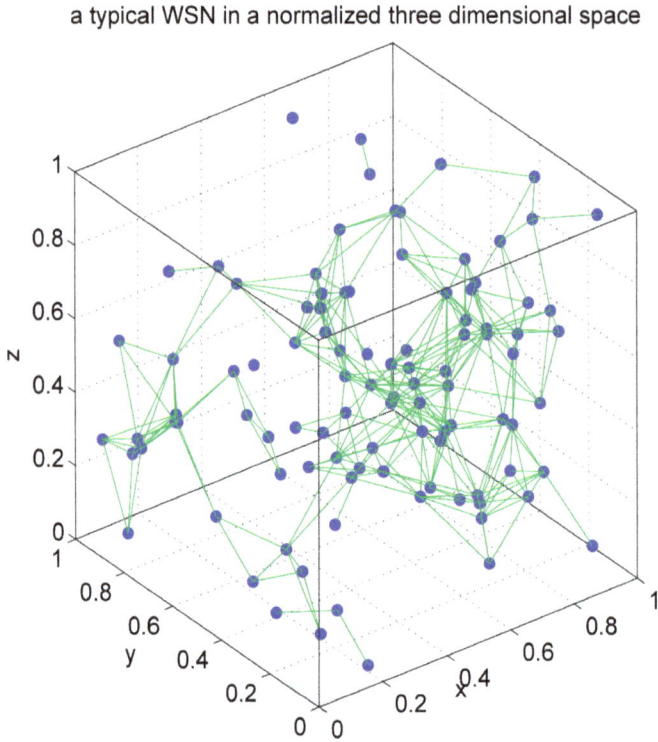

Figure 15. A typical WSN in normalized three dimensional space.

after normalization along three axial directions), such as building monitoring [35], underwater acoustic sensor networks[2], may be encountered. For example, the localization problem of sensors for building monitoring is actually defined in a three dimensional space since sensors are deployed in two dimensions on each floor and the whole network constructed by sensors on different floors forms a three dimensional one. Also, sensors in the underwater acoustic sensor networks are often deployed at different depth, with different longitude and latitude and thus form a three dimensional sensor network.

The presented two models in this chapter admit the higher dimensional localization problems as we did not specify the number of dimensions in the problem formulation and the model works for all possible dimensions.

As demonstrated in simulations, Model II outperforms Model I in the sense of estimation error for the cases with the same simulation setup. However, it is notable that Model II requires an extra dynamic model for the initialization, which is at the cost of implementation complexity and longer computation time. Fortunately, the dynamic models can be implemented with either digital or analog devices and thus the computation can be completed in a very short time. For example, the simulation examples for the two dimensional localization problem showed that it takes a time of 10^{-5} level for the dynamic models to converge. As to the implementation complexity, more hardware devices, such as summators, multipliers, dividers, integrators, etc, are needed to fabricate the analog circuit of Model II than that of Model I.

6. Conclusions and future work

In this chapter, we overviewed our recent work on range-free localization of sensors in WSNs via dynamic models. The range-free localization problem is formulated as two different optimization problems, each of which corresponds to a dynamic model, namely Model I and Model II, for the solution. Simulations in both one dimensional case and two dimensional case are performed and the two models are compared in both sceneries. The simulation results demonstrate effectiveness of the dynamic models.

Compared with conventional range-free localization algorithms, a prominent advantage of ODE based solutions is that the models are implementable by parallel hardware. As a promising direction for WSN localization, the following aspects are of fundamentally importance in both theory and practise and are open to all researchers,

• Is there better heuristic information applicable to improve the performance?

• Can Model II be modeled from a probabilistic perspective and therefore to evaluate its performance improvement relative to Model I in quantity?

• Recent advancement in ODE shows that a properly designed ODE model can receive finite time convergence to its steady state value. Can the results applied to the range-free WSN localization to gain a theoretically finite time convergence?

• Hardware implementation with VLSI, FPGA, etc. is a bright direction in practise.

• Can the presented models be used in range-base WSN localization, if not applicable directly, is the results helpful to supply a good "warm start", which accelerates the convergence?

• The presented models are essentially ODE models for solving nonlinear inequalities defined on a network. Are the models extendable to other network applications, such as robotic networks, smart grids, the internet of things, etc?

Acknowledgements

The authors would like to acknowledge the constant motivation by the fact presented in the movie Gattaca that never saving anything for the swim back enables Vincent to swim longer and faster than his genetically superior brother. The authors would like to acknowledge the support by the National Science Foundation of China under Grant NSFC: 61105090.

Author details

Shuai Li
Stevens Institute of Technology, USA

Yangming Li
Robot Sensor and Human-machine Interaction Lab, Institute of Intelligent Machines, Chinese Academy of Sciences, China

7. References

[1] Aalamifar, F., Vijay, G., Abedi Khoozani, P. & Ibnkahla, M. [2011]. Cognitive wireless sensor networks for highway safety, *Proceedings of the first ACM international symposium on Design and analysis of intelligent vehicular networks and applications*, DIVANet '11, ACM, New York, NY, USA, pp. 55–60.

[2] Akyildiz, I. F., Pompili, D. & Melodia, T. [2005]. Underwater acoustic sensor networks: Research challenges, *AD HOC NETWORKS (ELSEVIER* 3: 257–279.

[3] Baggio, A. [2005]. Wireless sensor networks in precision agriculture, *Precision Agriculture* pp. 1–22.

[4] Batalin, M., Sukhatme, G. & Hattig, M. [2004]. Mobile robot navigation using a sensor network, *Robotics and Automation, 2004. Proceedings. ICRA '04. 2004 IEEE International Conference on*, Vol. 1, pp. 636 – 641 Vol.1.

[5] Ceriotti, M., Corra, M., D'Orazio, L., Doriguzzi, R., Facchin, D., Guna, S., Jesi, G., Cigno, R., Mottola, L., Murphy, A., Pescalli, M., Picco, G., Pregnolato, D. & Torghele, C. [2011]. Is there light at the ends of the tunnel? wireless sensor networks for adaptive lighting in road tunnels, *Information Processing in Sensor Networks (IPSN), 2011 10th International Conference on*, pp. 187 –198.

[6] Chen, W., Liang, H., Mei, T., You, Z., Miao, S., Li, S., Zhou, Y. & Meng, M. Q.-H. [2007]. Design and implementation of wireless sensor network for robot navigation., *I. J. Information Acquisition* pp. 77–89.

[7] Hu, W., Bulusu, N., Chou, C. T., Jha, S., Taylor, A. & Tran, V. N. [2009]. Design and evaluation of a hybrid sensor network for cane toad monitoring, *ACM Trans. Sen. Netw.* 5(1): 4:1–4:28.

[8] Kaemarungsi, K. & Krishnamurthy, P. [2004]. Properties of indoor received signal strength for wlan location fingerprinting, *Mobile and Ubiquitous Systems: Networking and Services, 2004. MOBIQUITOUS 2004. The First Annual International Conference on*, pp. 14 – 23.

[9] Khanafer, M., Guennoun, M. & Mouftah, H. T. [2009]. Wsn architectures for intelligent transportation systems, *Proceedings of the 3rd international conference on New technologies, mobility and security*, NTMS'09, IEEE Press, Piscataway, NJ, USA, pp. 425–432.

[10] Li, S., Chen, S., Lou, Y., Lu, B. & Liang, Y. [2012]. A recurrent neural network for inter-localization of mobile phones, *IJCNN 2012*.

[11] Li, S., Liu, B., Chen, B. & Lou, Y. [2012]. Neural network based mobile phone localization using bluetooth connectivity, *NEURAL COMPUTING & APPLICATIONS* (0): –.

[12] Li, S., Meng, M. & Chen, W. [2007a]. Sp-nn: A novel neural network approach for path planning, *Robotics and Biomimetics, 2007. IEEE International Conference on*, pp. 1355 –1360.

[13] Li, S., Meng, M. & Liang, H. [2007]. A localization error estimation method based on maximum likelihood for wireless sensor networks, *Mechatronics and Automation, 2007. ICMA 2007. International Conference on*, pp. 348 –353.

[14] Li, S., Zhang, Z. & Ma, Y. [2007]. A sensor networks based method for detecting mine methane, *Information Acquisition, 2007. ICIA '07. International Conference on*, pp. 403 –407.

[15] Li, Y. [2010]. *Research On Robust Mapping Methods In Unstructured Environments*, PhD thesis, University of Science and Technology of China.

[16] Li, Y., Meng, M. & Chen, W. [2007b]. Data fusion based on rbf and nonparametric estimation for localization in wireless sensor networks, *Robotics and Biomimetics, 2007. ROBIO 2007. IEEE International Conference on*, pp. 1361 –1365.

[17] Li, Y., Meng, M., Li, S., Chen, W. & Liang, H. [2008]. Particle filtering for range-based localization in wireless sensor networks, *Intelligent Control and Automation, 2008. WCICA 2008. 7th World Congress on*, pp. 1629 –1634.

[18] Li, Y., Meng, M., Liang, H., Li, S. & Chen, W. [2008a]. Particle filtering for wsn aided slam, *Advanced Intelligent Mechatronics, 2008. AIM 2008. IEEE/ASME International Conference on*, IEEE, pp. 740–745.

[19] Li, Y., Meng, M. Q.-H., Liang, H., Li, S. & Chen, W. [2008b]. On wsn-aided simultaneous localization and mapping based on particle filtering, *Robot* **30**(5): 421–427.

[20] Milenkovi, A., Otto, C. & Jovanov, E. [2006]. Wireless sensor networks for personal health monitoring: Issues and an implementation, *Computer Communications (Special issue: Wireless Sensor Networks: Performance, Reliability, Security, and Beyond* **29**: 2521–2533.

[21] Mouftah, H. T., Khanafer, M., Guennoun, M. & Ave, K. E. [n.d.]. Wireless sensor network architectures for intelligent vehicular systems, pp. 1–7.

[22] Ogata, T., Nishide, S., Kozima, H., Komatani, K. & Okuno, H. [2010]. Inter-modality mapping in robot with recurrent neural network, *Pattern Recognition Letters* **31**(12): 1560–1569.

[23] Ogren, P., Fiorelli, E. & Leonard, N. [2004]. Cooperative control of mobile sensor networks:adaptive gradient climbing in a distributed environment, *Automatic Control, IEEE Transactions on* **49**(8): 1292 – 1302.

[24] Ortiz, C., Puig, J., Palau, C. & Esteve, M. [2007]. 3d wireless sensor network modeling and simulation, *Sensor Technologies and Applications, 2007. SensorComm 2007. International Conference on*, pp. 307 –312.

[25] S. Li, Y. L. & Liu, B. [2012]. Bluetooth aided mobile phone localization: a nonlinear neural circuit approach., *ACM Transactions on Embedded Computing Systems* **17**.

[26] Savvides, A., Srivastava, M., Girod, L. & Estrin, D. [2004]. Wireless sensor networks, Kluwer Academic Publishers, Norwell, MA, USA, chapter Localization in sensor networks, pp. 327–349.

[27] Sikka, P., Corke, P. I. & Overs, L. [2004]. Wireless sensor devices for animal tracking and control., *LCN'04*, pp. 446–454.

[28] Skowronski, M. & Harris, J. [2007]. Noise-robust automatic speech recognition using a predictive echo state network, *IEEE Transactions On Audio Speech And Language Processing* **15**(5): 1724–1730.

[29] Smith, K. [1999]. Neural networks for combinatorial optimization: a review of more than a decade of research, *INFORMS J. on Computing* **11**: 15–34.

[30] Taraktas, K., Ceylan, O. & Yagci, B. [2010]. Received signal strength technique performance in sensor network localization application, *Broadband, Wireless Computing, Communication and Applications (BWCCA), 2010 International Conference on*, pp. 357 –362.

[31] Tse, D. & Viswanath, P. [2005]. *Fundamentals of wireless communication*, Cambridge University Press, New York, NY, USA.

[32] VanZwol, T. R., Cheng, J. J. R. & Tadros, G. [2008]. Structural health monitoring of the golden gate bridge, *Canadian Journal of Civil Engineering* **35**(2): 179–189.

[33] Werner-Allen, G., Lorincz, K., Ruiz, M., Marcillo, O., Johnson, J., Lees, J. & Welsh, M. [2006]. Deploying a wireless sensor network on an active volcano, *Internet Computing, IEEE* **10**(2): 18 – 25.

[34] Xu, N., Rangwala, S., Chintalapudi, K. K., Ganesan, D., Broad, A., Govindan, R. & Estrin, D. [2004]. A wireless sensor network for structural monitoring, *Proceedings of the 2nd*

international conference on Embedded networked sensor systems, SenSys '04, ACM, New York, NY, USA, pp. 13–24.

[35] Zhou, J., Chen, Y., Leong, B. & Sundaramoorthy, P. S. [2010]. Practical 3d geographic routing for wireless sensor networks, *Proceedings of the 8th ACM Conference on Embedded Networked Sensor Systems*, SenSys '10, ACM, New York, NY, USA, pp. 337–350.

Permissions

The contributors of this book come from diverse backgrounds, making this book a truly international effort. This book will bring forth new frontiers with its revolutionizing research information and detailed analysis of the nascent developments around the world.

We would like to thank Bob tucker, for lending his expertise to make the book truly unique. He has played a crucial role in the development of this book. Without his invaluable contribution this book wouldn't have been possible. He has made vital efforts to compile up to date information on the varied aspects of this subject to make this book a valuable addition to the collection of many professionals and students.

This book was conceptualized with the vision of imparting up-to-date information and advanced data in this field. To ensure the same, a matchless editorial board was set up. Every individual on the board went through rigorous rounds of assessment to prove their worth. After which they invested a large part of their time researching and compiling the most relevant data for our readers. Conferences and sessions were held from time to time between the editorial board and the contributing authors to present the data in the most comprehensible form. The editorial team has worked tirelessly to provide valuable and valid information to help people across the globe.

Every chapter published in this book has been scrutinized by our experts. Their significance has been extensively debated. The topics covered herein carry significant findings which will fuel the growth of the discipline. They may even be implemented as practical applications or may be referred to as a beginning point for another development. Chapters in this book were first published by InTech; hereby published with permission under the Creative Commons Attribution License or equivalent.

The editorial board has been involved in producing this book since its inception. They have spent rigorous hours researching and exploring the diverse topics which have resulted in the successful publishing of this book. They have passed on their knowledge of decades through this book. To expedite this challenging task, the publisher supported the team at every step. A small team of assistant editors was also appointed to further simplify the editing procedure and attain best results for the readers.

Our editorial team has been hand-picked from every corner of the world. Their multi-ethnicity adds dynamic inputs to the discussions which result in innovative outcomes. These outcomes are then further discussed with the researchers and contributors who give their valuable feedback and opinion regarding the same. The feedback is then collaborated with the researches and they are edited in a comprehensive manner to aid the understanding of the subject.

Apart from the editorial board, the designing team has also invested a significant amount of their time in understanding the subject and creating the most relevant covers. They scrutinized every image to scout for the most suitable representation of the subject and create an appropriate cover for the book.

The publishing team has been involved in this book since its early stages. They were actively engaged in every process, be it collecting the data, connecting with the contributors or procuring relevant information. The team has been an ardent support to the editorial, designing and production team. Their endless efforts to recruit the best for this project, has resulted in the accomplishment of this book. They are a veteran in the field of academics and their pool of knowledge is as vast as their experience in printing. Their expertise and guidance has proved useful at every step. Their uncompromising quality standards have made this book an exceptional effort. Their encouragement from time to time has been an inspiration for everyone.

The publisher and the editorial board hope that this book will prove to be a valuable piece of knowledge for researchers, students, practitioners and scholars across the globe.

List of Contributors

M.A. Matin
Institut Teknologi Brunei, Brunei Darussalam

M.M. Islam
North South University, Dhaka, Bangladesh

Akshaye Dhawan
Department of Mathematics and Computer Science, Ursinus College, USA.

Wuyungerile Li, Ziyuan Pan and Takashi Watanabe
Shizuoka University, Japan

S. Chinnappen-Rimer
Department of Electrical and Electronic Engineering Science, University of Johannesburg, Johannesburg, South Africa

G. P. Hancke
Information Security Group, Royal Holloway, University of London, Egham, United Kingdom and Department of Electrical, Electronic and Computer Engineering, University of Pretoria, Pretoria, South Africa

Jan Nikodem, Marek Woda and Maciej Nikodem
Institute of Computer Science,
Automatic Control, and Robotics, Wroclaw University of Technology, Wroclaw, Poland

Mohamed M. A. Azim and Aly M. Al-Semary
Taibah University, Saudi Arabia

Alexander Klein
Network Architectures and Services - Institute for Informatics, Technical University Munich, 85748
Garching, Germany

Sheikh Omar M. and Mahmoud Samy A.
Faculty of Engineering and Design, Department of Systems and Computer Engineering, Carleton University, Ottawa, Canada

Elias Yaacoub and Adnan Abu-Dayya
QU Wireless Innovations Center (QUWIC), Doha, Qatar

A. R. Naseer
Principal and Professor of Computer Science & Engineering, Jyothishmathi Institute of Technology
& Science (JITS), Affiliated to Jawarharlal Nehru Technological University(JNTU) Hyderabad, India

Gustavo S. Quirino, Admilson R. L. Ribeiro and Edward David Moreno
Universidade Federal de Sergipe, Brasil

Shuai Li
Stevens Institute of Technology, USA

Yangming Li
Robot Sensor and Human-machine Interaction Lab, Institute of Intelligent Machines, Chinese
Academy of Sciences, China